INSTRUCTOR'S SOLUTION MANUAL

By

Willian B. Craine III

Andy Leung

Aaron Springford

Stats: Data and Models

Second Canadian Edition

Richard D. De Veaux, Williams College

Paul F. Velleman, Cornell University

David E. Bock, Cornell University

Augustin M. Vukov, University of Toronto

Augustine C. M. Wong, York University

PEARSON

Toronto

Acquisitions Editor: *David S. Le Gallais*
Marketing Manager: *Michelle Bish*
Program Manager: *Patricia Ciardullo*
Project Manager: *Marissa Lok*
Senior Developmental Editor: *John Polanszky*
Media Editor: *Ben Zaporozan*
Media Producer: *Kelli Cadet*
Production Services: *Kayci Wyatt, Electronic Publishing Services*
Cover Designer: *Anthony Leung*
Interior Designer: *Electronic Publishing Services*

Original edition published by Pearson Education, Inc., Upper Saddle River, New Jersey, USA. Copyright © 2013 Pearson Education, Inc. This edition is authorized for sale only in Canada.

If you purchased this book outside the United States or Canada, you should be aware that it has been imported without the approval of the publisher or the author.

1 2 3 4 5 6 7 8 9 10 [WC]

ISBN 978-0-321-99162-1

Table of Contents

Chapter 1: Stats Starts Here

1. **The news** Answers will vary.

3. **Orientation** *Who:* 40 undergraduate women. *What:* Whether or not the women could identify the sexual orientation of men based on a picture. *Why:* To see if ovulation affects a woman's ability to identify sexual orientation of a male. *How:* Showing very similar photos to the women, with half gay. *Variables:* Categorical variable: 'He's gay' or 'He's not gay'.

5. **Investments** *Who:* 48 China/India/Chindia funds listed at globeinvestor.com. *What:* 1 month, 1 year, and 5 year returns for each fund. *When:* The most recent periods of time. *Where:* globeinvestor.com website. *Why:* To compare investment returns for future investment decisions. *How:* globeinvestor.com uses reports from the fund companies. *Variables:* There are three variables, all of which are quantitative. 1 month return; 1 year return; 5 year return, annualized; all variables are measured as percentages.

7. **Air travel** *Who:* All airline flights in Canada. *What:* Type of aircraft, number of passengers, whether departures and arrivals were on schedule, and mechanical problems. *When:* This information is currently reported. *Where:* Canada. *Why:* This information is required by Transport Canada and the Canadian Transportation Agency. *How:* Data is collected from airline flight information. *Variables:* There are four variables. Type of aircraft, departure and arrival timeliness, and mechanical problems are categorical variables, and number of passengers is a quantitative variable.

9. **Cars** *Who:* Automobiles. *What:* Make, country of origin, type of vehicle, and age of vehicle (probably in years). *When:* Not specified. *Where:* A large university. *Why:* Not specified. *How:* A survey was taken in campus parking lots. *Variables:* There are four variables. Make, country of origin, and type of vehicle are categorical variables, and age of vehicle is a quantitative variable.

11. **Honesty** *Who:* Workers who buy coffee in an office. *What:* Amount of money contributed to the collection tray. *Where:* Newcastle. *Why:* To see if people behave more honestly when feeling watched. *How:* Counting money in the tray each week. *Variables:* Amount contributed (pounds) is a quantitative variable.

13. **Weighing bears** *Who:* 54 bears. *What:* Weight, neck size, length (no specified units), and sex. *When:* Not specified. *Where:* Not specified. *Why:* Since bears are difficult to weigh, the researchers hope to use the relationships between weight, neck size, length, and sex of bears to estimate the weight of bears, given the other more

observable features of the bear. *How*: Researchers collected data on 54 bears they were able to catch. *Variables*: There are four variables; weight, neck size, and length are quantitative variables, and sex is a categorical variable. No units are specified for the quantitative variables. *Concerns*: The researchers are (obviously!) only able to collect data from bears they were able to catch. This method is a good one, as long as the researchers believe the bears caught are representative of all bears, in regard to the relationships between weight, neck size, length, and sex.

15. **Tim Horton's doughnuts** *Who*: Doughnut types for sale at Tim Hortons. *What*: Various nutritional characteristics (see variables below). *When*: Not stated, but presumably the measurements were taken recently. *Where*: Tim Hortons Web site. *Why*: To help customers make good nutritional choices. *How*: Further research would be needed to learn how they made these measurements, but presumably at some specialized food analysis lab. *Variables*: There are eight variables, all quantitative: Number of calories (kca/s), amounts of trans fat (g), total fat (g), sodium (mg), sugar (g), protein (g), % daily value of iron (percentage), and % daily value of calcium (percentage). Units found by going to the Web site.

17. **Babies** *Who*: 882 births. *What*: Mother's age (in years), length of pregnancy (in weeks), type of birth (Caesarean, induced, or natural), level of prenatal care (none, minimal, or adequate), birth weight of baby (unit of measurement not specified, but probably grams), gender of baby (male or female), and baby's health problems (none, minor, major). *When*: 1998-2000. *Where*: Large city hospital. *Why*: Researchers were investigating the impact of prenatal care on newborn health. *How*: It appears that they kept track of all births in the form of hospital records, although it is not specifically stated. *Variables:* There are three quantitative variables: mother's age, length of pregnancy, and birth weight of baby. There are four categorical variables: type of birth, level of prenatal care, gender of baby, and baby's health problems.

19. **Herbal medicine.** *Who*: Experiment volunteers. *What*: Herbal cold remedy or sugar solution, and cold severity. *When*: Not specified. *Where*: Major pharmaceutical firm. *Why*: Scientists were testing the efficacy of an herbal compound on the severity of the common cold. *How*: The scientists set up a controlled experiment. *Variables*: There are two variables. Type of treatment (herbal or sugar solution) is categorical, and severity rating (on a scale from 0 to 5) is quantitative. *Concerns*: The severity of a cold seems subjective and difficult to quantify. Also, the scientists may feel pressure to report negative findings about the herbal product.

21. **Streams** *Who:* Streams. *What:* Name of stream, substrate of the stream (limestone, shale, or mixed), acidity of the water (measured in pH), temperature (in degrees Celsius), and BCI (unknown units). *When:* Not specified. *Where:* Northern Ontario. *Why:* Research is conducted for an ecology class. *How:* Not specified. *Variables:* There are five variables. Name and substrate of the stream are categorical variables, and acidity (pH), temperature (in degrees Celsius), and BCI are quantitative variables.

23. **Refrigerators** *Who:* 41 models of refrigerators. *What:* Brand, cost (probably in dollars), size (in cu. ft.), type, estimated annual energy cost (probably in dollars), overall rating, and repair history (in percent requiring repair over the past five years). *When:* 2002-2006. *Where:* United States. *Why:* The information was compiled to provide information to the readers of *Consumer Reports. How:* Not specified. *Variables:* There are seven variables. Brand, type, and overall rating are categorical variables. Cost, size (cu. ft.), estimated energy cost, and repair history (percentage) are quantitative variables.

25. **Kentucky Derby 2012** *Who:* Kentucky Derby races. *What:* Year, winner, jockey, trainer, owner, and time (in minutes, seconds, and hundredths of a second). *When:* 1875 to 2012. *Where:* Churchill Downs, Louisville, Kentucky. *Why:* It is interesting to examine the trends in the Kentucky Derby. *How:* Official statistics are kept for the race each year. *Variables:* There are six variables. Date, Winner, jockey, trainer, and owner are categorical variables. Duration is quantitative variables.

1. **Graphs in the news** Answers will vary.

3. **Tables in the news** Answers will vary.

5. **Forest fires 2010** The relative frequency distribution is shown below:

Cause of fire	Percentage
Lightning	46.94
Human activities	51.64
Unknown	1.42

Causes for forest fires are about equally split between human activities and lightning. Only 1.42% of forest fires are due to unknown causes.
(Example: 46.94% = 0.4694 = 3279/6986)

7. **Teen smokers** According to the Monitoring the Future study, teen smoking brand preferences differ somewhat by region. Although Marlboro is the most popular brand in each region, with about 58% of teen smokers preferring this brand in each region, teen smokers from the South prefer Newports at a higher percentage than teen smokers from the West, with 22.5% of teen smokers preferring this brand, compared to only 10.1% in the South. Teen smokers in the West are also more likely to have no particular brand than teen smokers in the South. 12.9% of teen smokers in the West have no particular brand, compared to only 6.7% in the South. Both regions have about 9% of teen smokers that prefer one of over 20 other brands.

9. **Oil spills as of 2010**
   a) Grounding, accounting for 160 spills, is the most frequent cause of oil spillage for these 460 spills. A substantial number of spills, 132, were caused by collision. Less prevalent causes of oil spillage in descending order of frequency were loading/discharging, other/unknown causes, fire/explosions, and hull failures.
   b) If being able to differentiate between these close counts is required, use the bar chart. Since each spill only has one cause, the pie chart is also acceptable as a display, but it's difficult to tell whether, for example, there is a greater percentage of spills caused by fire/explosions or hull failure. If you want to showcase the causes of oil spills as a fraction of all 460 spills, use the pie chart.

11. **Global warming** Perhaps the most obvious error is that the percentages in the pie chart only add up to 93%, when they should, of course, add up to 100%. Furthermore, the three-dimensional perspective view distorts the regions in the graph, violating the area principle. The regions corresponding to No Solid Evidence and Due to Human Activity should be roughly the same size, at 32% and 34% of respondents, respectively. However, the 32% region looks bigger, and the angle for the 34% region makes it look only slightly bigger than the 18% region. Always use simple, two-dimensional graphs.

13. **Complications**
    a) A bar chart is the proper display for these data. A pie chart is not appropriate since these are counts, not fractions of a whole.

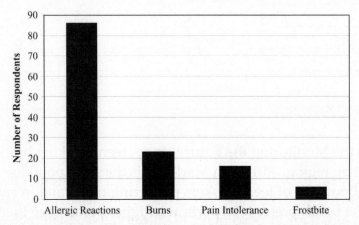

    b) The *Who* for these data is athletic trainers who used cryotherapy, which should be a cause for concern. A trainer who treated many patients with cryotherapy would be more likely to have seen complications than one who used cryotherapy rarely. We would prefer a study in which the *Who* referred to patients so we could assess the risks of each complication.

15. **Spatial distribution**
    a) The relative frequency distribution of quadrant location is given below. Not all proportions are equal. In particular, the relative frequency for Quadrant 4 is approximately twice the other frequencies.

Quadrant	Quadrant 1	Quadrant 2	Quadrant 3	Quadrant 4
Relative Frequency	0.18	0.21	0.22	0.39

    b) The relative frequency distribution of quadrant location is given below. There seems to have some similarity with that in part a. For example, Quadrant 4 has the highest relative frequency and Quadrant 1 has the lowest.

Quadrant	Quadrant 1	Quadrant 2	Quadrant 3	Quadrant 4
Relative Frequency	0.12	0.24	0.28	0.36

17. Politics revisited

a) The females in this course were 45.5% Liberal, 46.8% Moderate, and 7.8% Conservative. (They don't add up to exactly 100% due to the roundoffs.)

b) The males in this course were 43.5% Liberal, 38.3% Moderate, and 18.3% Conservative.

c) A segmented bar chart comparing the distributions is below.

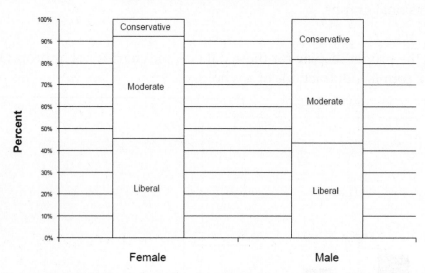

d) Politics and sex do not appear to be independent in this course. Although the percentage of liberals was roughly the same for each sex, females had a greater percentage of moderates and a lower percentage of conservatives than males.

19. Canadian languages 2011

a) 22,564,665 Canadians speak English only. 22,564,665/33,121,175 total Canadians ≈ 68.1%

b) 4,165,015 Canadians speak French only and 5,795,575 speak both French and English, for a total of 9,960,590 French speakers. 9,960,590/33,121,175 total Canadians ≈ 30.1%

c) 4,047,175 French and 3,328,725 French and English speakers yield a total of 7,375,900 French speakers in Quebec. 7,375,900/7,815,955 Quebec residents ≈ 94.4%

d) 7,375,900 Quebec residents speak French and 9,960,590 Canadians speak French. The percentage of French-speaking Canadians who live in Quebec is 7,375,900/9,960,590 ≈ 74.1%

e) If language knowledge were independent of province, we would expect the percentage of French-speaking residents of Quebec to be the same as the overall percentage of Canadians who speak French. Since 30.1% of all Canadians speak French while 94.4% of residents of Quebec speak French, there is evidence of an association between language knowledge and province.

21. Weather forecasts

a) The table shows the marginal totals. It rained on 34 of 365 days, or 9.3% of the days.

b) Rain was predicted on 90 of 365 days. $90/365 \approx$ 24.7% of the days.

		Actual Weather		
		Rain	No Rain	**Total**
Forecast	Rain	27	63	90
	No Rain	7	268	275
	Total	34	331	365

c) The forecast of rain was correct on 27 of the days it actually rained and the forecast of No Rain was correct on 268 of the days it didn't rain. So, the forecast was correct a total of 295 times. $295/365 \approx 80.8\%$ of the days.

d) On rainy days, rain had been predicted 27 out of 34 times (79.4%). On days when it did not rain, forecasters were correct in their predictions 268 out of 331 times (81.0%). These two percentages are very close. There is no evidence of an association between the type of weather and the ability of the forecasters to make an accurate prediction.

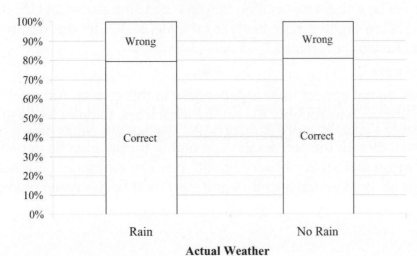

Weather Forecast Accuracy

23. Driver's licenses 2011

a) There are 10.0 million drivers under 20 and a total of 208.3 million drivers in the U.S. That's about 4.8% of U.S. drivers under 20.

b) There are 103.5 million males out of 208.4 million total U.S. drivers, or about 49.7%.

c) Each age category appears to have about 50% male and 50% female drivers. The segmented bar chart shows a pattern in the deviations from 50%. At younger ages, males form the slight majority of drivers. The percentage of male drivers continues to shrink until, at around age 45, female drivers hold a slight majority. This continues into the 85 and over category.

Registered U.S. Drivers by Age and Gender

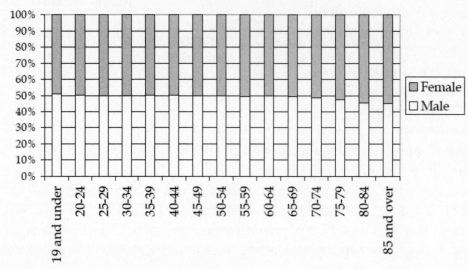

Age in Years

d) There appears to be a slight association between age and gender of U.S. drivers. Younger drivers are slightly more likely to be male, and older drivers are slightly more likely to be female.

25. Anorexia

These data provide no evidence that Prozac might be helpful in treating anorexia. About 71% of the patients who took Prozac were diagnosed as "Healthy." Even though the percentage was higher for the placebo patients, this does not mean that Prozac is hurting patients. The difference between 71% and 73% is not likely to be statistically significant (will be discussed in later chapters).

27. Smoking gene?

a) The marginal distribution of genotype is given below:

Genotype	Marginal percentage
GG	42.71
GT	45.08
TT	12.21

b) The conditional distributions of genotype for the four categories of smokers are given in columns 2–5 of the table below.

| Genotype | Cigarettes per day | | | | All |
	1–10	11–20	21–30	31 and more	
GG	48.06	42.60	38.32	36.75	42.71
GT	42.96	44.75	47.39	48.28	45.08
TT	8.99	12.65	14.29	14.98	12.21
All	100.00	100.00	100.00	100.00	100.00

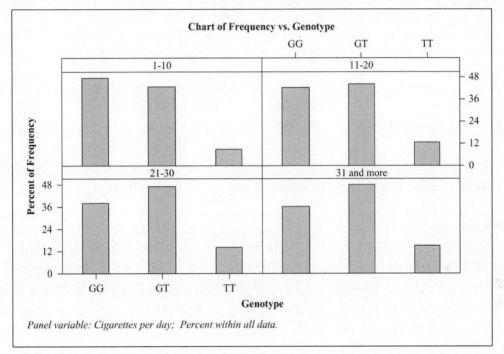

Panel variable: Cigarettes per day; Percent within all data.

c) Though not a very noticeable difference, the percentages of smokers with genotype GT (also TT) are slightly higher among heavy smokers. However, this is only an observed association. This does not prove that presence of the T allele increases susceptibility to nicotine addiction. We cannot conclude that this increase was caused by the presence of the T allele. There can be many factors associated with the presence of the T allele, and some of these factors might be the reason for the increase in susceptibility to nicotine addiction.

29. **Antidepressants and bone fractures** These data provide evidence that taking a certain class of antidepressants (SSRI) might be associated with a greater risk of bone fractures. Approximately 10% of the patients taking this class of antidepressants experience bone fractures. This is compared to only approximately 5% in the group that were not taking the antidepressants.

31. Cell phones

a) The two-way table and the conditional distributions (percentages) of 'car accident' (crash or non-crash) for cell phone owners and non-cell phone owners are given below. The proportion of crashes is higher for cell phone owners than for non-cell phone owners.

	Cell phone ownership		
Car accident	Cell phone owner	Non-cell phone owner	All
Crash	20	10	30
Non-crash	58	92	150
All	78	102	180

	Cell phone ownership		
Car accident	Cell phone owner	Non-cell phone owner	All
Crash	25.64	9.80	16.67
Non-crash	74.36	90.20	83.33
All	100.00	100.00	100.00

b) On the basis of this study, we cannot conclude that the use of a cell phone increases the risk of a car accident. This is only an observed association between cell phone ownership and the risk of car accidents. We cannot conclude that the higher proportion of accidents was caused by the use of a cell phone. There can be lots of other factors common to cell phone owners, and some of those factors can be the reason for the accidents.

33. Blood pressure

a) The marginal distribution of blood pressure for the employees of the company is the total column of the table,

Blood pressure	under 30	30 - 49	over 50	Total
low	27	37	31	95
normal	48	91	93	232
high	23	51	73	147
Total	98	179	197	474

converted to percentages. 20% low, 49% normal, and 31% high blood pressure.

b) The conditional distribution of blood pressure within each age category is: Under 30: 28% low, 49% normal, 23% high 30–49: 21% low, 51% normal, 28% high Over 50: 16% low, 47% normal, 37% high.

c) A segmented bar chart of the conditional distributions of blood pressure by age category is at the right.

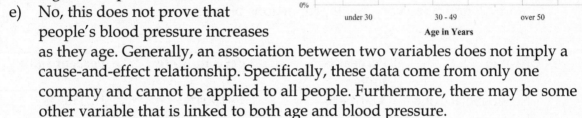

Blood Pressure of Employees

d) In this company, as age increases, the percentage of employees with low blood pressure decreases, and the percentage of employees with high blood pressure increases.

e) No, this does not prove that people's blood pressure increases as they age. Generally, an association between two variables does not imply a cause-and-effect relationship. Specifically, these data come from only one company and cannot be applied to all people. Furthermore, there may be some other variable that is linked to both age and blood pressure.

35. **Aboriginal identity 2011**

a) The second column includes some individuals in the 3rd, 4th, and 5th columns, so it is not a standard contingency table.

b) Use the label "Aboriginal population not included in columns 3, 4, and 5" (or call them "other Aboriginals"). Use the value in second column minus those in the third, fourth and fifth columns as the new value.

c) There are 27 070 Canadians who are Inuit from Nunavut. The Canadian population is 32 852 320, so the proportion of Canadians who are Inuit from Nunavut is 27 070/32 852 320 = 0.00082 = 0.08% (approx.).

d) There are 27 070 Canadians who are Inuit from Nunavut. The total Canadian Aboriginal population is 1 400 685, so the proportion of Canadian Aboriginals who are Inuit from Nunavut is 27 070/1 400 685 = 0.0193 = 1.93% (approx.).

e) The total Canadian Aboriginal population is 1 400 685 and of them 59 440 are Inuit. So 59 440/1 400 685 = 0.0424 = 4.24% of Canadian Aboriginals are Inuit.

f) The total Canadian Aboriginal population is 1 400 685 and of them 27 360 are from Nunavut. So 27 360/1 400 685 = 0.0195 = 1.95% of Canadian Aboriginals are from Nunavut.

g) The total population in Nunavut is 31 700 and 27 360 of them are Inuit. So 27 070/31 700 = 0.8539 = 85.39% of the people from Nunavut are Inuit.

h) There are 27 360 Nunavut Aboriginals and 27 070 of them are Inuit. So 27 070/27 360 = 0.9894 = 98.94% of Nunavut Aboriginals are Inuit.

i) The total Inuit population is 59 440 and of them 27 070 are from Nunavut. So 27 070/59 440 = 0.4554 = 45.54% of Inuit live in Nunavut.

j)   The total number of Ontario Aboriginals is 301 430, and 301 430 – 201 105 – 86 015 – 3360 = 10 950 of them are other Aboriginals (i.e., other than Inuit, Metis, or N.A. Indian) and so 10 950/301 430 = 0.0363 = 3.63% of Ontario Aboriginals could not be simply classified as Inuit, Metis, or N.A. Indian.

k)   A table of percentages of total provincial population for each Aboriginal identity group (Inuit, Metis, N.A. Indian) for Newfoundland, Ontario, Saskatchewan, and Alberta is given below. The second table is a bit easier if using MINITAB. The side-by-side bar charts below show that Saskatchewan has the highest proportion of N.A. Indian and Metis. Ontario, Saskatchewan, and Alberta have very small proportions of Inuit.

Region	Percent N.A. Indian	Percent Metis	Percent Inuit
Newfoundland and Labrador	3.80764%	1.51004%	1.23504%
Ontario	1.58954%	0.67986%	0.02656%
Saskatchewan	10.23137%	5.19945%	0.02875%
Alberta	3.26992%	2.71498%	0.05563%

Group	Newfoundland and Labrador	Ontario	Saskatchewan	Alberta
N.A. Indian	3.80764%	1.58954%	10.23137%	3.26992%
Metis	1.51004%	0.67986%	5.19945%	2.71498%
Inuit	1.23504%	0.02656%	0.02875%	0.05563%

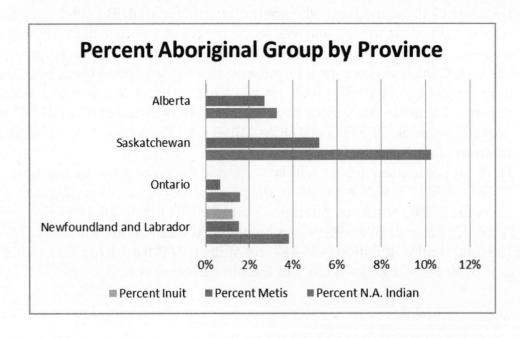

37. Hospitals

a) The marginal totals have been added to the table:

		Discharge delayed		
		Large Hospital	**Small Hospital**	**Total**
Procedure	**Major surgery**	120 of 800	10 of 50	130 of 850
	Minor surgery	10 of 200	20 of 250	30 of 450
	Total	130 of 1000	30 of 300	160 of 1300

160 of 1300, or about 12.3%, of the patients had a delayed discharge.

b) Yes. Major surgery patients were delayed 130 of 850 times, or about 15.3% of the time. Minor surgery patients were delayed 30 of 450 times, or about 6.7% of the time.

c) Large Hospital had a delay rate of 130 of 1000, or 13%. Small Hospital had a delay rate of 30 of 300, or 10%. The small hospital has the lower overall rate of delayed discharge.

d) Large Hospital: Major Surgery 15% delayed and Minor Surgery 5% delayed. Small Hospital: Major Surgery 20% delayed and Minor Surgery 8% delayed. Even though small hospital had the lower overall rate of delayed discharge, the large hospital had a lower rate of delayed discharge for each type of surgery.

e) No. While the overall rate of delayed discharge is lower for the small hospital, the large hospital did better with *both* major surgery and minor surgery.

f) The small hospital performs a higher percentage of minor surgeries than major surgeries. 250 of 300 surgeries at the small hospital were minor (83%). Only 200 of the large hospital's 1000 surgeries were minor (20%). Minor surgery had a lower delay rate than major surgery (6.7% to 15.3%), so the small hospital's overall rate was artificially inflated. Simply put, it is a mistake to look at the overall percentages. The real truth is found by looking at the rates after the information is broken down by type of surgery, since the delay rates for each type of surgery are so different. The larger hospital is the better hospital when comparing discharge delay rates.

39. Graduate admissions

a) 1284 applicants were admitted out of a total of 3014 applicants. 1284/3014 = 42.6%

Program	Males Accepted (of applicants)	Females Accepted (of applicants)	Total
1	511 of 825	89 of 108	600 of 933
2	352 of 560	17 of 25	369 of 585
3	137 of 407	132 of 375	269 of 782
4	22 of 373	24 of 341	46 of 714
Total	1022 of 2165	262 of 849	1284 of 3014

b)   1022 of 2165 (47.2%) males were admitted. 262 of 849 (30.9%) females were admitted.

c)   Since there are four comparisons to make, the table at the right organizes the percentages of males and females accepted in each program. Females are accepted at a higher rate in every program.

Program	Males	Females
1	61.9%	82.4%
2	62.9%	68.0%
3	33.7%	35.2%
4	5.9%	7%

d)   The comparison of acceptance rate within each program is most valid. The overall percentage is an unfair average. It fails to take the different numbers of applicants and different acceptance rates of each program. Women tended to apply to the programs in which gaining acceptance was difficult for everyone. This is an example of Simpson's Paradox.

Chapter 3: Displaying and Summarizing Quantitative Data

1. **Histogram** Answers will vary.

3. **In the news** Answers will vary.

5. **Thinking about shape**
   a) The distribution of the number of speeding tickets each student in final year of university has ever had is likely to be unimodal and skewed to the right. Most students will have very few speeding tickets (maybe 0 or 1), but a small percentage of students will likely have comparatively many (3 or more) tickets.
   b) The distribution of player's scores at the U.S. Open Golf Tournament would most likely be unimodal and slightly skewed to the right. The best golf players in the game will likely have around the same average score, but some golfers might be off their game and score 15 strokes above the mean. (Remember that high scores are undesirable in the game of golf!)
   c) The weights of female babies in a particular hospital over the course of a year will likely have a distribution that is unimodal and symmetric. Most newborns have about the same weight, with some babies weighing more and less than this average. There may be slight skew to the left, since there seems to be a greater likelihood of premature birth (and low birth weight) than post-term birth (and high birth weight).
   d) The distribution of the length of the average hair on the heads of students in a large class would likely be bimodal and skewed to the right. The average hair length of the males would be at one mode, and the average hair length of the females would be at the other mode, since women typically have longer hair than men. The distribution would be skewed to the right, since it is not possible to have hair length less than zero, but it is possible to have a variety of lengths of longer hair.

7. **Sugar in cereals**
   a) The distribution of the sugar content of breakfast cereals is bimodal, with a cluster of cereals with sugar content around 10% sugar and another cluster of cereals around 45% sugar. The lower cluster shows a bit of skew to the right. Most cereals in the lower cluster have between 0% and 10% sugar. The upper cluster is symmetric, with centre around 45% sugar.
   b) There are two different types of breakfast cereals, those for children and those for adults. The children's cereals are likely to have higher sugar contents, to make them taste better (to kids, anyway!). Adult cereals often advertise low sugar content.

9. Test scores

a) The number of students scoring 40 or higher is approximately $16 + 5 + 3 + 1 = 25$ (these are only approximate heights of the bars in the histogram after 40) and the percentage = $25/110 = 22.7$ percent.

b) The number of students scoring between 25 and 35 is the sum of the heights of the two bars between 25 and 35. This is approximately $21 + 31 = 52$, and so the percentage is $52/110 = 47.3$ percent.

c) The distribution is symmetric. The centre (the median) is around 32. The scores range from 0–60, but the few scores close to zero may be outliers (with a gap of just five we might not be able to conclude as a clear outlier, but they are somewhat unusual compared to the rest of the scores).

11. Election 2011

a) The distribution is right skewed and so it is logical to expect the mean to be greater than the median. The median is the $(308+1)/2$th = 154.5th value in the ordered data set. This value must be in the fourth interval from the left of the histogram, i.e., the median must be in the interval 0.45–0.55.

b) The distribution is right skewed, and could be bimodal. The median is between 0.45 and 0.55. The values of the percentage of rejected ballots range from 0–2.5 (approx.). The largest value could be an outlier, so might warrant further investigation. The apparent bimodality might also warrant further investigation.

13. Summaries

a) The mean price of the electric smoothtop ranges is $1001.50.

b) To find the median and the quartiles, first order the list.
565 750 850 900 1000 1050 1050 1200 1250 1400
The median price of the electric smoothtop ranges is $1025.
Quartile 1 = $850 and Quartile 3 = $1200.

c) The range of the distribution of prices is Max – Min = $1400 – $565 = $835.
The IQR = Q3 – Q1 = $1200 – $850 = $350.

15. Mistake

a) As long as the boss's true salary of $200 000 is still above the median, the median will be correct. The mean will be too large, since the total of all the salaries will decrease by $2 000 000 – $200 000 = $1 800 000, once the mistake is corrected.

b) The range will likely be too large. The boss's salary is probably the maximum, and a lower maximum would lead to a smaller range. The IQR will likely be unaffected, since the new maximum has no effect on the quartiles. The standard deviation will be too large, because the $2 000 000 salary will have a large squared deviation from the mean.

17. Standard deviation I

a) Set 2 has the greater standard deviation. Both sets have the same mean (6), but set two has values that are generally farther away from the mean.
SD(Set 1) = 2.24 SD(Set 2) = 3.16

b) Set 2 has the greater standard deviation. Both sets have the same mean (15), maximum (20), and minimum (10), but 11 and 19 are farther from the mean than 14 and 16.
SD(Set 1) = 3.61 SD(Set 2) = 4.53

c) The standard deviations are the same. Set 2 is simply Set 1 + 80. Although the measures of centre and position change, the spread is exactly the same.
SD(Set 1) = 4.24 SD(Set 2) = 4.24

19. Payroll

a) The mean salary is $\dfrac{1200 + 700 + 6(400) + 4(500)}{12} = 525$

The median salary is the middle of the ordered list:
400 400 400 400 400 400 500 500 500 500 700 1200
The median is $450.

b) Only two employees, the supervisor and the inventory manager, earn more than the mean wage.

c) The median better describes the wage of the typical worker. The mean is affected by the two higher salaries.

d) The IQR is the better measure of spread for the payroll distribution. The standard deviation and the range are both affected by the two higher salaries.

21. Alberta casinos 2013   The stem and leaf plot, a dotplot, and the five-number summary (plus mean) for these data are shown below. The distribution looks slightly right-skewed (mean larger than median). The median number of slot machines is 428. The interquartile range is 731 – 299 = 432.

Stem and leaf plot:
```
 1 | 7
 2 | 0035
 3 | 0033
 4 | 0224
 5 | 3
 6 | 000
 7 | 0678
 8 | 66
 9 |
10 | 0
```
(6|0 means 600 slot machines)

Dotplot:

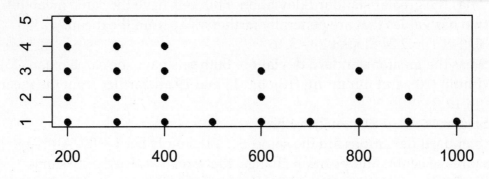

Numbers of slot machines

Descriptive Statistics: numbers of slot machines

Min.	1st Qu.	Median	3rd Qu.	Max.
170	299	428	731	1000

23. How tall? The histogram shows some low outliers in the distribution of height estimates. These are probably poor estimates and will pull the mean down. The median is likely to give a better estimate of the professor's true height.

25. World Series champs

The distribution of the number of homeruns hit by Joe Carter during the 1983–1998 seasons is skewed to the left, with a typical number of homeruns per season in the high 20s to low 30s. The season in which Joe hit no homeruns looks to be an outlier. With the exception of this no-homerun season, Joe's total number of homeruns per season was between 13 and 35. The median is 27 homeruns.

27. Hurricanes 2010

a) A dotplot of the number of hurricanes each year from 1944 through 2010 is shown below. Each dot represents a year in which there were that many hurricanes.

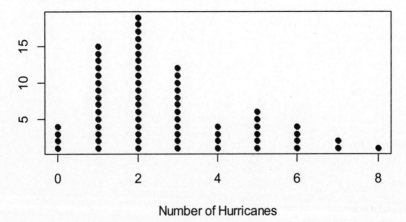

Number of Hurricanes

b) The distribution of the number of hurricanes per year is unimodal and skewed to the right, with centre around 2 hurricanes per year. The number of hurricanes per year ranges from 0 to 8. There are no outliers. There may be a second mode at 5 hurricanes per year, but since there were only 6 years in which 5 hurricanes occurred, this may simply be natural variability.

29. World Series champs again

a) This is not a histogram. The horizontal axis should be the number of homeruns per year, split into bins of a convenient width. The vertical axis should show the frequency; that is, the number of years in which Carter hit a number of home runs within the interval of each bin. The display shown is a bar chart/time series plot hybrid that simply displays the data table visually, in time order. It is of no use in describing the shape, center, spread, or unusual features of the distribution of home runs hit per year by Carter.

b) The histogram is shown below.

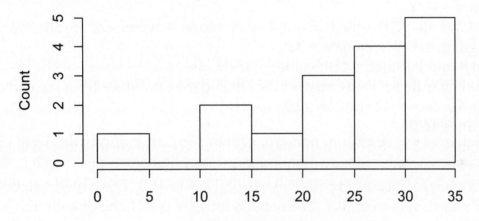

31. Acid rain The distribution of the pH readings of water samples in Allegheny County, Pennsylvania, is bimodal. A roughly uniform cluster is centred on a pH of 4.4. This cluster ranges from pH of 4.1 to 4.9. Another smaller, tightly packed cluster is centred on a pH of 5.6. Two readings in the middle seem to belong to neither cluster.

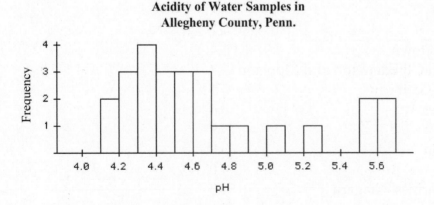

33. Housing price boom

a) The stem and leaf plot for rounded data is shown below:

```
-0 | 1
 0 | 112334444
 0 | 55566
 1 | 234
 1 | 5
 2 |
 2 | 8
 3 |
 3 |
 4 |
 4 | 6
```

Percent Increase (rounded)

(4|6 means 46 percent)

b) Minimum = –1.0

The first quartile (Q1) = median of lower 10 values = average of the 5th and the 6th values in the ordered data = 3.0

Median = middle value = 11th value = 5.0

The third quartile (Q3) = average of the 5th and the 6th values from the end of the ordered data = (12+13)/2 = 12.5

Maximum = 46.0

c) The mean is 8.62 (or 8.57 with the rounded data). Yes, the mean must be larger than the median because the distribution is right-skewed.

d) The percentage increases in price range from –1% to 46%. The largest value 46 (possibly 28 also) is an outlier. The median increase is 5%. The distribution is right-skewed.

e) The data sorted by the percentage increase is shown below. This shows large increases in the Prairie and Atlantic regions.

Data Display

Row	Metropolitan Area	Percentage Increase
1	Windsor	–0.6
2	Victoria	1.2
3	Charlottetown	1.4
4	Saint John, Fredericton and Moncton	2.4
5	Ottawa–Gatineau	3.1
6	Kitchener	3.4
7	Québec	3.9
8	Hamilton	3.9
9	London	4.0
10	St. Catharines–Niagara	4.3

11	Montréal	4.5
12	Toronto and Oshawa	4.5
13	Calgary	5.3
14	Vancouver	6.1
15	Greater Sudbury and Thunder Bay	6.3
16	St. John's	12.0
17	Halifax	12.8
18	Edmonton	13.5
19	Winnipeg	15.0
20	Regina	27.8
21	Saskatoon	46.2

35. Trimmed mean

a) The sum of the 10 values given is 64, so the mean is 64/10 = 6.4
The data arranged in increasing order: 1, 5, 6, 6, 7, 7, 8, 8, 8, 8. The median is 7 (the average of the 5th and the 6th values).
There are 10 values in the data set and 10% of 10 is 1. To calculate the 10% trimmed mean, delete the smallest and the largest value in the sorted data set and calculate the average of the remaining values. The 10% trimmed mean is 55/8 = 6.875.

b) The data of Exercise 34, arranged in increasing order:
1 1 2 2 2 2 2 2 2 3 3 3 3 3 3 4 4 4 4 4 4 4 5 5 5 5 6 7 7 8 8 8 10 10 10 16. There are 36 values in the data set and 5% of 36 is 1.8, or approximately 2. Hence, to calculate the 5% trimmed mean, delete the two smallest and two largest values in the sorted data set and calculate the average of the remaining values. The 5% trimmed mean is 144/32 = 4.5. The mean is 4.78, and the median is 4.0.

c) The data in part (a) is left skewed. As a result, the median is bigger than the trimmed mean, which is bigger than the mean. The data in part (b) is right skewed, so the median is smaller than the trimmed mean, which is smaller than the mean.

37. Zip codes Even though zip codes are numbers, they are not quantitative in nature. Zip codes are categories. A histogram is not an appropriate display for categorical data. The histogram the Holes-R-Us staff member displayed doesn't take into account that some 5-digit numbers do not correspond to zip codes or that zip codes falling into the same classes may not even represent similar cities or towns. The summary statistics are meaningless for the same reason. The employee could design a better display by constructing a bar chart that groups together zip codes representing areas with similar demographics and geographic locations.

39. Golf courses

a)

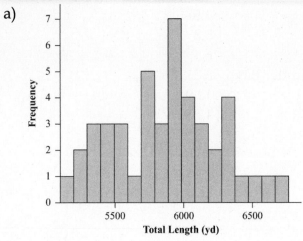

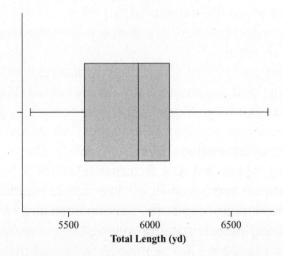

b) In any distribution, 50% of scores lie between Quartile 1 and Quartile 3. In this case, Quartile 1 = 5585.75 yards and Quartile 3 = 6131 yards.

c) The distribution of golf course lengths appears roughly symmetric, so the mean and standard deviation are the preferred measures of centre and spread.

d) The distribution of the lengths of all the golf courses in Vermont is roughly unimodal and symmetric. The mean length of the golf courses is approximately 5900 yards. Vermont has golf courses anywhere from 5185 yards to 6796 yards long. There are no outliers in the distribution.

e) (Mean – Std Dev, Mean + Std Dev) ➔ (5892.91 – 386.59, 5892.91 + 386.59) ➔ (5506.32, 6279.5). There are 28 observations in this range. Or 28/45 = 62% are in this range.

(Mean – 2Std Dev, Mean + 2Std Dev) ➔ (5892.91 – 2(386.59), 5892.91 + 2(386.59)) ➔ (5119.73, 6666.09). There are 44 observations in this range. Or 44/45 = 98% are in this range.

These are reasonably close to what we expect according to the empirical rule for roughly bell-shaped distributions.

41. Math scores 2009

a) The 5-number summary of the national averages is given below:
   Descriptive Statistics: Ave Score
   Min.   1st Qu.   Median   3rd Qu.   Max.
   419 487 496.5 514 546
   The IQR, mean, and the standard deviation of the national averages is given below:
   Descriptive Statistics: Ave Score
   Variable   Mean   StDev   IQR
   Avg. Score   495.7   29.66   27
   The distribution of average scores appears to be left-skewed. The long left tail pulls the mean down toward the smaller values (in fact, three of these small values are smaller than Q1 – 1.5 × IQR, and are thus suspect outliers). The low outliers attract the mean toward them, whereas the median is resistant to outliers. (Alternatively, one might say that there are several very small scores, and putting them aside, there is some slight right-skewness)

b) Since there are outliers and asymmetry in the data set, the 5-number summary is better than the mean and standard deviation which are not resistant and are more useful for symmetric distributions.

c) Thirty-four countries participated in the program. The highest national average is 546 and the lowest is 419. The median national average is 496.5. The interquartile range is 27. Twenty-five percent (i.e., 9 countries) of the participating countries had a national average 487 or below and at least 25% of the countries had a national average of 514 or above. Canada's national average (which is 527) is in the top 25% of all participating countries, more specifically, the 5th highest of all participating countries. The United States' national average (which is 487) is the 25th highest of all participating countries.

d) Using the empirical rule, the middle 68% of the students have their scores within one standard deviation from the mean, i.e., 527 – 88 = 439 to 527 + 88 = 615.
   The middle 95% of the students have their scores within two standard deviations from the mean, i.e., 527 – 2 × 88 = 351 to 527 + 2 × 86 = 703.
   Just about all of the students will have their scores within three standard deviations from the mean, i.e., 527 – 3 × 88 = 263 to 527 + 3 × 88 = 791.
   Using this information we can fill in the blanks as shown below:
   > About two-thirds of students scored between 439 and 615.
   > Only about 5% scored less than 351 or more than 703.
   > Only a real math genius could have scored above 791.

43.   First Nations 2010

a) A histogram (or a stemplot) is an appropriate graphical display. Both these displays are given below. The distribution is right skewed. This means the portion of large registry groups is relatively small. There are some (one or two) outliers (indicated by the gaps in the histogram or stemplot). The median size is 717. The interquartile range is 1220.

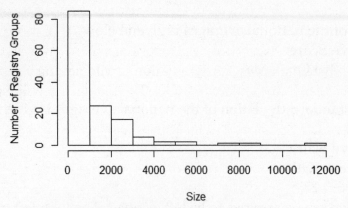

```
 0 | 01112222222233333333444444444444444
 0 | 55555555555556666666666677777777777777788888899
 1 | 000000111122222344
 1 | 56677889999
 2 | 01112233344
 2 | 5567
 3 | 0234
 3 | 8
 4 | 033
 4 |
 5 |
 5 | 57
 6 |
 6 |
 7 | 4
 7 |
 8 | 0
 8 |
 9 |
 9 |
10 |
10 |
11 | 2
   (7 | 4 means 7400)
```

The 5-number summary is:

Minimum	Q1	Median	Q3	Maximum
41	452	717	1672	11202

b) If we calculate the band sizes, most band sizes will be same as the registry group sizes since only Six Nations of the Grand River in Ontario consists of more than one registry group. This very large band, consisting of 13 registry groups, will increase considerably the mean and standard deviation, while having a much smaller effect on the more resistant median and IQR. The histogram will have a bigger gap due to this very large band.

45. Bi-lingual ni-lingual 2011

a) The histogram of the proportion of city residents who are bilingual is shown below. The highest proportion of bilinguals is in Montreal with more than 50% bilinguals, and the lowest proportion is in Brantford with less than 5% bilinguals. The median is about 8%. The distribution of the proportion of city residents who are bilingual is right skewed. This means a relatively smaller number of cities with high proportion of bilinguals and more cities with a low proportion of bilinguals. The distribution of proportion of bilinguals appears bimodal it looks like the distribution of a few cities with a large proportion of bilinguals has a distinct shape compared to the other cities. Summary statistics are given below. About 25% of the cities have less than 7% bilinguals.

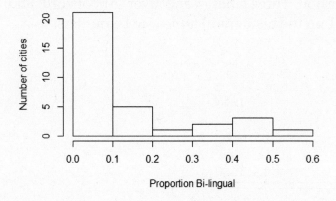

Descriptive Statistics: Proportion Bi-lingual

Mean	StDev	Min.	Q1	Median	Q3	Max.
0.153	0.145	0.042	0.066	0.077	0.173	0.539

b) The dotplot below shows the cities in Quebec and New Brunswick in red. They have relatively high proportions of bilinguals.

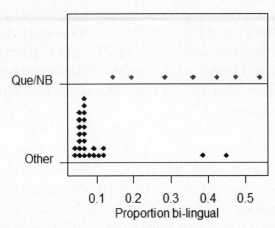

c)   The histogram of the proportion of city residents who speak neither English nor French is shown below. The highest proportion of residents who speak neither English nor French is in Vancouver with more than 5%, and the lowest proportion is in Saguenay with about 0.03%. The median is about 0.69%. The distribution of the proportion of city residents who speak neither English nor French is right skewed. This means there is a relatively smaller number of cities with a high proportion of residents who speak neither English nor French and more cities with a low proportion. Three cities (Vancouver, Abbotsford, and Toronto, the bars above 0.031 on the histogram) appear to be outliers.

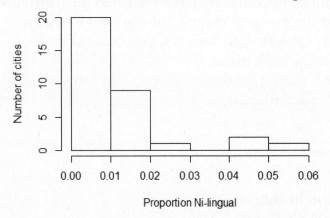

Descriptive Statistics: Proportion Ni-lingual

Mean	StDev	Minimum	Q1	Median	Q3	Maximum
0.0111	0.0135	0.0003	0.0026	0.0069	0.0144	0.0560

47. Elections 2011 again

a) The closeness of the mean and median suggests that the distribution of the percentage of voter turnout in the 2011 election is approximately symmetric.

b) The distribution of the percentage of voter turnout in the 2011 election is approximately symmetric, with mean 61.068 and standard deviation 5.772.

c) Mean + 2 Stdev = 61.068 + 2 × 5.772 = 72.612
Mean − 2 Stdev = 61.068 − 2 × 5.772 = 49.524
From the histogram there appears to be about 12.5 observations below 50 and 6.5 observations above 72.5 (assuming uniform distribution within bins). This is about 19/308 = 6.2%. Thus, about 93.8% of the observations are within 2 standard deviations from the mean. This is pretty close to 95% as we would expect according to the empirical rule for bell-shaped distributions. The exact answer is 292/308 or 94.8%.

d) The overall percentage won't be exactly the same as the mean of the percentages as the number of eligible voters is not exactly the same in each electoral district. However, since the number of eligible voters in each riding is pretty close, we should expect the overall percentage to be fairly close to 61.07%.

49. Run times continued

a) The mean will be smaller because it depends on the sum of all of the values. The standard deviation will be smaller because we have truncated the distribution at 32. The range will be smaller for the same reason. The statistics based on quartiles of the data (median, Q1, Q3, IQR) won't change because the distribution was truncated to the right of Q3.

b) The spread of the distribution will be relatively smaller because we have added observations near the middle. Thus, we expect the standard deviation and IQR to be smaller. We should only see a small change to the mean and median because the added observations are near the middle of the distribution. The range will not change.

c) The spread of the distribution will be relatively larger because we have added observations at about 2 minutes left and right of the mean. Thus, we expect the standard deviation and IQR to be bigger. Because the additional observations were added symmetrically to either side of the mean, there should be little change to the mean or median, and the range will remain unchanged.

d) Subtracting one minute from each time shifts the entire distribution to the left by one minute. The mean and median should decrease by one minute. However, there has been no change to the shape of the distribution, so the standard deviation, IQR, and range will all remain the same.

e) The values added at 35 are further from the mean than the values at 29.5, so the mean will increase. The overall spread of the distribution has increased, so the standard deviation will increase. The added observations at 35 minutes lie above Q3, so the IQR will increase. Because the same number of observations was added on either side of the median, its value won't change. The minimum and maximum values have not changed, so the range remains unchanged as well.

f) We have moved some values outside of the Q1 to Q3 interval further away from the centre of the distribution. Q1 and Q3 remain unchanged, so the IQR remains unchanged. Because the adjustment is symmetric, the mean and median are unchanged. The range remains the same because the minimum and maximum are unchanged. The standard deviation increases because the spread of the distribution has increased.

51. Grouped data

a) The class midpoints and the mean and the standard deviation of the midpoints are shown below:

Height (class midpoint)

61	61	61	61	61	61	61	61	61	61	61	61	61	61	61
61	61	64	64	64	64	64	64	64	64	64	64	64	64	64
64	64	64	64	64	64	64	64	64	64	64	64	64	64	64
64	64	64	64	67	67	67	67	67	67	67	67	67	67	67
67	67	67	67	67	67	67	67	67	67	67	67	67	67	67
67	67	67	67	67	67	67	67	67	67	67	70	70	70	70
70	70	70	70	70	70	70	70	70	70	70	70	70	70	70
70	70	70	70	70	73	73	73	73	73	73	73	73	73	73
73	73	73	73	73	76	76	76	76	76					

Descriptive Statistics: Height (class midpoint)

Variable	N	Mean	StDev
Height (class mid)	130	67.069	3.996

The actual heights and the mean and the standard deviation of the actual values are given below. The mean and the standard deviation of the actual data are very close to those calculated using the midpoints. For grouped data, we assume that all the values in a class are equal to the midpoint. This assumption is usually reasonable unless the class width is big.

Height (actual)

60	60	61	61	61	61	61	61	62	62	62	62	62	62	62
62	62	63	63	63	63	63	63	63	64	64	64	64	64	65
65	65	65	65	65	65	65	65	65	65	65	65	65	65	65
65	65	65	65	66	66	66	66	66	66	66	66	66	66	66
66	66	66	66	66	66	66	67	67	67	67	67	67	67	68
68	68	68	68	68	68	68	68	68	68	68	69	69	69	69
69	70	70	70	70	70	70	70	70	70	70	70	71	71	71
71	71	71	71	71	72	72	72	72	72	72	72	72	72	73
73	73	73	74	74	75	75	75	75	76					

Descriptive Statistics: Height (actual)

Variable	N	Mean	StDev
Height (actual)	130	67.115	3.792

b)

$$\text{Mean} = \frac{\sum_i f_i m_i}{\sum_i f_i}, \quad Variance = \frac{\sum_i f_i(m_i - \bar{X})^2}{\sum_i f_i - 1}$$

53. Unequal bin widths

a) The last wide bin now has a rather big rectangle above it, with height equal to about 13. The number of run times exceeding 32.5 appears to have grown!

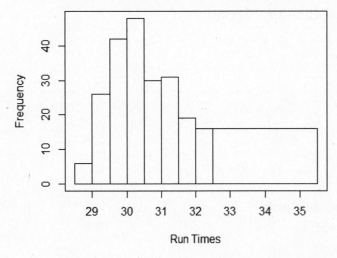

b) Now the rectangle height over the last bin equals the average of the 6 rectangle heights in the original histogram, i.e. the area of this rectangle equals the total area of the 6 rectangles it replaced. We have been true to the Area Principle, and kept area proportional to frequency. (Since area is height multiplied by width, the rectangle height must be proportional to frequency divided by bin width)

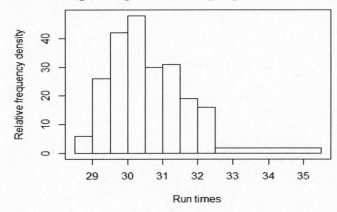

55. Rock concert accidents

a) The histogram and boxplot of the distribution of "crowd crush" victims' ages both show that a typical crowd crush victim was approximately 18–20 years of age, that the range of ages is 36 years, and that there are two outliers, one victim at age 36–38 and another victim at age 46–48. In addition, both show some right-skewness.

b) This histogram shows that there may have been two modes in the distribution of ages of "crowd crush" victims, one at 18–20 years of age and another at 22–24 years of age. Boxplots, in general, can show symmetry and skewness, but not features of shape like bimodality or uniformity.

c) Median is the better measure of centre, since the distribution of ages is right skewed and has outliers. Median is more resistant to outliers than the mean.

d) IQR is a better measure of spread, since the distribution of ages has outliers. IQR is more resistant to outliers than the standard deviation.

Chapter 4: Understanding and Comparing Distributions

1. **In the news** Answers will vary.

3. **Graduation?**
   a) The distribution of the percent of incoming freshman who graduate on time is roughly symmetric. The mean and the median are reasonably close to one another, and the quartiles are approximately the same distance from the median.
   b) Upper Fence: $Q3 + 1.5(IQR) = 74.75 + 1.5(74.75 - 59.15)$

   $$= 74.75 + 23.4$$

   $$= 98.15$$

   Lower Fence: $Q1 - 1.5(IQR) = 59.15 - 1.5(74.75 - 59.15)$

   $$= 59.15 - 23.4$$

   $$= 35.75$$

   Since the maximum value of the distribution of the percent of incoming freshmen who graduate on time is 87.4% and the upper fence is 98.15%, there are no high outliers. Likewise, since the minimum is 43.2% and the lower fence is 35.75%, there are no low outliers. Since the minimum and maximum percentages are within the fences, all percentages must be within the fences.

   c) A boxplot of the distribution of the percent of incoming freshmen who graduate on time is at the right.

   d) The distribution of the percent of incoming freshmen who graduate on time is roughly symmetric, with mean of approximately 68% of freshmen graduating on time. Universities surveyed had between 43.2% and 87.4% of students graduating on time, with the middle 50% of universities reporting between 59.15% and 74.75% graduating on time.

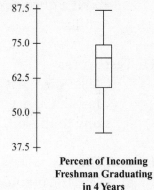

Percent of Incoming Freshman Graduating in 4 Years

5. **Dots to boxes**

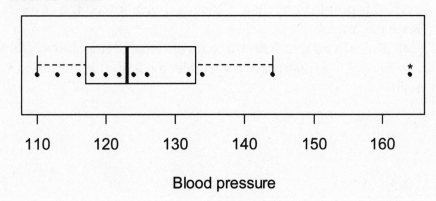

Blood pressure

7. **Hospital stays**
    a) The histograms of male and female hospital stay durations would be easier to compare if they were constructed with the same scale, perhaps from 0 to 20 days.
    b) The distribution of hospital stays for men is skewed to the right, with many men having very short stays of about 1 or 2 days. The distribution tapers off to a maximum stay of approximately 25 days. The distribution of hospital stays for women is skewed to the right, with a mode at approximately 5 days, and tapering off to a maximum stay of approximately 22 days. Typically, hospital stays for women are longer than those for men.
    c) The peak in the distribution of women's hospital stays can be explained by childbirth. This time in the hospital increases the length of a typical stay for women, and not for men.

9. **Women's basketball**
    a) Both women have a median score of about 17 points per game, but Yan is much more consistent. Her IQR is about 2 points, while Bini's is over 10.
    b) If the coach wants a consistent performer, she should take Yan. She'll almost certainly deliver somewhere between 15 and 20 points. But, if she wants to take a chance and needs a "big game," she should take Bini. Bini scores over 24 points about a quarter of the time. On the other hand, she scores under 11 points about as often.

11. **Test scores**
    Class A is Class 1. The median is 60, but has less spread than Class B, which is Class 2. Class C is Class 3, since its median is higher, which corresponds to the skew to the left.

13. **Caffeine**
    a) *Who:* 45 volunteers. *What:* Level of caffeine consumption and memory test score. *When:* Not specified. *Where:* Not specified. *Why:* The student researchers want to see the possible effects of caffeine on memory. *How:* It appears that the researchers imposed the treatment of level of caffeine consumption in an experiment. However, this point is not clear. Perhaps they allowed the subjects to choose their own level of caffeine.
    b) Caffeine level is a categorical variable with three levels: no caffeine, low caffeine, and high caffeine. Test score is a quantitative variable, measured in number of items recalled correctly.

c)

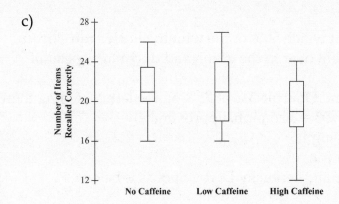

d) The groups consuming no caffeine and low caffeine had comparable memory test scores. A typical score from these groups was around 21. However, the scores of the group consuming no caffeine were more consistent, with a smaller range and smaller interquartile range than the scores of the group consuming low caffeine. The group consuming high caffeine had lower memory scores in general, with a median score of about 19. No one in the high caffeine group scored above 24, but 25% of each of the other groups scored above 24.

15. Population growth 2010

a) Comparative boxplots below:

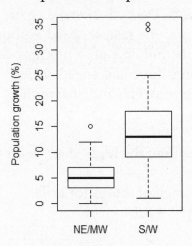

b) The distribution of population growth in NE/MW states is unimodal, symmetric, and tightly clustered around 5% growth. The distribution of population growth in S/W states is much more spread out, with most states having population growth between 5% and 25%. A typical state had about 15% growth. There were two outliers, with 34% and 35% growth, respectively. Generally, the growth rates in the S/W states were higher and more variable than the rates in the NE/MW states.

17. Derby speeds 2011

   a) The median speed is the speed at which 50% of the winning horses ran slower. Find 50% on the left, move straight over to the graph and down to a speed of about 36 mph.

   b) Quartile 1 is at 25% on the left and Quartile 3 is at 75% on the left. Matching these to the ogive, Q1 = 34.5 mph and Q3 = 36.5 mph, approximately.

   c) Range = Max – Min = 38 – 31 = 7 mph
      IQR = Q3 – Q1 = 36.5 – 34.5 = 2 mph

   d) An approximate boxplot of winning Kentucky Derby Speeds is below:

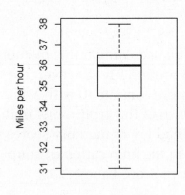

Derby speeds

   e) The distribution of winning speeds in the Kentucky Derby is skewed to the left. The lowest winning speed is just under 31 mph and the fastest speed is about 38 mph. The median speed is approximately 36 mph, and 75% of winning speeds are above 34.5 mph. Only a few percent of winners have had speeds below 33 mph. The middle 50% of winning speeds are between 34.5 and 36.5 mph.

19. Reading scores

   a) The highest score for boys was 6, which is higher than the highest score for girls, 5.9.

   b) The range of scores for boys is greater than the range of scores for girls.
      Range = Max – Min, Range(Boys) = 4, Range(Girls) = 3.1

   c) The girls had the greater IQR.
      IQR = Q3 – Q1, IQR(Boys) = 4.9 – 3.9 = 1, IQR(Girls) = 5.2 – 3.8 = 1.4

   d) The distribution of boys' scores is more skewed. The quartiles are not the same distance from the median. In the distribution of girls' scores, Q1 is 0.7 units below the median, while Q3 is 0.7 units above the median.

   e) Overall, the girls did better on the reading test. The median, 4.5, was higher than the median for the boys, 4.3. Additionally, the upper quartile score was higher for girls than boys, 5.2 compared to 4.9. The girls' lower quartile score was slightly lower than the boys' lower quartile score, 3.8 compared to 3.9.

   f) The overall mean is calculated by weighting each mean by the number of students. 14 (4.2) + 11 (4.6)/25 = 4.38

21. PISA Math by gender 2009

a) The median for males is slightly higher than that for females. There are two low outliers in each boxplot. The median of the differences between the mean scores for males and females is about 12. All but one country has the mean for males greater than that for females. Only one country has a mean score for females greater than that for males (Sweden).

b) In the boxplots below, Canada, USA, and Mexico are indicated by the symbols 'x', 'U', 'M', respectively. In the boxplots for males, Israel, Turkey, Chile, and Mexico are outliers because their mean scores are unusually low. In the boxplot for females, Mexico and Chile are outliers because their mean scores are unusually low.

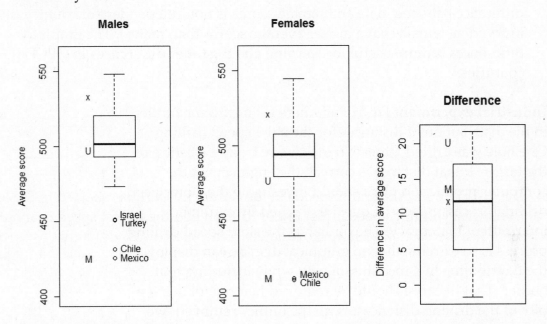

c) The 5-number summaries for the male scores, female scores, and differences are given below.

Summary statistics:

Column	Min	Q1	Median	Q3	Max
Male	425.4	492.9	501.3	520.4	547.8
Female	410.4	479.2	494.1	507.5	544.5
Difference	–1.75	5.00	11.8	17.1	21.8

For the female average scores,
Q1 = 479.2
Q3 = 507.5
IQR = Q3– Q1 = 28.3
Q1 – 1.5*IQR = 436.8
The average female score for Mexico is 411.8, which is less than Q1 – 1.5 * IQR = 436.8 and is thus an outlier.

d) The separate boxplots for males and females show that some countries have particularly low average scores (the outliers Israel, Chile, Turkey, and Mexico) compared with the average scores from the rest of the countries. The boxplot for differences shows that although the average female scores are lower, the difference between male and female scores is not independent of country (in fact, in Sweden, females have higher average scores than males). The graph for differences is more useful for learning about gender differences in OECD countries.

23. Industrial experiment First of all, there is an extreme outlier in the distribution of distances for the slow speed drilling. One hole was drilled almost an inch away from the centre of the target! If that distance is correct, the engineers at the computer production plant should investigate the slow speed drilling process closely. It may be plagued by extreme, intermittent inaccuracy. The outlier in the slow speed drilling process is so extreme that no graphical display can display the distribution in a meaningful way while including that outlier. That distance should be removed before looking at a plot of the drilling distances. With the outlier removed, we can determine that the slow drilling process is more accurate.

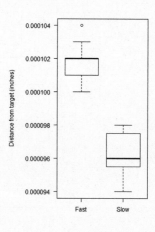

The greatest distance from the target for the slow drilling process, 0.000098 inches, is still more accurate than the smallest distance for the fast drilling process, 0.000100 inches.

25. Crime comparisons
a) The side-by-side boxplots for Canada and the U.S. are given below. Homicide, aggravated assaults, and robberies are higher in the U.S.. Break and enter is higher in Canada. Theft and motor vehicle theft is about the same in Canada and the U.S.. The pattern appears to be similar in the two countries, for example, homicide is the least frequent among the six types in both countries and theft is the most frequent.

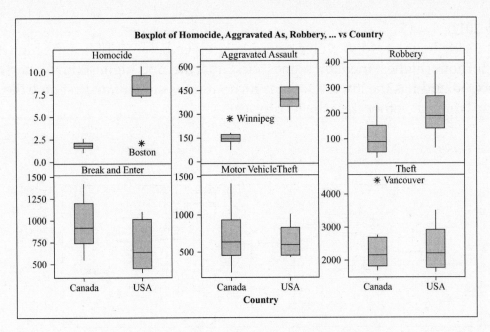

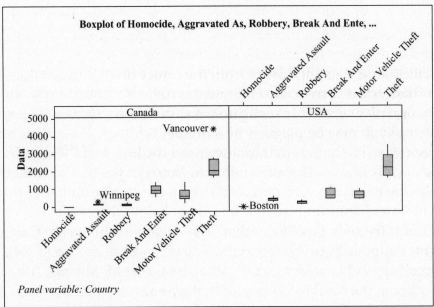

b) Winnipeg has aggravated assaults in the U.S. range, and Boston has homicides in the Canadian range. Vancouver is a high outlier in Theft.

c) Population in the city and poverty can be lurking variables.

27. Elections 2011

a) Side-by-side boxplots for % voter turnout are given below. PEI has the highest voter turnout (highest median, highest first quartile, and highest third quartile). Newfoundland has the lowest % voter turnout (lowest median, first quartile, and the third quartile, other than the territories).

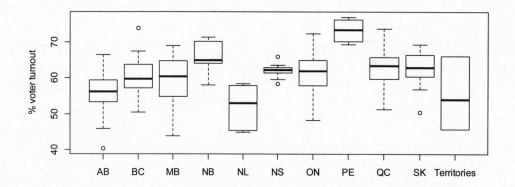

b) The low outlier in Alberta would be an outlier in any province or territory. The rest of the outliers are outliers within their own province or territories, but they overlap with the distributions of voter turnout in other provinces so in some sense they are not outliers compared to all Canadian ridings. Also, some appear to barely have exceeded the 1.5 IQR distance from the box, and those in Nova Scotia hardly appear unusually far from the median, but the box is strangely narrow, likely due to chance variation considering the small number of ridings.

c) The boxplot of Canadian ridings (overall) is given below. There are outliers: Cardigan in PEI is the only positive outlier. The negative outliers are Calgary East, Nunavut, Random-Burin-St. George's, Bonavista-Gander-Grand Falls-Windsor, Churchill, and Fort McMurray-Athabasca. Fort McMurray-Athabasca is the only outlier in the combined boxplot that was also an outlier within its province/territory. Cardigan is a country-level outlier but is not an outlier within PEI because of the overall high voter turnout in PEI. With the exception of Calgary East, the negative outliers all seem to be in relatively remote areas of the country, which could account for the low turnout.

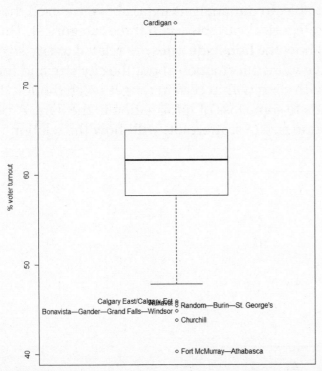

d) Manitoba has a left-skewed distribution (the median is closer to the third quartile than to the first and the upper whisker is shorter compared to the lower whisker), therefore the mean must be smaller than the median. BC has a right-skewed distribution (the median is closer to the first quartile than to the third, and the most extreme value is larger than the median), therefore the mean must be larger than the median.

29. Homicides 2011

a) The side-by-side boxplots of homicide rates are shown below. Large cities have the highest median homicide rate. Medium-sized cities have the least variable homicide rate. There are two large outliers.

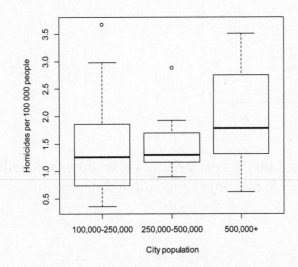

b) City size is a quantitative variable. In our analysis in (a) above we used the variable city size to create a categorical variable (with three categories). This helps us get some idea about how the homicide rates are related to city size, but the actual city size values have more information about the city size and forming categories combining cities with sizes within certain ranges (even though the actual sizes are different) leads to some loss of information in the data. A plot of homicide rate versus population size (a scatterplot) will show the relation between the two variables.

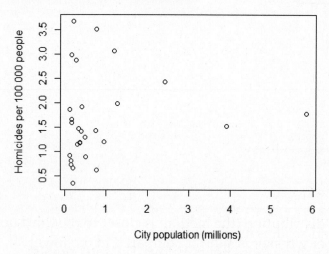

c) The side-by-side boxplots of homicide rates are shown below. Quebec has the lowest homicide rate and the West has the highest.

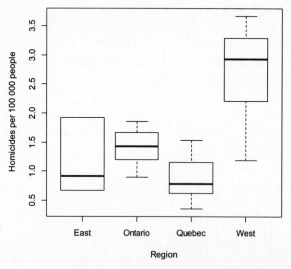

d) If we have the annual data for the 11-year period, we could examine the variation or change in homicide rate over time with a timeplot.

31. He shoots he scores the hat-trick

a) **Descriptive Statistics: points scored**

Variable	Minimum	Q1	Median	Q3	Maximum
points scored	48.0	98.3	145.5	192.8	215.0

b)

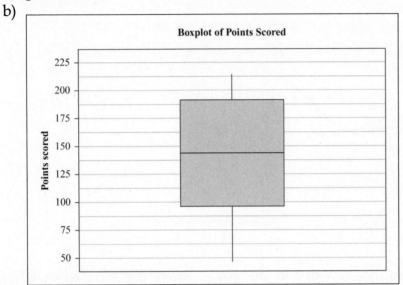

c) 48 > Q1 – 1.5 IQR = –43.45 and so 48 is not an outlier.

d) The plot of the total points versus year is shown below. The total points decreases with time. The boxplot does not show the relationship with time.

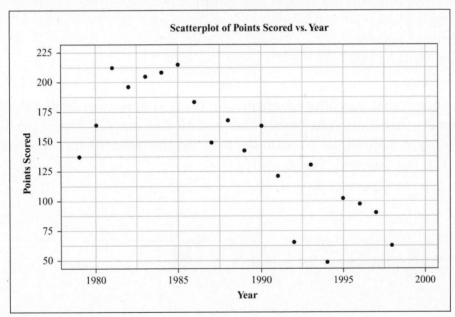

33. Hurricanes 2010 again The smoothed timeplot is below. It appears that the number of hurricanes per year had a lull during the period from the mid-1960s to about 1990, and has increased fairly rapidly since. The number of hurricanes is at least as high now as it was at the beginning of the time period. This information was obscured in the previous analysis using only a dotplot.

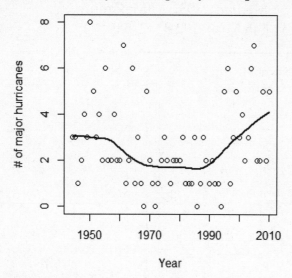

35. Tsunami 2013

a) The distribution is unimodal with a mode near 7. The results are slightly left skewed with a few small observations near 3.

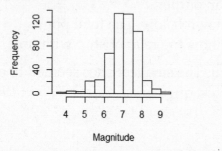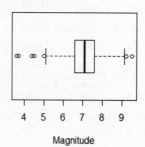

b) It doesn't appear that magnitude depends on time. That is, over time it doesn't appear that the distribution has really changed all that much. There is perhaps a slight increase in magnitudes. We do see that the positive outliers occurred either early in the time series or recently.

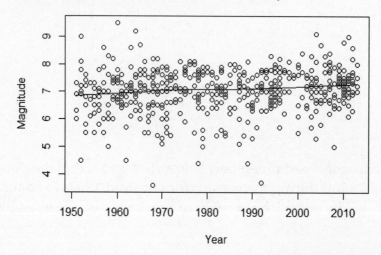

c) There are several countries with only 1 observation and thus only appear as a single point. There are also too many countries. We could group countries together in similar regions.

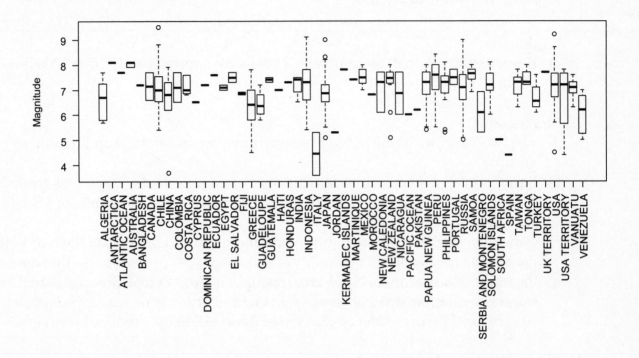

d) The quakes seemed to be the biggest in Peru. The smallest ones seemed to occur in Greece. The three countries with the biggest quakes were Chile (9.5 in 1960), USA (9.2 in 1964) and Indonesia (9.1 in 2004). The smallest were in Greece (4.5 in 1962) and USA (4.5 in 1952).

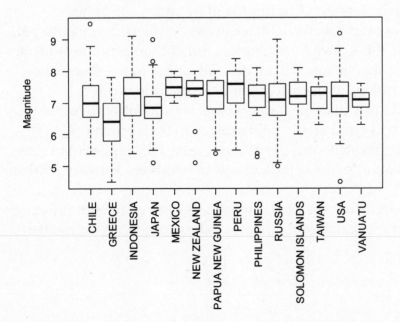

37. Assets again

a) The distribution of logarithm of assets is preferable, because it is roughly unimodal and symmetric. The distribution of the square root of assets is still skewed right, with outliers.

b) If $\sqrt{Assets} = 50$, then the company's assets are approximately $50^2 = 2500$ million dollars.

c) If $\log(Assets) = 3$, then the company's assets are approximately $10^3 = 1000$ million dollars.

39. Stereograms

a) The two variables discussed in the description are fusion time and treatment group.

b) Fusion time is a quantitative variable, measured in seconds. Treatment group is a categorical variable, with subjects either receiving verbal clues only, or visual and verbal clues.

c) Generally, the Visual/Verbal group had shorter fusion times than the No/Verbal group. The median for the Visual/Verbal group was approximately the same as the lower quartile for the No/Verbal group. The No/Verbal Group also had an extreme outlier, with at least one subject whose fusion time was approximately 50 seconds. There is evidence that visual information may reduce fusion time.

41. First Nations 2010 by province

a) The size of the largest band in Ontario is about 24 000. The 4th largest is about 7500. The median band size is slightly higher in Saskatchewan, but it is hard to read the difference. Boxplots do not show the mean. Both the means must be greater than the median since distributions are right skewed (and there are large outliers).

b) The readability of the boxplots could be improved by adding gridlines.

c) Both the distributions are right skewed. Many suspect outliers may just be part of the long upper tail of this skewed distribution, rather than genuine outliers.

d) Median and IQR do not change much since deleting one extreme observation does not change the positions of the median and the quartile much.

e) The mean band size is bigger than the median band size in both provinces because the distribution of band size is right skewed in both provinces.

f) Seventy-five percent of Saskatchewan bands are at least 1500 (approx.) in size (the first quartile). The middle half of Saskatchewan bands are between 1500 and 2500 (approx.) in size.

g) The distribution of the transformed sizes is much less skewed in both provinces. The median band size in Saskatchewan appears to be about twice the median band size in Ontario.

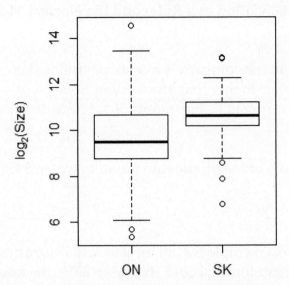

h) Six Nations is still an outlier, but it is much less extreme (less unusual) than it was before the transformation. There are additional outliers, most notably the Lucky Man band (size = 110) in North Central Saskatchewan.

Chapter 5: The Standard Deviation as a Ruler and the Normal Model

1. **Payroll**
   a) The distribution of salaries in the company's weekly payroll is skewed to the right. The mean salary, $700, is higher than the median, $500.
   b) The IQR, $600, measures the spread of the middle 50% of the distribution of salaries.

   $$Q3 - Q1 = IQR$$
   $$Q3 = Q1 + IQR$$
   $$Q3 = \$350 + \$600$$
   $$Q3 = \$950$$

   50% of the salaries are found between $350 and $950.

   c) If a $50 raise were given to each employee, all measures of centre or position would increase by $50. The minimum would change to $350, the mean would change to $750, the median would change to $550, and the first quartile would change to $400. Measures of spread would not change. The entire distribution is simply shifted up $50. The range would remain at $1200, the IQR would remain at $600, and the standard deviation would remain at $400.
   d) If a 10% raise were given to each employee, all measures of centre, position, and spread would increase by 10%.
      Minimum = $330   Mean = $770   Median = $550   Range = $1320
      IQR = $660   First Quartile = $385   St. Dev. = $440

3. **SAT or ACT?** Measures of centre and position (lowest score, top 25% above, mean, and median) will be multiplied by 40 and increased by 150 in the conversion from ACT to SAT by the rule of thumb. Measures of spread (standard deviation and IQR) will only be affected by the multiplication.
   Lowest score = 910   Mean = 1230   Standard deviation = 120
   Top 25% above = 1350   Median = 1270   IQR = 240

5. **Temperatures** In January, with mean temperature 2°C and standard deviation in temperature 6°C, a high temperature of 13°C is less than 2 standard deviations above the mean. In July, with mean temperature 24°C and standard deviation 5°C, a high temperature of 13° is more than two standard deviations below the mean. A high temperature of 13° is less likely to happen in July, when 13°C is farther away from the mean.

7. **Final exams**
   a) Anna's average is (83 + 83)/2 = 83. Megan's average is (77 + 95)/2 = 86. Only Megan qualifies for language honours, with an average higher than 85.
   b) On the French exam, the mean was 81 and the standard deviation was 5. Anna's score of 83 was 2 points, or 0.4 standard deviations, above the mean. Megan's score of 77 was 4 points, or 0.8 standard deviations below the mean.

On the Spanish exam, the mean was 74 and the standard deviation was 15. Anna's score of 83 was 9 points, or 0.6 standard deviations, above the mean. Megan's score of 95 was 21 points, or 1.4 standard deviations, above the mean.

Measuring their performance in standard deviations is the only fair way in which to compare the performance of the two women on the test.

Anna scored 0.4 standard deviations above the mean in French and 0.6 standard deviations above the mean in Spanish, for a total of 1.0 standard deviation above the mean.

Megan scored 0.8 standard deviations below the mean in French and 1.4 standard deviations above the mean in Spanish, for a total of only 0.6 standard deviations above the mean.

Anna did better overall, but Megan had the higher average. This is because Megan did very well on the test with the higher standard deviation, where it was comparatively easy to do well.

9. **Cattle**
   a) A steer weighing 1000 pounds would be about 1.81 standard deviations below the mean weight. $z = \dfrac{y - \mu}{\sigma} = \dfrac{1000 - 1152}{84} \approx -1.81$
   b) A steer weighing 1000 pounds is more unusual. Its z-score of –1.81 is further from 0 than the 1250 pound steer's z-score of 1.17.

11. **More cattle**
   a) The new mean would be 1152 – 1000 = 152 pounds. The standard deviation would not be affected by subtracting 1000 pounds from each weight. It would still be 84 pounds.
   b) The mean selling price of the cattle would be 0.40(1152) = $460.80. The standard deviation of the selling prices would be 0.40(84) = $33.60.

13. **Cattle, part III** Generally, the minimum and the median would be affected by the multiplication and subtraction. The standard deviation and the IQR would only be affected by the multiplication.
   Minimum = 0.40(980) – 20 = $372.00   Median = 0.40(1140) – 20 = $436
   Standard deviation = 0.40(84) = $33.60   IQR = 0.40(102) = $40.80

15. **Professors** The standard deviation of the distribution of years of teaching experience for college professors must be 6 years. College professors can have between 0 and 40 (or possibly 50) years of experience. A workable standard deviation would cover most of that range of values with ±3 standard deviations around the mean. If the standard deviation were 6 months ($\frac{1}{2}$ year), some professors would have years of experience 10 or 20 standard deviations away from the mean, whatever it is. That isn't possible. If the standard deviation were 16 years, ±2 standard deviations would be a range of 64 years. That's way too high. The only

reasonable choice is a standard deviation of 6 years in the distribution of years of experience.

17. **Trees**
    a) The Normal model for the distribution of tree diameters is below.

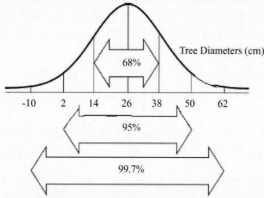

    b) Approximately 95% of the trees are expected to have diameters between 2 cm and 50 cm.
    c) Approximately 2.5% of the trees are expected to have diameters less than 2 cm.
    d) Approximately 34% of the trees are expected to have diameters between 14 inches and 26 cm
    e) Approximately 16% of the trees are expected to have diameters over 38 cm.

19. **Trees, part II** The use of the Normal model requires a distribution that is unimodal and symmetric. The distribution of tree diameters is neither unimodal nor symmetric, so use of the Normal model is not appropriate.

21. **Winter Olympics 2010 downhill**
    a) The 2010 Winter Olympics downhill times have a mean of 117.34 seconds and standard deviation of 2.456 seconds. 114.875 seconds is one standard deviation below the mean. If the Normal model is appropriate, 16% of the times should be below 99.7 seconds.
    b) 8 out of 59 times (13.45%) are below 114.875 seconds.
    c) The percentages in parts a) and b) do not agree more closely because the Normal model is not appropriate in this situation.
    d) The histogram of 2010 Winter Olympic Downhill times is below. The Normal model is not appropriate for the distribution of times because the distribution is not symmetric.

2010 Men's Alpine Downhill

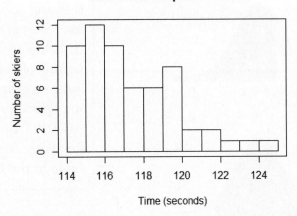

23. TV watching

a) Approximately 16% of the university students are expected to watch less than 1 standard deviation below the mean number of hours of TV.

b) The distribution of the number of hours of TV watched per week has mean 3.66 hours and standard deviation 4.93 hours. According to the Normal model, students who watch fewer than 1 standard deviation below the mean number of hours of TV are expected to watch less than –1.27 hours of TV per week. Of course, it is impossible to watch less than 0 hours of TV, let alone less than –1.27 hours.

c) The distribution of the number of hours of TV watched per week by university students is skewed heavily to the right. Use of the Normal model is not appropriate for this distribution, since it is not unimodal and symmetric.

25. Normal models

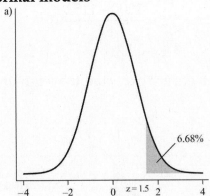

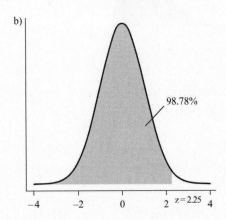

27.   More Normal models

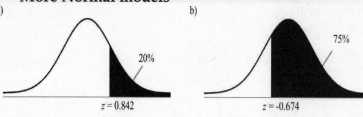

Note that z values in this and future exercises often are to three decimal places (from software), and hence the answers may differ very slightly from answers using the z-table in the textbook.

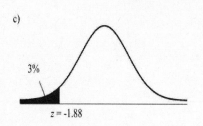

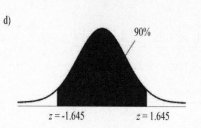

c)

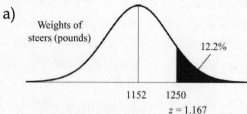

d)

29. Normal cattle

a)

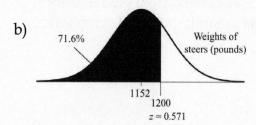

$$z = \frac{y - \mu}{\sigma}$$

$$z = \frac{1250 - 1152}{84}$$

$$z \approx 1.167$$

According to the Normal model, 12.2% of steers are expected to weigh over 1250 pounds.

b)

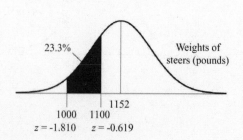

$$z = \frac{y - \mu}{\sigma}$$

$$z = \frac{1200 - 1152}{84}$$

$$z \approx 0.571$$

According to the Normal model, 71.6% of steers are expected to weigh under 1200 pounds.

c)

$$z = \frac{y - \mu}{\sigma}$$

$$z = \frac{1000 - 1152}{84}$$

$$z \approx -1.810$$

$$z = \frac{y - \mu}{\sigma}$$

$$z = \frac{1100 - 1152}{84}$$

$$z \approx -0.619$$

According to the Normal model, 23.3% of steers are expected to weigh between 1000 and 1100 pounds.

31.   **Cattle, finis**

a)

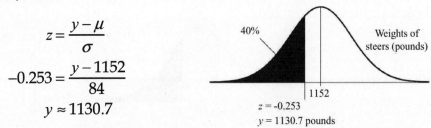

$$z = \frac{y - \mu}{\sigma}$$

$$-0.253 = \frac{y - 1152}{84}$$

$$y \approx 1130.7$$

According to the Normal model, the weight at the 40th percentile is 1130.7 pounds. This means that 40% of steers are expected to weigh less than 1130.7 pounds.

b)

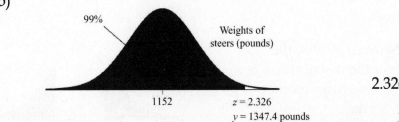

$$z = \frac{y - \mu}{\sigma}$$

$$2.326 = \frac{y - 1152}{84}$$

$$y \approx 1347.4$$

According to the Normal model, the weight at the 99th percentile is 1347.4 pounds. This means that 99% of steers are expected to weigh less than 1347.4 pounds.

c)

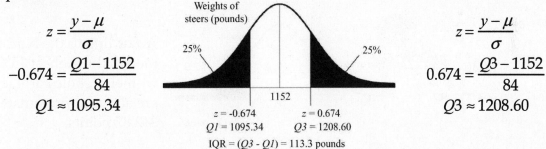

$$z = \frac{y - \mu}{\sigma}$$

$$-0.674 = \frac{Q1 - 1152}{84}$$

$$Q1 \approx 1095.34$$

$$z = \frac{y - \mu}{\sigma}$$

$$0.674 = \frac{Q3 - 1152}{84}$$

$$Q3 \approx 1208.60$$

According to the Normal model, the IQR of the distribution of weights of Angus steers is about 113.3 pounds.

33. Cholesterol

a) The Normal model for cholesterol levels of adult American women is at the right.

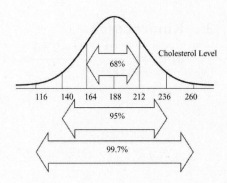

b) According to the Normal model, 30.85% of American women are expected to have cholesterol levels over 200.

$$z = \frac{y - \mu}{\sigma}$$

$$z = \frac{200 - 188}{24}$$

$$z = 0.5$$

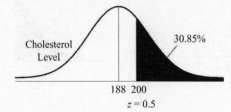

c) According to the Normal model, 17.00% of American women are expected to have cholesterol levels between 150 and 170.

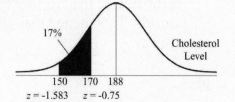

d) According to the Normal model, the interquartile range of the distribution of cholesterol levels of American women is approximately 32.38 points.

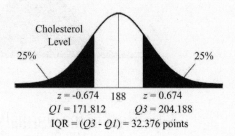

e) $$z = \frac{y - \mu}{\sigma}$$

$$1.036 = \frac{y - 188}{24}$$

$$y = 212.87$$

According to the Normal model, the highest 15% of women's cholesterol levels are above approximately 212.87 points.

35. Kindergarten

a) $$z = \frac{y - \mu}{\sigma} = \frac{91 - 97.0}{4.9} = -1.22$$

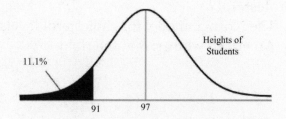

According to the Normal model, approximately 11.1% of kindergarten kids are expected to be less than 91 cm tall.

b) $z = \dfrac{y - \mu}{\sigma} \Rightarrow \dfrac{y_1 - 97}{4.9} = -1.282 \Rightarrow y = 90.72$ and

$z = \dfrac{y - \mu}{\sigma} \Rightarrow \dfrac{y_2 - 97}{4.9} = 1.282 \Rightarrow y = 103.28$

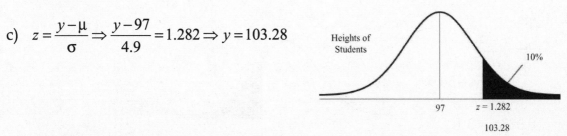

According to the Normal model, the middle 80% of kindergarten kids are expected to be between 90.72 and 103.28 cm tall. (The appropriate values of $z = \pm 1.282$ are found by using right and left tail percentages of 10% of the Normal model.)

c) $z = \dfrac{y - \mu}{\sigma} \Rightarrow \dfrac{y - 97}{4.9} = 1.282 \Rightarrow y = 103.28$

According to the Normal model, the tallest 10% of kindergarteners are expected to be at least 103.28 cm tall.

37. Undercover?
Assuming that heights are nearly Normal, we can model the heights of Greek males with $N(167.8, \ 7)$. According to the model, about 1% of Greek men are expected to be taller than the average Dutch man. He wouldn't stand out radically, but he'd probably have a hard time keeping a "low profile"!

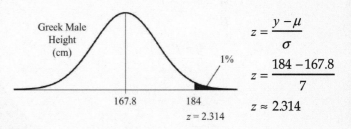

39. Helmet sizes
a) The smallest size is 56.0 – 2.96 × 1.8 = 50.672 cm and the biggest is 56.0 + 2.96 × 1.8 = 61.328 cm. (NOTE: It's okay to use 3 instead of 2.96 in these calculations.)
b) In order to fit 60% of the heads, the helmet should be in the range of 56.0 – 0.84 × 1.8 = 54.488 cm and 56.0 + 0.84 × 1.8 = 57.512 cm.
c) z values for 51 cm and 54 cm are (51 – 56)/1.8 = –2.78 and (54 – 56)/1.8 = –1.11 respectively. The proportion between for 51 cm and 54 cm is 0.1335 – 0.0027 = 0.1308 = 13.08%.
d) The smallest of the biggest 15% of heads is 56.0 + 1.04 × 1.8 = 57.872 cm.
e) Circumferences less than or equal to = 56 – 2.05 × 1.8 = 52.31 cm.

41. **TOEFL scores again**
    a) Note that 600 = 540 + 60, i.e., the score 600 is one standard deviation above the mean. Using the 68–95–99.7 rule, the proportion of students scoring above 600 is 16% (i.e., (100 – 68)/2).
    b) If the distribution is normal, the 10th and the 90th percentile are 540 – 1.28 × 60 = 463.2 and 540 + 1.28 × 60 = 616.8, not very different from the given values.
    c) For a normal distribution with mean 540 and standard deviation 60, the lower and the upper quartiles are 640 – 0.67 × 60 = 499.8 and 640 + 0.67 × 60 = 580.2 and the IQR is 580.2 – 499.8 = 80.4. This is not that far from the given value (i.e., 84).
    d) The z-value is (677 – 540)/60 = 2.28 and so the proportion scoring above 667 = 1 – 0.9887 = 0.0113 = 1.13%.

43. **Eggs**

    a) $z = \dfrac{y - \mu}{\sigma}$

    $0.583 = \dfrac{54 - 50.9}{\sigma}$

    $0.583\sigma = 3.1$

    $\sigma = 5.317$

    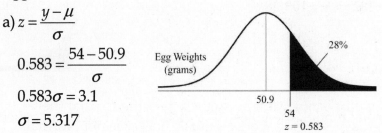

    According to the Normal model, the standard deviation of the egg weights for young hens is expected to be 5.3 grams.

    b) $z = \dfrac{y - \mu}{\sigma}$

    $-2.054 = \dfrac{54 - 67.1}{\sigma}$

    $-2.054\sigma = -13.1$

    $\sigma = 6.377$

    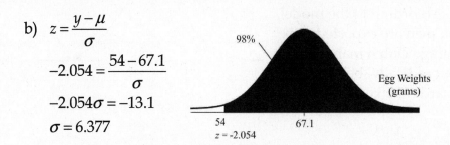

    According to the Normal model, the standard deviation of the egg weights for older hens is expected to be 6.4 grams.

    c) The younger hens lay eggs that have more consistent weights than the eggs laid by the older hens. The standard deviation of the weights of eggs laid by the younger hens is lower than the standard deviation of the weights of eggs laid by the older hens.

d) A good way to visualize the solution to this problem is to look at the distance between 54 and 70 grams in two different scales. First, 54 and 70 grams are 16 grams apart. When measured in standard deviations, the respective z-scores, –1.405 and 1.175, are 2.580 standard deviations apart. So, 16 grams must be the same as 2.580 standard deviations.

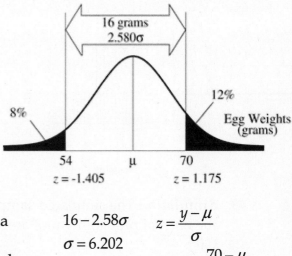

According to the Normal model, the mean weight of the eggs is 62.7 grams, with a standard deviation of 6.2 grams. (Note that the first equation above could also have simply stated that 54 standardized equals –1.405, very similar to the second equation above. Now, solve your two equations in two unknowns.)

$$16 - 2.58\sigma$$
$$\sigma = 6.202$$

$$z = \frac{y - \mu}{\sigma}$$

$$1.175 = \frac{70 - \mu}{6.202}$$
$$\mu = 62.713$$

45. **Music library**
   a) The Normal probability plot is not straight, so there is evidence that the distribution of the lengths of songs in Corey's music library is not Normal.
   b) The distribution of the lengths of songs in Corey's music library appears to be skewed to the right. The Normal probability plot shows that the longer songs in Corey's library are much longer than the lengths predicted by the Normal model. The song lengths are much longer than their scores would predict for a Normal model.

47. **Tsunamis 2013 again** The histogram, boxplot, and NPP are shown below. All three plots show several low outliers. Apart from these outliers, the histogram looks close to symmetric (maybe slightly left skewed). The boxplot has similar information. Other than the low outliers, the median line is close to the middle of the two quartiles and the two whiskers are of approximately equal length.

   The normal probability plot shows a slight curvature (to the right). The many low outliers make the distribution look slightly left skewed.

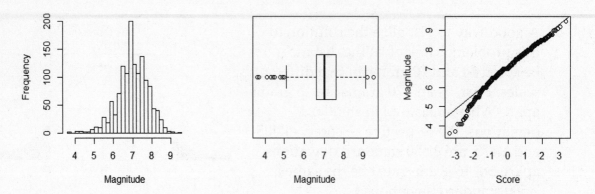

49. **Simulation** The simulated sample from the normal distribution, the histogram, and the normal probability plot are shown below:

C1

–1.75644	0.77867	–1.07557	–0.44651	1.29822	–1.47837	–1.00798
–0.32509	0.58245	–1.92505	0.66664	–0.72072	1.54382	0.18497
0.14243	2.06861	0.46728	–2.43506	–0.55715	–1.74849	0.18366
--0.29445	–1.95336	–0.75296	–0.67982	–0.48151	–0.49129	1.84136
–1.09386	–1.17643					

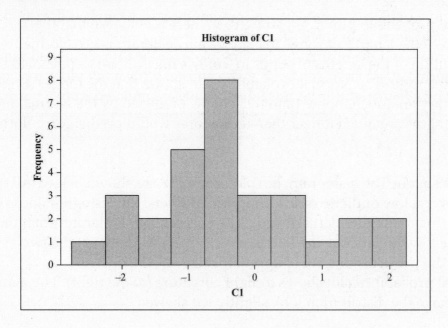

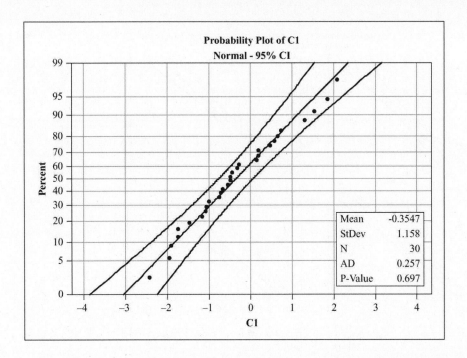

a) The histogram is close to a bell shape and the normal probability plot is close to a straight line. Usually it is easier to judge straightness. It can be difficult to judge *bell-shapedness* with small samples.

b) The histograms and the normal probability plots for the four simulated samples from the normal (0, 1) are given below. The normal probability plots are close to straight lines. We expect this because the data were simulated from a normal distribution. The shapes of the histograms vary from sample to sample even though the data are from a normal distribution. So it is easier to see the essential pattern through the noise or random jitter in the NPP.

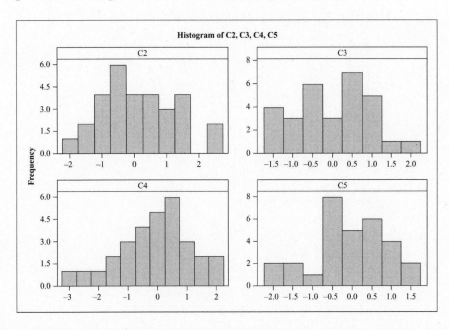

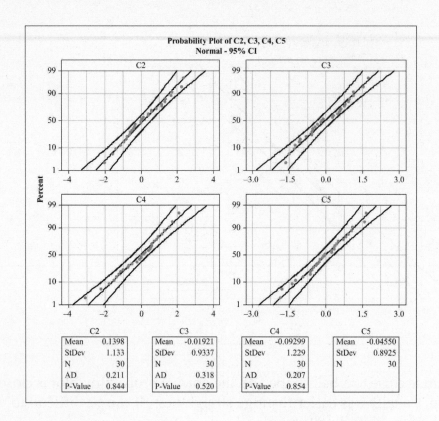

51. Some more simulations

a) The histogram and the normal probability plot from a simulated sample of 100 observations from a normal (0, 1) distribution are shown below. The NPP is close to a straight line. The histogram is close to a bell shape, but this can vary, for example, if we double the number of bins (as shown below) it looks less bell-shaped. So it is easier to judge straightness of the NPP.

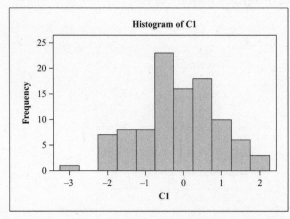

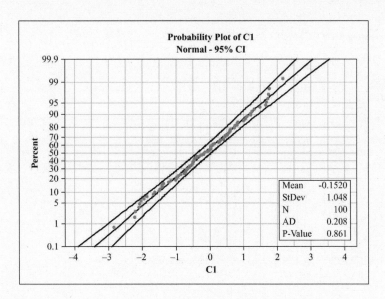

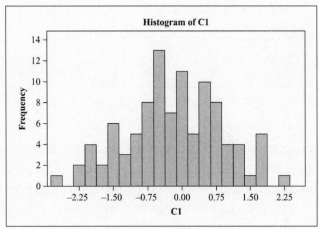

b) The histograms and the normal probability plots for the four simulated samples of 100 observations from the normal (0, 1) are given below. The normal probability plots are close to straight lines. We expect this because the data were simulated from a normal distribution. The shapes of the histograms are also close to bell shapes. Random variation from sample to sample is clear in the histograms.

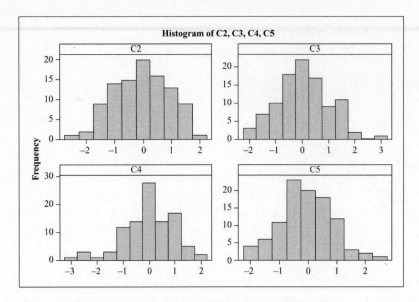

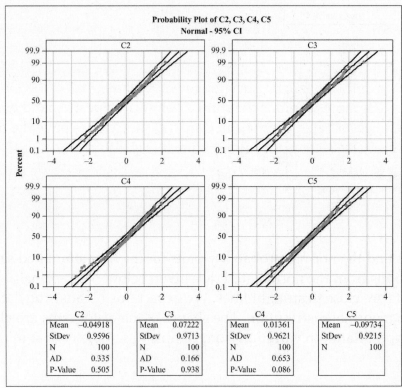

Part I Review: Exploring and Understanding Data

1. **Homicide across Canada 2011**
   a) A histogram for the homicide rates is shown below:

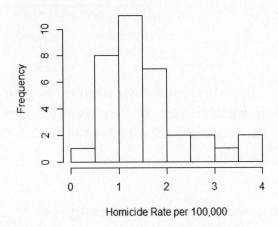

   The homicide rates have median 1.36, quartiles Q1 = 0.92 and Q3 = 1.92, and interquartile range 1.00. The distribution is skewed to the right.

   b) Boxplots for the geographic regions are shown below. It appears that there are substantial differences between regions in Canada. The Prairies have the highest homicide rates, and Quebec has the lowest homicide rates.

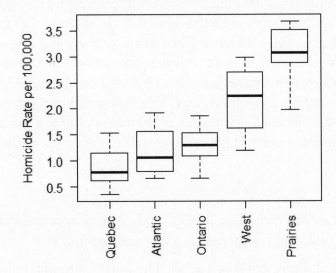

3. **Singers**
    a) The two statistics could be the same if there were many sopranos of that height.
    b) The distribution of heights of each voice part is roughly symmetric. The basses and tenors are generally taller than the altos and sopranos, with the basses being slightly taller than the tenors. The sopranos and altos have about the same median height. Heights of basses and sopranos are more consistent than altos and tenors.

5. **Beanstalks**
    a) The greater standard deviation for the distribution of men's heights means that their heights are more variable than the heights of women.
    b) The z-score for women to qualify is 2.4 compared with 1.75 for men, so it is harder for women to qualify.

7. **University survey**
    a) *Who* – Local residents near a Canadian university. *What* – Age, whether or not the respondent attended college or university, and whether or not the respondent had a favourable opinion of a Canadian university. *When* – Not specified. *Where* – Region around a Canadian university. *Why* – The information will be included in a report to the university's directors. *How* – 850 local residents were surveyed by phone.
    b) There is one quantitative variable, age, probably measured in years. There are two categorical variables, college or university attendance (yes or no), and opinion of the university (favourable or unfavourable).
    c) There are several problems with the design of the survey. No mention is made of a random selection of residents. Furthermore, there may be a non-response bias present. People with an unfavourable opinion of the university may hang up as soon as the staff member identifies himself or herself. Also, response bias may be introduced by the interviewer. The responses of the residents may be influenced by the fact that employees of the university are asking the questions. There may be greater percentage of favourable responses to the survey than truly exist.

9. **Fraud detection**
    a) Even though they are numbers, the SIC code is a categorical variable. A histogram is a quantitative display, so it is not appropriate.
    b) The Normal model will not work at all. The Normal model is for modelling distributions of unimodal and symmetric quantitative variables. SIC code is a categorical variable.

11. Cramming

a) Comparitive boxplots of the distributions of Friday and Monday scores are shown at the right.

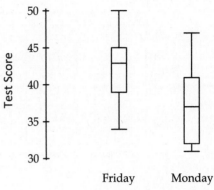

b) The distribution of scores on Friday was generally higher by about 5 points. Students fared worse on Monday after preparing for the test on Friday. The spreads are about the same, but the scores on Monday are slightly skewed to the right.

c) A histogram of the distribution of change (Friday-Monday) in test score is shown at the right.

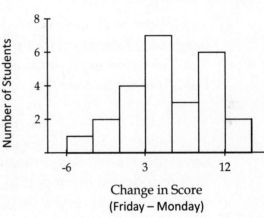

d) The distribution of changes in score is roughly unimodal and symmetric, and is centred near 4 points. Changes ranged from a student who scored 5 points higher on Monday, to two students who each scored 14 points higher on Friday. Only three students did better on Monday.

13. Off or on reserve 2011

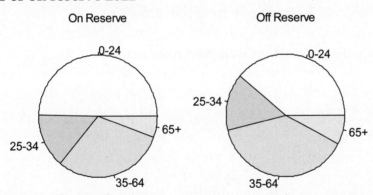

a) There is a higher proportion living on reserves in the 0–24 age group and a lower proportion living on reserves in the 35–64 age group.

b) Side by side barplots for the four conditional residence distributions are shown below. For the youngest age group, there are many more living on reserve. As the age groups increase in age, the percentage living on reserve declines.

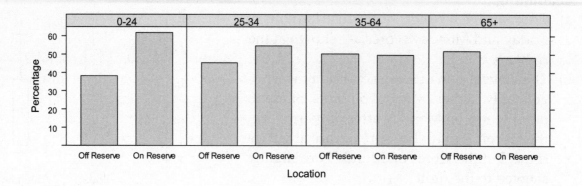

c) The residence variable is categorical. The age variable is quantitative, but has been made categorical for this display.

d) Yes. The following mean ages will occur if only 1/8 of the males live on reserves, but 3/4 of the females live on reserves.

Residency	Male	Female
On	30	40
Off	50	60
Overall	47.5	45

For the males, the overall mean is:

$$\frac{n_1\bar{x}_1 + n_2\bar{x}_2}{n_1 + n_2} = p\bar{x}_1 + (1 - p)\bar{x}_2 = \left(\frac{1}{8}\right)30 + \left(\frac{7}{8}\right)50 = 47.5$$

where $p = \dfrac{n_1}{n_1 + n_2}$ is the proportion living on reserve.

For the females: $\dfrac{n_1\bar{x}_1 + n_2\bar{x}_2}{n_1 + n_2} = p\bar{x}_1 + (1 - p)\bar{x}_2 = \left(\dfrac{3}{4}\right)40 + \left(\dfrac{1}{4}\right)60 = 45.$

15. Hard water

a) The variables in this study are both quantitative. Annual mortality rate for males is measured in deaths per 100 000. Calcium concentration is measured in parts per million.

b) The distribution of calcium concentration is skewed right, with many towns having concentrations below 25 ppm. The rest of the towns have calcium concentrations that are distributed in a fairly uniform pattern from 25 ppm to 100 ppm, tapering off to a maximum concentration around 150 ppm. The distribution of mortality rates is unimodal and symmetric, with centre approximately 1500 deaths per 100 000. The distribution has a range of 1000 deaths per 100 000, from 1000 to 2000 deaths per 100 000.

17. Seasons
a) The two histograms have different horizontal and vertical scales. This makes a quick comparison impossible.
b) The centre of the distribution of average temperatures in January is around 2 degrees, compared to a centre of the distribution of July temperatures around 23. The January distribution is also much more spread out than the July distribution. The range is over 34 degrees in January, compared to a range of over 14 in July. The distribution of average temperature in January is skewed slightly to the right, while the distribution of average temperature in July is roughly symmetric.
c) The distribution of difference in average temperature (July – January) for 60 large U.S. cities is slightly skewed to the left, with median at approximately 23 degrees. There are several low outliers: cities with very little difference between their average July and January temperatures. The single high outlier is a city with a large difference in average temperature between July and January. The middle 50% of differences are between approximately 21 and 25 degrees.

19. Old Faithful?
a) The distribution of duration of the 222 eruptions is bimodal, with modes at approximately 2 minutes and 4.5 minutes. The distribution is fairly symmetric around each mode.
b) The bimodal shape of the distribution of duration of the 222 eruptions suggests that there may be two distinct groups of eruption durations. The usual summary statistics would try to summarize these two groups as a single group, which wouldn't make sense. However, it could be useful to report the two modes, the range, and perhaps the lowest dip point between the modes.
c) The intervals between eruptions are generally longer for long eruptions than the intervals for short eruptions. Over 75% of the short eruptions had intervals of approximately 60 minutes or less, while almost all of the long eruptions had intervals of more than 60 minutes.

21. Liberty's nose
a) The distribution of the ratio of arm length to nose length of 18 girls in a statistics class is unimodal and roughly symmetric, with centre around 15. There is one low outlier, a ratio of 11.8. A boxplot is shown at the right. A histogram or stemplot is also an appropriate display.

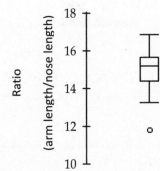

b) In the presence of an outlier, the 5-number summary is the appropriate choice for summary statistics. The 5-number summary is 11.8, 14.4, 15.25, 15.7, 16.9. The IQR is 1.3.

c) The ratio of 9.3 for the Statue of Liberty is very low, well below the lowest ratio in the statistics class, 11.8, which is already a low outlier. Compared to the girls in the statistics class, the Statue of Liberty's nose is very long in relation to her arm.

23. **Simpson's paradox** Overall, the follow-up group was insured only 11.1% of the time as compared to 16.6% for the not-traced group. At first, it appears that the group is associated with presence of health insurance. But for blacks, the follow-up group was quite close (actually slightly higher) in terms of being insured: 8.9% to 8.7%. The same is true for whites. The follow-up group was insured 83.3% of the time, compared to 82.5% of the not-traced group. When broken down by race, we see that the group is not associated with the presence of health insurance for either race. This demonstrates Simpson's paradox, because the overall percentages lead us to believe that there is an association between health insurance and group, but we see the truth when we examine the situation more carefully.

25. **Be quick!**

a) The Normal model for the distribution of reaction times is shown at the right.

b) The distribution of reaction times is unimodal and symmetric, with mean 1.5 seconds, and standard deviation 0.18 seconds. According to the Normal model, 95% of drivers are expected to have reaction times between 1.14 seconds and 1.86 seconds.

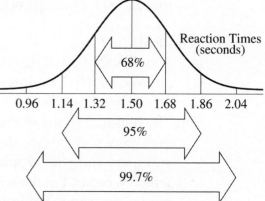

c) According to the Normal model, 8.24% of drivers are expected to have reaction times below 1.25 seconds.

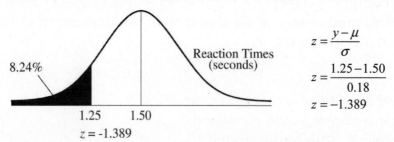

$$z = \frac{y - \mu}{\sigma}$$

$$z = \frac{1.25 - 1.50}{0.18}$$

$$z = -1.389$$

d) According to the Normal model, 24.13% of drivers are expected to have reaction times between 1.6 seconds and 1.8 seconds.

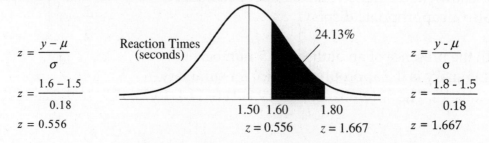

$$z = \frac{y - \mu}{\sigma}$$

$$z = \frac{1.6 - 1.5}{0.18}$$

$$z = 0.556$$

$$z = \frac{y - \mu}{\sigma}$$

$$z = \frac{1.8 - 1.5}{0.18}$$

$$z = 1.667$$

e) According to the Normal model, the interquartile range of the distribution of reaction times is expected to be 0.24 seconds.

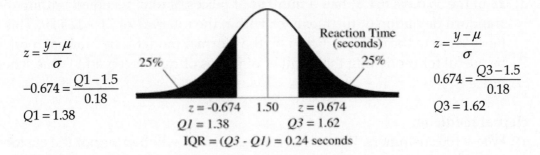

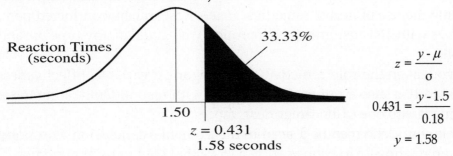

f) According to the Normal model, the slowest 1/3 of all drivers is expected to have reaction times of 1.58 seconds or more. (Remember that a high reaction time is a SLOW reaction time!)

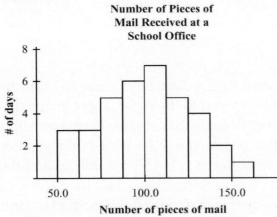

27. **Mail**

a) A histogram of the number of pieces of mail received at a school office is shown below.

b) Since the distribution of the number of pieces of mail is unimodal and symmetric, the mean and standard deviation are appropriate measures of centre and spread. The mean number of pieces of mail is 100.25, and the standard deviation is 25.54 pieces.

c) The distribution of the number of pieces of mail received at the school office is unimodal and symmetric, with mean 100.25 and standard deviation 25.54. The lowest number of pieces of mail received in a day was 52 and the highest was 151.

d) 23 of the 36 days (64%) had a number of pieces of mail received within one standard deviation of the mean, or within the interval 74.71 – 125.79. This is fairly close to the 68% predicted by the Normal model. The Normal model may be useful for modelling the number of pieces of mail received by this school office.

29. Herbal medicine

a) *Who* – 100 customers. *What* – Researchers asked whether or not the customer had taken the cold remedy and had customers rate the effectiveness of the remedy on a scale from 1 to 10. *When* – Not specified. *Where* – Store where natural health products are sold. *Why* – The researchers were from the Herbal Medicine Council, which sounds suspiciously like a group that might be promoting the use of herbal remedies. *How* – Researchers conducted personal interviews with 100 customers. No mention was made of any type of random selection.

b) "Have you taken the cold remedy?" is a categorical variable. Effectiveness on a scale of 1 to 10 is also a categorical variable, with respondents rating the remedy by placing it into one of 10 categories.

c) Very little confidence can be placed in the Council's conclusions. Respondents were people who already shopped in a store that sold natural remedies. They may be pre-disposed to thinking that the remedy was effective. Furthermore, no attempt was made to randomly select respondents in a representative manner. Finally, the Herbal Medicine Council has an interest in the success of the remedy.

31. Engines

a) The count of cars is 38.

b) The mean displacement is higher than the median displacement, indicating a distribution of displacements that is skewed to the right. There are likely to be several very large engines in a group that consists of mainly smaller engines.

c) Since the distribution is skewed, the median and IQR are useful measures of centre and spread. The median displacement is 148.5 cubic inches and the IQR is 126 cubic inches.

d) Your neighbour's car has an engine that is bigger than the median engine, but 227 cubic inches is smaller than the third quartile of 231, meaning that at least 25% of cars have a bigger engine than your neighbour's car. Don't be impressed!

e) Using the Outlier Rule (more than 1.5 IQRs beyond the quartiles) to find the fences:
Upper Fence: Q3 + 1.5(IQR) = 231 + 1.5(126) = 420 cubic inches.
Lower Fence: Q1 – 1.5(IQR) = 105 – 1.5(126) = –84 cubic inches.

Since there are certainly no engines with negative displacements, there are no low outliers. Q1 + Range = 105 + 275 = 380 cubic inches. This means that the maximum must be less than 380 cubic inches. Therefore, there are no high outliers (engines over 420 cubic inches).

f) It is not reasonable to expect 68% of the car engines to measure within one standard deviation of the mean. The distribution of engine displacements is skewed to the right, so the Normal model is not appropriate.

g) Multiplying each of the engine displacements by 16.4 to convert cubic inches to cubic centimetres would affect measures of position and spread. All of the summary statistics (except the count!) could be converted to cubic centimetres by multiplying each by 16.4.

33. Toronto students

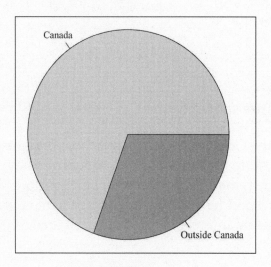

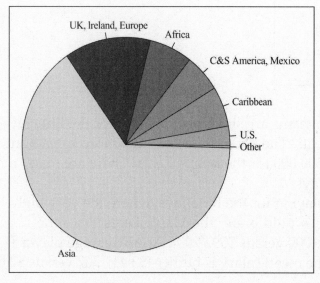

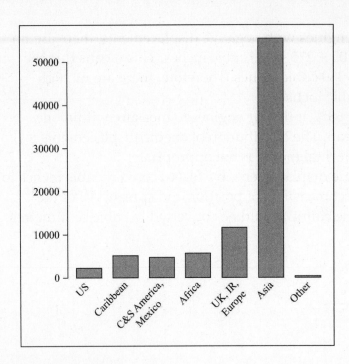

35. Toronto teams

a) Side-by-side boxplots for the two teams are shown below:

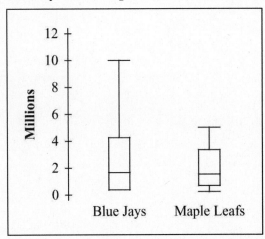

> The median salaries for the two teams are similar, $1 675 000 for the Blue Jays
> and $1 725 000 for the Maple Leafs. The Blue Jay salaries are more spread out,
> with IQR $3 888 500 versus $2 500 000 for the Maple Leafs. There are many more
> very high salaries for the Blue Jays.

b) As a superstar I would prefer playing for the Blue Jays, where the very high
    salaries occur. As a new player I would prefer the Maple Leafs, since the
    minimum salary is higher ($500 000 versus $392 200). As an average player I
    would prefer the Blue Jays as the mean salary is higher ($3 417 367 versus $2 099
    545).

c) As an owner, I would prefer the salary structure of the Maple Leafs as the total
    salary budget is much smaller ($46 190 000 versus $82 016 800).

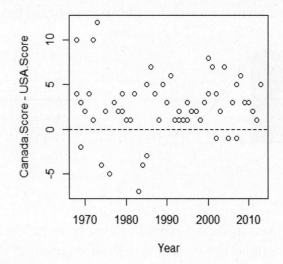

Year

37. Some assembly required

a) According to the Normal model, the standard deviation is 0.43 hours.

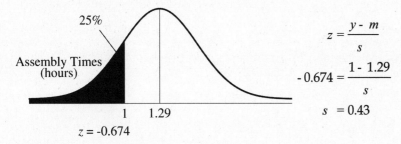

$$z = \frac{y - m}{s}$$

$$-0.674 = \frac{1 - 1.29}{s}$$

$$s = 0.43$$

b) According to the Normal model, the company would need to claim that the desk takes "less than 1.40 hours to assemble," not the catchiest of slogans!

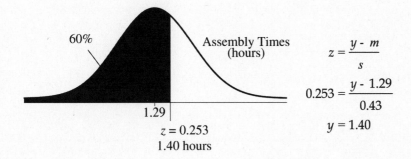

$$z = \frac{y - m}{s}$$

$$0.253 = \frac{y - 1.29}{0.43}$$

$$y = 1.40$$

c) According to the Normal model, the company would have to lower the mean assembly time to 0.89 hour (53.4 minutes).

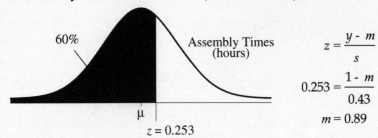

$$z = \frac{y - m}{s}$$

$$0.253 = \frac{1 - m}{0.43}$$

$$m = 0.89$$

d) The new instructions and part-labelling may have helped lower the mean, but it also may have changed the standard deviation, making the assembly times more consistent as well as lower.

Chapter 6: Scatterplots, Association, and Correlation

1. **Association**
   a) Either weight in grams or weight in ounces could be the explanatory or response variable. Greater weights in grams correspond with greater weights in ounces. The association between weight of apples in grams and weight of apples in ounces would be positive, straight, and perfect. Each apple's weight would simply be measured in two different scales. The points would line up perfectly.
   b) Circumference is the explanatory variable, and weight is the response variable, since one-dimensional circumference explains three-dimensional volume (and therefore weight). For apples of roughly the same size, the association would be positive, straight, and moderately strong. If the sample of apples contained very small and very large apples, the association's true curved form would become apparent.
   c) There would be no association between shoe size and GPA of university freshmen.
   d) Number of kilometres driven is the explanatory variable, and litres remaining in the tank is the response variable. The greater the number of kilometres driven, the less gasoline there is in the tank. If a sample of different cars is used, the association is negative, straight, and moderate. If the data is gathered on different trips with the same car, the association would be strong.

3. **Association III**
   a) Altitude is the explanatory variable, and temperature is the response variable. As you climb higher, the temperature drops. The association is negative, straight, and strong.
   b) At first, it appears that there should be no association between ice cream sales and air conditioner sales. When the lurking variable of temperature is considered, the association becomes more apparent. When the temperature is high, ice cream sales tend to increase. Also, when the temperature is high, air conditioner sales tend to increase. Therefore, there is likely to be an increase in the sales of air conditioners whenever there is an increase in the sales of ice cream. The association is positive, straight, and moderate. Either one of the variables could be used as the explanatory variable.
   c) Age is the explanatory variable, and grip strength is the response variable. The association is neither negative nor positive, but is curved and moderate in strength due to the variability in grip strength among people in general. The very young would have low grip strength, and grip strength would increase as age increased. After reaching a maximum (at whatever age physical prowess peaks), grip strength would decline again, with the elderly having low grip strengths.
   d) Blood alcohol content is the explanatory variable, and reaction time is the response variable. As blood alcohol level increase, so does the time it takes to react to a stimulus. The association is positive, probably curved, and strong. The

scatterplot would probably be almost linear for low concentrations of alcohol in the blood, and then begin to rise dramatically, with longer and longer reaction times for each incremental increase in blood alcohol content.

5. **Scatterplots**
   a) None of the scatterplots show little or no association, although #4 is very weak.
   b) #3 and #4 show negative association. Increases in one variable are generally related to decreases in the other variable.
   c) #2, #3, and #4 show a straight association.
   d) #2 shows a moderately strong association.
   e) #1 and #3 show a very strong association. #1 shows a curved association and #3 shows a straight association.

7. **Performance IQ scores versus brain size** The scatterplot of IQ scores versus Brain Sizes is scattered, with no apparent pattern. There appears to be little or no association between the IQ scores and brain sizes displayed in this scatterplot.

9. **Firing pottery**
   a) A histogram of the number of broken pieces is shown at the right. 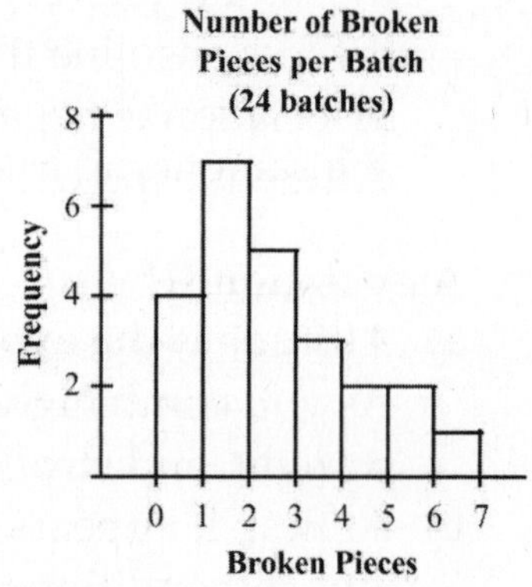
   b) The distribution of the number of broken pieces per batch of pottery is skewed right, centred around 1 broken piece per batch. Batches had from 0 and 6 broken pieces. The scatterplot does not show the centre or skewness of the distribution.
   c) The scatterplot shows that the number of broken pieces increases as the batch number increases. If the eight daily batches are numbered sequentially, this indicates that batches fired later in the day generally have more broken pieces. This information is not visible in the histogram.
   d) Answers will vary.

11. **Matching**
   a) 0.006   b) 0.777
   c) –0.923   d) –0.487

13. **Correlation facts**
   a) True.
   b) False. Correlation will remain the same
   c) False. Correlation has no units.

15. **Roller coasters**
    a) It is appropriate to calculate correlation. Both height of the drop and speed are quantitative variables, the scatterplot shows an association that is straight enough, and there are not outliers.
    b) There is a strong, positive, straight association between drop and speed; the greater the height of the initial drop, the higher the top speed.

17. **Lunchtime**
    a) The correlation between time toddlers spent at the table and the number of calories consumed by the toddlers is $r = -0.649$.
    b) If time spent at the table were recorded in hours instead of minutes, the correlation would not change at all. Correlation is calculated from z-scores, and since these unitless measures of position are unaffected by changes in scale, correlation is likewise unaffected.
    c) The correlation between time spent at the table and calorie consumption for toddlers is –0.649, a moderate, negative correlation. Toddlers who spent more time at the table tended to consume fewer calories.
    d) The analyst's remark is speculative. There are many possible explanations for the behaviour of the toddlers. The data merely show us an association between time spent at the table and calories consumed by toddlers, and association is not the same thing as a cause-and-effect relationship.

19. **Fuel economy 2010**
    a) The scatterplot of gas mileage versus horsepower is shown below:

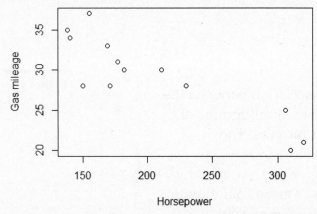

    b) There is a strong, negative, straight association between horsepower and mileage of the selected vehicles. There don't appear to be any outliers. All of the cars seem to fit the same pattern. Cars with more horsepower tend to have lower mileage. [NOTE: The plot seems to be missing two of the 15 models, but in fact there are two pairs of identical points, and our software unfortunately does not distinguish the overlapping points from single points.]
    c) Since the relationship is linear, with no outliers, correlation is an appropriate measure of strength. The correlation between horsepower and mileage of the selected vehicles is –0.909.

d) There is a strong linear relationship in the negative direction between horsepower and highway gas mileage. Lower fuel efficiency is associated with higher horsepower.

e) The scatterplot of fuel consumption versus horsepower is shown below. There is a strong, positive, straight association between fuel consumption and horsepower — more powerful cars burn more fuel. The correlation between the fuel consumption and horsepower is stronger than the correlation between mileage and horsepower, and the direction of association is reversed (correlation = 0.921). The strength of correlation changed here because the transformation from mileage to consumption is non-linear. The two presentations of the data are in some sense equivalent; mileage focuses on the distance travelled using a fixed amount of fuel, and consumption focuses on the amount of fuel needed to travel a certain distance. Both plots are roughly linear, but I might prefer the transformed scale, since the correlation is slightly higher.

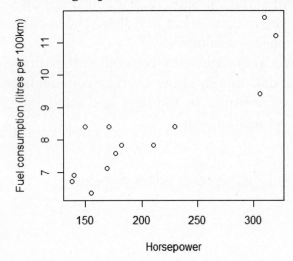

21. Burgers There is no apparent association between the number of grams of fat and the number of milligrams of sodium in several brands of fast food burgers. The correlation is only $r = 0.199$, which is close to zero, an indication of no association. One burger had a much lower fat content than the other burgers, at 19 grams of fat, with 920 milligrams of sodium. Without this (comparatively) low fat burger, the correlation would have been even lower (and negative).

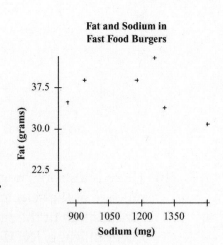

23. **Attendance 2010**
    a) Number of runs scored and attendance are quantitative variables, the relationship between them appears to be straight, and there are no outliers, so calculating a correlation is appropriate.
    b) The association between attendance and runs scored is positive, straight, and moderate in strength. Generally, as the number of runs scored increases, so does attendance.
    c) There is evidence of an association between attendance and runs scored, but a cause-and effect relationship between the two is not implied. There may be lurking variables that can account for the increases in each. For example, perhaps winning teams score more runs and also have higher attendance. We don't have any basis to make a claim of causation.

25. **Politics** The candidate might mean that there is an association between television watching and crime. The term correlation is reserved for describing linear associations between quantitative variables. We don't know what type of variables "television watching" and "crime" are, but they seem categorical. Even if the variables are quantitative (hours of TV watched per week, and number of crimes committed, for example), we aren't sure that the relationship is linear. The politician also seems to be implying a cause-and-effect relationship between television watching and crime. Association of any kind does not imply causation.

27. **Height and reading**
    a) Actually, this *does* mean that taller children in elementary school are better readers. However, this does *not* mean that height causes good reading ability.
    b) Older children are generally both taller and are better readers. Age is the lurking variable.

29. **Hard water** It is not appropriate to summarize the strength of the association between water hardness and pH with a correlation, since the association is curved and not straight enough.

31. **Correlation errors**
    a) If the association between GDP and infant mortality is linear, a correlation of –0.772 shows a moderate, negative association. Generally, as GDP increases, infant mortality rate decreases.
    b) Continent is a categorical variable. Correlation measures the strength of linear associations between quantitative variables.

33. **Baldness and heart disease** Even though the variables baldness and heart disease were assigned numerical values, they are categorical. Correlation is only an appropriate measure of the strength of linear association between quantitative variables. Their conclusion is meaningless.

35. Thrills 2011 The scatterplot below shows that the association between duration and length is straight, positive, and moderate, with no outliers. Generally, rides on coasters with a greater length tend to last longer. The correlation between length and duration is 0.698, indicating a positive moderate association.

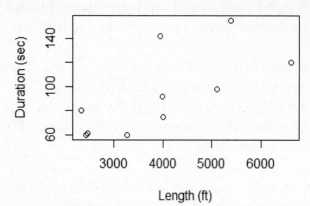

37. Thrills III

a) We would expect that as one variable (e.g., length of ride) increases, the rank will improve, which means it will decrease. The idea here is that more is better.

b) Drop has the strongest correlation ($r = -0.193$), but even that correlation is very weak. The scatterplot shows no apparent association. The number one ranked coaster, Bizarro, has a fairly typical drop. Other factors appear to influence the rank of coaster more than any of the ones measured in this data set.

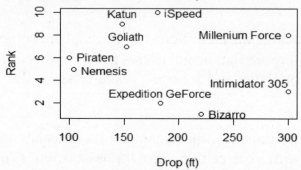

c) Other variables may account for the ranking. For example, other quantitative variables such as number of loops and number of corkscrews, categorical variables such as whether the coaster is made of wood or steel and whether or not there are tunnels may all have an effect on the rank.

39. Does my bum look big in this?

a) All possible correlations are given below:

Correlations: Year, Bust size (in), Waist size (in), Hips size (in)

	Year	Bust size	Waist size
Bust size	1.000		
Waist size	0.992	0.992	
Hips size	0.936	0.936	0.929

b) The plots for body size measurements against each other and against time are shown below. Most statistical software packages give a plot of all possible pairs of variables known as a scatterplot matrix that looks like the second plot below. All these relationships are linear, strong, and positive in direction, i.e., one increases as the other increases. There are no clear outliers, and the relationship between bust size and year is the strongest.

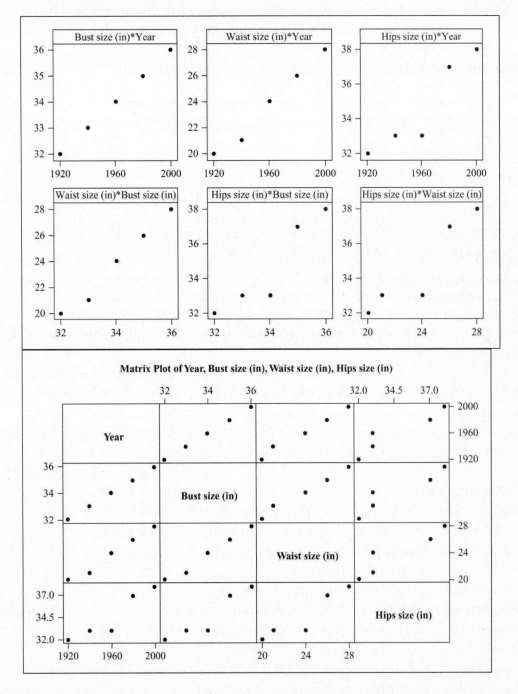

c) The correlations between body measurements are not really surprising as we expect some proportionality between them (otherwise women's body shape is changing over time). The positive correlation with time can happen, for example, if we are getting bigger (increasing weight) over time due to various reasons (maybe lack of activity because most of the work we did in the good old days has been taken over by machines).

41. Census at school

a) The pair-wise correlations between the variables are given below. The correlation between armspan and height is the strongest with $r = 0.781$.

Correlations: age, height, armspan, wristbone, middlefinger, foot

	age	height	armspan	wristbone	middlefinger
height	0.272				
	0.004				
armspan	0.219	0.781			
	0.021	0.000			
wristbone	−0.008	0.397	0.449		
	0.937	0.000	0.000		
middlefinger	0.059	0.560	0.565	0.360	
	0.537	0.000	0.000	0.000	
foot	0.188	0.542	0.548	0.436	0.461
	0.049	0.000	0.000	0.000	0.000

Cell Contents: Pearson correlation
          P-Value

b) Disregarding gender, the form of the relation is linear, positive (i.e., one increases as the other increases), and moderate in strength, with no very unusual outliers.

c) The relationship is similar for both genders, but it looks likes the points for females (the red points) are closer to a straight line than those for males (the blue points). This means the relationship between arm span and height is stronger for females than for males. If we omitted the information about gender in the plot above, it would become a *lurking* variable.

43. Planets (more or less)

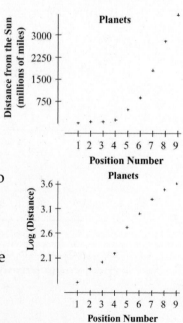

a) The association between Position Number of each planet and Distance from the Sun (in millions of miles) is very strong, positive, and curved. The scatterplot is shown at the right.

b) The relationship between Position Number and Distance from the Sun is not linear. Correlation is a measure of the degree of *linear* association between two variables.

c) The scatterplot of the logarithm of Distance versus Position Number (shown at the right) still shows a strong, positive relationship, but it is straighter than the previous scatterplot. It still shows a curve in the scatterplot, but it is straight enough that correlation may now be used as an appropriate measure of the strength of the relationship between logarithm of Distance and Position Number, which will in turn give an indication of the strength of the association.

45. Mandarin or Cantonese 2011

a) Correlation between the two mother tongue counts is 0.995. This is very close to 1. This is not surprising if we imagine that the ratio of Mandarin to Cantonese speakers is (nearly) the same in all communities (or for reasons related to the answer in part d.).

b) The plot of Cantonese speaker count versus Mandarin speaker count is shown below. Most points are concentrated to a small region and just two points are far away from them but in the same direction. This can make the correlation very high (just like when you have only two points the correlation is exactly 1) and correlation is not an appropriate measure in situations like this.

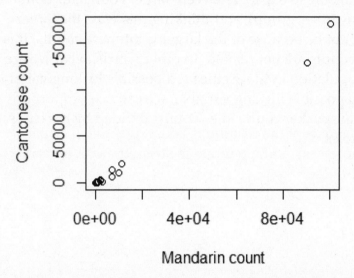

c) The plot of log(Cantonese) versus log(Mandarin) is shown below. Yes, now there is a concentration of points toward smaller communities. The correlation between log(Cantonese) and log(Mandarin) is 0.985. The correlation is now appropriately measured.

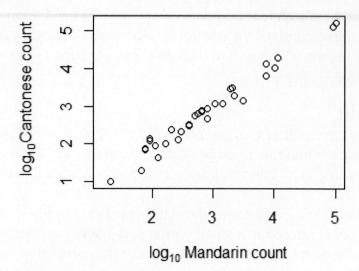

d) In larger cities, it is reasonable to expect relatively larger counts for both these groups (as well as for any other group), and thinking this way, the larger Mandarin counts might not be because of the larger Cantonese counts. It is even possible that these two counts are not related. In other words, what we are suggesting is that the population in those cities is a possible lurking variable. If we think population is a possible lurking variable, we can use proportions of Cantonese and Mandarin speaker counts (i.e., counts divided by the city's population) as the variables.

Chapter 7: Linear Regression

1. Regression equations

$\bar{x}$	s_x	$\bar{y}$	s_y	r	$\hat{y} = b_0 + b_1 x$
a) 10	2	20	3	0.5	$\hat{y} = 12.5 + .75\,x$
b) 2	0.06	7.2	1.2	−0.4	$\hat{y} = 23.2 - 8\,x$
c) 12	6	152	30	−0.8	$\hat{y} = 200 - 4\,x$
d) 2.5	1.2	25	100	0.6	$\hat{y} = -100 + 50\,x$

a)

$$b_1 = \frac{rs_y}{s_x}$$
$$b_1 = \frac{(0.5)(3)}{2}$$
$$b_1 = 0.75$$

$$\hat{y} = b_0 + b_1 x$$
$$\bar{y} = b_0 + b_1 \bar{x}$$
$$20 = b_0 + 0.75(10)$$
$$b_0 = 12.5$$

b)

$$b_1 = \frac{rs_y}{s_x}$$
$$b_1 = \frac{(-0.4)(1.2)}{0.06}$$
$$b_1 = -8$$

$$\hat{y} = b_0 + b_1 x$$
$$\bar{y} = b_0 + b_1 \bar{x}$$
$$7.2 = b_0 - 8(2)$$
$$b_0 = 23.2$$

c)

$$\hat{y} = b_0 + b_1 x$$
$$\bar{y} = b_0 + b_1 \bar{x}$$
$$\bar{y} = 200 - 4(12)$$
$$\bar{y} = 152$$

$$b_1 = \frac{rs_y}{s_x}$$
$$-4 = \frac{-0.8 s_y}{6}$$
$$s_y = 30$$

d)

$$\hat{y} = b_0 + b_1 x$$
$$\bar{y} = b_0 + b_1 \bar{x}$$
$$\bar{y} = -100 + 50(2.5)$$
$$\bar{y} = 25$$

$$b_1 = \frac{rs_y}{s_x}$$
$$50 = \frac{r(100)}{1.2}$$
$$r = 0.6$$

3. Residuals
a) The scattered residuals plot indicates an appropriate linear model.
b) The curved pattern in the residuals plot indicates that the linear model is not appropriate. The relationship is not linear.
c) The fanned pattern indicates that the linear model is not appropriate. The model's predicting power decreases as the values of the explanatory variable increase.

5. Least squares
If the four x-values are plugged into $\hat{y} = 7 + 1.1x$, the four predicted values are $\hat{y} = 18, 29, 51$, and 62, respectively. The four residuals are –8, 21, –31, and 18. The squared residuals are 64, 441, 961, and 324, respectively. The sum of the squared residuals is 1790. Least squares means that no other line has a sum lower than 1790.

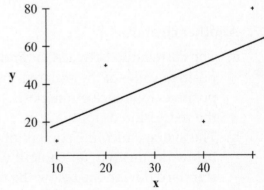

7.   **Real estate**
     a) The response variable is the asking price (in thousands of dollars) and the explanatory variable is the size (in square feet).
     b) Thousands of dollars per square feet.
     c) The price usually increases as the size increases, so the slope must be positive.

9.   **Real estate again** 70.2% of the variability in *Price* is explained by the linear regression of *Price* on *size*.

11. **Real estate redux**
     a) Correlation = $\sqrt{R^2} = \sqrt{0.702} = 0.84$ (assuming that the relationship will be positive as expected).
     b) We would predict the price to be 0.84 standard deviations above the average price.
     c) We would predict the price to be 1.68 standard deviations below the average price.

13. **More real estate**
     a) For every square foot increase in size, the *price* is expected to increase by 0.37 thousand dollars (i.e., 0.37 × 1000 = $370).
     b) The predicted *Asking Price* = 49.30 + 0.37 × 1000 = $419.3 thousand dollars.
     c) The predicted price is 49.30 + 0.37 × 1200 = 493.3 thousand dollars and asking price = 493 300 – 6000 = 487 300. The value 6000 is the residual for a 1200-square-foot condo with asking price $487 300.

15. **Cigarettes**
     a) A linear model is probably appropriate. The residuals plot shows some initially low points, but there is no clear curvature.
     b) 92.4% of the variability in nicotine level is explained by variability in tar content. (In other words, 92.4% of the variability in nicotine level is explained by the linear model.)

17. **Another cigarette**
     a) The correlation between tar and nicotine is $r = \sqrt{R^2} = \sqrt{0.924} = 0.961$. The positive value of the square root is used, since the relationship is believed to be positive. Evidence of the positive relationship is the positive coefficient of tar in the regression output.
     b) The average nicotine content of cigarettes that are two standard deviations below the mean in tar content would be expected to be about 1.922 (2 × 0.961) standard deviations below the mean nicotine content.

c) Cigarettes that are one standard deviation above average in nicotine content are expected to be about 0.961 standard deviations (in other words, r standard deviations) above the mean tar content.

19. **Last cigarette**
    a) $\widehat{Nicotine} = 0.15403 + 0.065052(Tar)$ is the equation of the regression line that predicts nicotine content from tar content of cigarettes.
    b)

    $\widehat{Nicotine} = 0.15403 + 0.065052(Tar)$ The model predicts that a cigarette with 4 mg of tar will have about 0.414 mg of nicotine.

    $\widehat{Nicotine} = 0.15403 + 0.065052(4)$

    $\widehat{Nicotine} = 0.414$

    c) For each additional mg of tar, the model predicts an increase of 0.065 mg of nicotine.
    d) The model predicts that a cigarette with no tar would have 0.154 mg of nicotine.
    e)

    $\widehat{Nicotine} = 0.15403 + 0.065052(Tar)$ The model predicts that a cigarette with 7 mg of tar will have 0.6094 mg of nicotine. If the residual is –0.5, the cigarette actually had 0.1094 mg of nicotine.

    $\widehat{Nicotine} = 0.15403 + 0.065052(7)$

    $\widehat{Nicotine} = 0.6094$

21. **What slope?** The only slope that makes sense is 300 pounds per foot. 30 pounds per foot is too small. For example, a Honda Civic is about 14 feet long, and a Cadillac DeVille is about 17 feet long. If the slope of the regression line were 30 pounds per foot, the Cadillac would be predicted to outweigh the Civic by only 90 pounds! (The real difference is about 1500 pounds.) Similarly, 3 pounds per foot is too small. A slope of 3000 pounds per foot would predict a weight difference of 9000 pounds (4.5 tons) between Civic and DeVille. The only answer that is even reasonable is 300 pounds per foot, which predicts a difference of 900 pounds. This isn't very close to the actual difference of 1500 pounds, but at least it is in the right ballpark.

23. **Misinterpretations**
    a) R^2 is an indication of the strength of the model, not the appropriateness of the model. A scattered residuals plot is the indicator of an appropriate model.
    b) Regression models give predictions, not actual values. The student should have said, "The model predicts that a bird 30 centimetres tall is expected to have a wingspan of 34 centimetres."

25. **ESP**
    a) First, since no one has ESP, you must have scored 2 standard deviations above the mean by chance. On your next attempt, you are unlikely to duplicate the

extraordinary event of scoring 2 standard deviations above the mean. You will likely "regress" toward the mean on your second try, getting a lower score. If you want to impress your friend, don't take the test again. Let your friend think you can read his mind!

b) Your friend doesn't have ESP, either. No one does. Your friend will likely "regress" toward the mean score on his second attempt as well, meaning his score will probably go up. If the goal is to get a higher score, your friend should try again.

27. SAT scores

a) The association between SAT Math scores and SAT Verbal Scores is linear, moderate in strength, and positive. Students with high SAT Math scores typically have high SAT Verbal scores.

b) One student got a 500 Verbal and 800 Math. That set of scores doesn't seem to fit the pattern.

c) $r = 0.685$ indicates a moderate, positive association between SAT Math and SAT Verbal, but only because the scatterplot shows a linear relationship. Students who score one standard deviation above the mean in SAT Math are expected to score 0.685 standard deviations above the mean in SAT Verbal. Additionally, R^2 = $(0.685)^2 = 0.469225$, so 46.9% of the variability in math score is explained by variability in verbal score.

d) The scatterplot of verbal and math scores shows a relationship that is straight enough, so a linear model is appropriate.

$$b_1 = \frac{rs_{Math}}{s_{Verbal}}$$

$$b_1 = \frac{(0.685)(96.1)}{99.5}$$

$$b_1 = 0.661593$$

$$\hat{y} = b_0 + b_1 x$$

$$\bar{y} = b_0 + b_1 \bar{x}$$

$$612.2 = b_0 + 0.661593(596.3)$$

$$b_0 = 217.692$$

The equation of the least squares regression line for predicting SAT Math score from SAT Verbal score is

$$\widehat{Math} = 217.692 + 0.662(Verbal)$$

e) For each additional point in verbal score, the model predicts an increase of 0.662 points in math score. A more meaningful interpretation might be scaled up. For each additional 10 points in verbal score, the model predicts an increase of 6.62 points in math score.

f)

$$\widehat{Math} = 217.692 + 0.662(Verbal)$$

$$\widehat{Math} = 217.692 + 0.662(500)$$

$$\widehat{Math} = 548.692$$

According to the model, a student with a verbal score of 500 is expected to have a math score of 548.692.

g)

$$\widehat{Math} = 217.692 + 0.662(Verbal)$$

$$\widehat{Math} = 217.692 + 0.662(800)$$

$$\widehat{Math} = 747.292$$

According to the model, a student with a verbal score of 800 is expected to have a math score of 747.292. She actually scored 800 on math, so her *residual* was 800 – 747.292 = 52.708 points.

29. SAT, take 2

a) $r = 0.685$. The correlation between SAT Math and SAT Verbal is a unitless measure of the degree of linear association between the two variables. It doesn't depend on the order in which you are making predictions.

b) The scatterplot of verbal and math scores shows a relationship that is straight enough, so a linear model is appropriate.

$$b_1 = \frac{rs_{Verbal}}{s_{Math}}$$

$$b_1 = \frac{(0.685)(99.5)}{96.1}$$

$$b_1 = 0.709235$$

$$\hat{y} = b_0 + b_1 x$$

$$\bar{y} = b_0 + b_1 \bar{x}$$

$$596.3 = b_0 + 0.709235(612.2)$$

$$b_0 = 162.106$$

The equation of the least squares regression line for predicting SAT Verbal score from SAT Math score is:

$$\widehat{Verbal} = 162.106 + 0.709(Math)$$

c) A positive residual means that the student's actual verbal score was higher than the score the model predicted for someone with the same math score.

d)

$$\widehat{Verbal} = 162.106 + 0.709(Math)$$

$$\widehat{Verbal} = 162.106 + 0.709(500)$$

$$\widehat{Verbal} = 516.606$$

According to the model, a person with a math score of 500 is expected to have a verbal score of 516.606 points.

e)

$$\widehat{Math} = 217.692 + 0.662(Verbal)$$

$$\widehat{Math} = 217.692 + 0.662(516.606)$$

$$\widehat{Math} = 559.685$$

According to the model, a person with a verbal score of 516.606 is expected to have a math score of 559.685 points.

f) The prediction in part e) does not cycle back to 500 points because the regression equation used to predict math from verbal is a different equation than the regression equation used to predict verbal from math. One was generated by minimizing squared residuals in the verbal direction, the other was generated by minimizing squared residuals in the math direction. If a math score is one standard deviation above the mean, its predicted verbal score regresses toward the mean. The same is true for a verbal score used to predict a math score.

31. Used cars 2013

a)

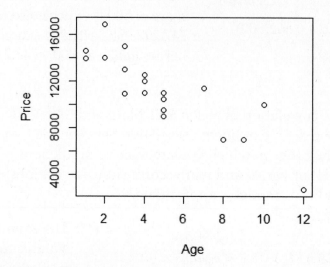

b) There is a strong, negative, linear association between price and age of used Toyota Corollas. A few observations are somewhat unusual, but not very unusual.

c) The scatterplot provides evidence that the relationship is straight enough. A linear model will likely be appropriate.

d) Since $R^2 = 0.748$, simply take the square root to find $|r| = \sqrt{0.748} = 0.865$. Since association between age and price is negative, $r = \sqrt{0.748} = -0.865$.

e) 74.8% of the variability in price of a used Toyota Corolla can be accounted for by variability in the age of the car.

f) The relationship is not perfect. Other factors, such as options, condition, and mileage, explain the rest of the variability in price.

33. More used cars 2013

a) The scatterplot from **Used cars 2013** shows that the relationship is straight, so the linear model is appropriate. The regression equation to predict the price of a used Toyota Corolla from its age is: Price = $15,729 – $935.9 × Age
The regression output is below:
Coefficients:

	Estimate	Std. Error	t value	Pr(>|t|)	
(Intercept)	15729.0	723.2	21.749	2.26e – 14	***
Age	–935.9	127.9	–7.315	8.56e – 07	***

b) According to the model, for each additional year in age, the car is expected to drop $936 in price.

c) The model predicts that a new Toyota Corolla (0 years old) will cost $15 729.

d) Using the model for price as a function of age, the average six year old Corolla has a price of $15 729 – $935.9 × 6 = $10 114, which might be appropriate.

e) Buy the car with the negative residual. Its actual price is lower than predicted.

f) According to the linear model, the average price for a five-year-old Corolla is $15 729 – $935.9 × 5 = $11 050, so the residual is $9000 – $11 050 = –$2050.

g) The model would not be useful for predicting the price of a 17-year-old Corolla. The oldest car in the data is only 12 years old. Predicting a price after 17 years would be an extrapolation. In fact, the predicted price using the model would be negative(!) $15 729 – $935.9 × 17 = –$181.

35. Burgers

a) The scatterplot of calories vs. fat content in fast food hamburgers is at the right. The relationship appears linear, so a linear model is appropriate.

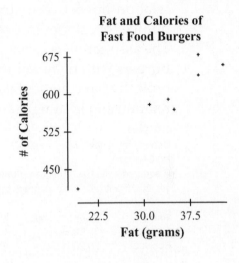

Fat and Calories of Fast Food Burgers

Dependent variable is: **Calories**
No Selector
R squared = 92.3%   R squared (adjusted) = 90.7%
s = 27.33 with 7 - 2 = 5 degrees of freedom

Source	Sum of Squares	df	Mean Square	F-ratio
Regression	44664.3	1	44664.3	59.8
Residual	3735.73	5	747.146	

Variable	Coefficient	s.e. of Coeff	t-ratio	prob
Constant	210.954	50.10	4.21	0.0084
Fat	11.0555	1.430	7.73	0.0006

b) From the computer regression output, R^2 = 92.3%. 92.3% of the variability in the number of calories can be explained by the variability in the number of grams of fat in a fast food burger.

c) From the computer regression output, the regression equation that predicts the number of calories in a fast food burger from its fat content is:

$$\widehat{Calories} = 210.954 + 11.0555(Fat)$$

d) The residuals plot at the right shows no pattern. The linear model appears to be appropriate.

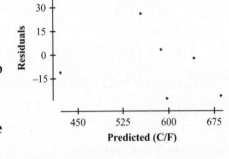

e) The model predicts that a fat-free burger would have 210.954 calories. Since there are no data values close to 0, this is an extrapolation outside the data and isn't of much use.

f) For each additional gram of fat in a burger, the model predicts an increase of 11.056 calories.

g) $\widehat{Calories} = 210.954 + 11.056(Fat) = 210.954 + (11.0555(28)) = 520.508$

The model predicts a burger with 28 grams of fat will have 520.508 calories. If the residual is +133, the actual number of calories is 520.508 + 133 ≈ 653.508 calories.

37. A second helping of burgers

a) The model from **Burgers** was for predicting number of calories from number of grams of fat. In order to predict grams of fat from the number of calories, a new linear model needs to be generated.

b) The scatterplot at the right shows the relationship between number fat grams and number of calories in a set of fast food burgers. The association is strong, positive, and linear. Burgers with higher numbers of calories typically have higher fat contents. The relationship is straight enough to apply a linear model.

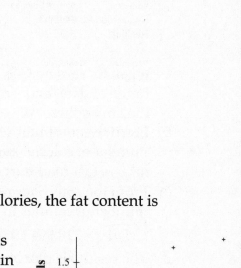

Calories and Fat in Fast Food Burgers

Dependent variable is: **Fat**
No Selector
R squared = 92.3%   R squared (adjusted) = 90.7%
s = 2.375 with 7 - 2 = 5 degrees of freedom

Source	Sum of Squares	df	Mean Square	F-ratio
Regression	337.223	1	337.223	59.8
Residual	28.2054	5	5.64109	

Variable	Coefficient	s.e. of Coeff	t-ratio	prob
Constant	-14.9622	6.433	-2.33	0.0675
Calories	0.083471	0.0108	7.73	0.0006

The linear model for predicting fat from calories is:

$$\widehat{Fat} = -14.9622 + 0.083471(Calories)$$

The model predicts that for every additional 100 calories, the fat content is expected to increase by about 8.3 grams.

The residuals plot shows no pattern, so the model is appropriate. $R^2 = 92.3\%$, so 92.3% of the variability in fat content can be explained by the model.

$$\widehat{Fat} = -14.9622 + 0.083471(Calories)$$

$$\widehat{Fat} = -14.9622 + 0.083471(600)$$

$$\widehat{Fat} \approx 35.1$$

According to the model, a burger with 600 calories is expected to have 35.1 grams of fat.

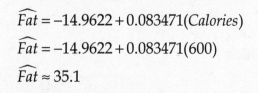

39. New York bridges

a) Overall, the model predicts the condition score of new bridges to be 4.95, a little below the cutoff of 5. Of course, looking at the scatterplot, we can see that the cluster of new bridges all have scores above 5. Still, the model is not a very encouraging one in regards to the conditions of New York City bridges.

b) According to the model, the condition of the bridges in New York City is decreasing by an average of 0.0048 per year.

c) We shouldn't place too much faith in the model. First of all, the scatterplot shows some curvature, so the linear model may not be appropriate. Even if it is appropriate, R^2 is very low, and the standard deviation of the residuals, 0.6708, is quite high in relation to the scope of the data values themselves. This association is very weak.

41. Climate change 2011

a) The correlation between CO_2 level and mean temperature is
$$r = \sqrt{R^2} = \sqrt{0.84} = 0.9165.$$

b) 91.58% of the variability in mean temperature can be accounted for by variability in CO_2 level.

c) Since the scatterplot of CO_2 level and mean temperature shows a relationship that is straight enough, use of the linear model is appropriate. The linear regression model that predicts mean temperature from CO_2 level is:
$$\widehat{MeanTemp} = 11.0276 + 0.0089CO_2$$

d) The model predicts that an increase in CO_2 level of 1 ppm is associated with an increase of 0.0089 °C in mean temperature.

e) According to the model, the mean temperature is predicted to be 11.0276 °C when there is no CO_2 in the atmosphere. This is an extrapolation outside of the range of data, and isn't very meaningful in context, since there is always CO_2 in the atmosphere. We want to use this model to study the change in CO_2 level and how it relates to the change in temperature.

f) The residuals plot shows no apparent patterns. The linear model appears to be an appropriate one.

g) $\widehat{MeanTemp} = 11.0276 + 0.0089CO_2 = 11.0276 + 0.089(400) = 14.59.$ According to the model, the temperature is predicted to be 14.59°C when the CO_2 level is 400 ppm.

43. Body fat

a) The scatterplot of % body fat and weight of 20 male subjects, at the right, shows a strong, positive, linear association. Generally, as a subject's weight increases, so does % body fat. The association is straight enough to justify the use of the linear model.

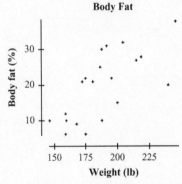

Weight and Body Fat

The linear model that predicts % body fat from weight is:

$$\widehat{\%Fat} = -27.3763 + 0.249874(Weight)$$

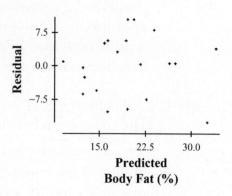

b) The residuals plot, at the right, shows no apparent pattern. The linear model is appropriate.

c) According to the model, for each additional pound of weight, body fat is expected to increase by about 0.25%.

d) Only 48.5% of the variability in % body fat can be accounted for by the model. The model is not expected to make predictions that are accurate.

e)

$$\widehat{\%Fat} = -27.3763 + 0.249874(Weight)$$

$$\widehat{\%Fat} = -27.3763 + 0.249874(190)$$

$$\widehat{\%Fat} = 20.09976$$

According to the model, the predicted body fat for a 190-pound man is 20.09976%. The residual is $21 - 20.09976 \approx 0.9\%$.

45. Heptathlon 2012

a) Both high jump height and 800 metre time are quantitative variables, the association is straight enough to use linear regression. We exclude the high leverage point Chantae McMillan.

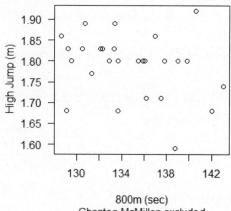

800m (sec)
Chantae McMillan excluded

```
Coefficients:
             Estimate Std. Error  t value Pr(>|t|)
(Intercept)  2.420905   0.463133    5.227  1.85e-05 ***
X800m       -0.004690   0.003437   -1.364    0.184
---
Signif. codes:  0 '***' 0.001 '**' 0.01 '*' 0.05 '.' 0.1 ' ' 1

Residual standard error: 0.07288 on 26 degrees of freedom
Multiple R-squared: 0.06682,    Adjusted R-squared: 0.03092
F-statistic: 1.862 on 1 and 26 DF,  p-value: 0.1841
```

The regression equation to predict high jump from 800 m results is:

$$\widehat{HighJump} = 2.42 - 0.004690(800 \text{ } m \text{ } time)$$

According to the model, the predicted high jump decreases by an average of 0.004690 metres for each additional second in 800 metre time.

b) $R^2 = 0.06682$. This means that 6.682% of the variability in high jump height is accounted for by the variability in 800 metre time.

c) Yes, good high jumpers tend to be fast runners. The slope of the association is negative. Faster runners tend to jump higher, as well. However, this is a very weak relationship, as evident in the graph and the small value of R^2.

d) The residuals plot is fairly patternless. The scatterplot shows a slight tendency for less variation in high jump height among the slower runners than the faster runners. Overall, the linear model is appropriate.

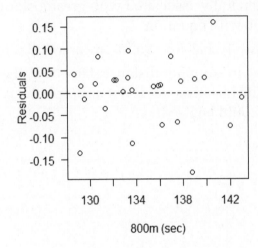

e) The linear model is not particularly useful for predicting high jump performance. First of all, 6.7% of the variability in high jump height is accounted for by the variability in 800 metre time, leaving 93.3% of the variability unaccounted for. Secondly, the residual standard deviation is 0.07288 metres, which is not much smaller than the standard deviation of all high jumps, 0.074 metres. Predictions are not likely to be accurate.

47. Hard water

a) There is a fairly strong, negative, linear relationship between calcium concentration (in ppm) in the water and mortality rate (in deaths per 100 000). Towns with higher calcium concentrations tended to have lower mortality rates.

b) The linear regression model that predicts mortality rate from calcium concentration is *Mortality* = 1676 - 3.23(*Calcium*).

c) The model predicts a decrease of 3.23 deaths per 100 000 for each additional ppm of calcium in the water. For towns with no calcium in the water, the model predicts a mortality rate of 1676 deaths per 100 000 people.

d) Exeter had 2348.6 more deaths per 100 000 people than the model predicts.

e)

$$\widehat{Mortality} = 1676 - 3.23(Calcium)$$

$$\widehat{Mortality} = 1676 - 3.23(100)$$

$$\widehat{Mortality} = 1353$$

The town of Derby is predicted to have a mortality rate of 1353 deaths per 100 000 people.

f)   43% of the variability in mortality rate can be explained by variability in calcium concentration.

49. Brakes

a)   The association between speed and stopping distance is strong, positive, and appears straight. Higher speeds are generally associated with greater stopping distances. The linear regression model, with equation

$\overline{Stopping\ Distance} = -20.667 + 1.151(Speed)$, has $R^2 = 96.7\%$, meaning that the model explains 96.7% of the variability in stopping distance. However, the residuals plot has a curved pattern. The linear model is not appropriate. A model using re-expressed variables should be used.

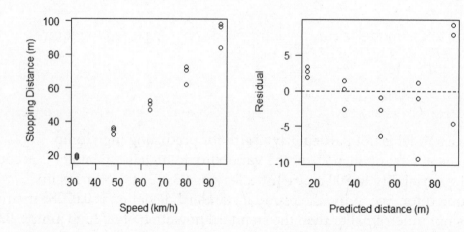

b)   Stopping distances appear to be relatively higher for higher speeds. This increase in the rate of change might be able to be straightened by taking the square root of the response variable, stopping distance. The scatterplot of $\sqrt{Distance}$ versus Speed seems like it might be a bit straighter.

c)   The model for the re-expressed data is

$\overline{\sqrt{Stopping\ distance}} = 1.777 + 0.08226(Speed)$. The residuals plot is better than the linear fit before re-expressing the data, and shows no pattern. $R^2 = 98.4\%$, so 98.4% of the variability in the square root of the stopping distance can be explained by the model.

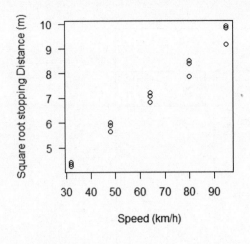 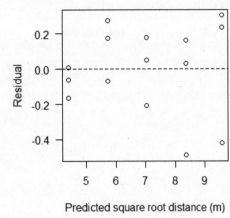

d) $\overline{\sqrt{Stopping\,distance}} = 1.777 + 0.08226(Speed)$

$\overline{\sqrt{Stopping\,distance}} = 1.777 + 0.08226(90)$

$\overline{\sqrt{Stopping\,distance}} = 9.1804$

$\overline{\sqrt{Stopping\,distance}} = 84.28\,m$

According to the model, a car travelling 90 km/h is expected to require approximately 84.3 metres to come to a stop.

51. Immigrant commuters

a) There is a strong positive linear relationship between the immigrant percentage and the Canadian born percentage. There are no outliers.

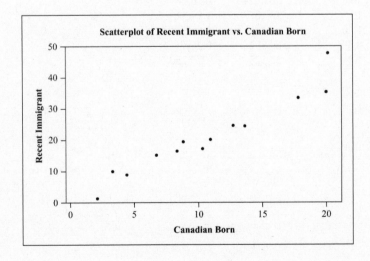

b) The fitted line is shown below.

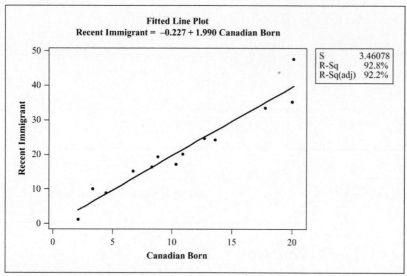

Fitted Line Plot
Recent Immigrant = −0.227 + 1.990 Canadian Born

S	3.46078
R-Sq	92.8%
R-Sq(adj)	92.2%

c)   1.99%. (This is the slope.)

d)   Montreal has a large residual (shown in the table and scatterplot below) compared to other observations. The predicted value for Montreal is 39.8, but the actual value is 47.8.

CMA	Canadian Born	Recent Immigrant	Fits	Residuals
Montréal	20.1	47.8	39.7781	8.02194
Toronto	20.0	35.4	39.5790	-4.17903
Ottawa–Hull	17.8	33.5	35.2004	-1.70035
Calgary	12.7	24.6	25.0498	-0.44977
Winnipeg	13.6	24.3	26.8411	-2.54105
Vancouver	10.9	20.1	21.4672	-1.36722
Edmonton	8.8	19.3	17.2876	2.01243
Victoria	10.3	17.2	20.2730	-3.07303
Hamilton	8.3	16.4	16.2924	0.10759
London	6.7	15.1	13.1079	1.99208
Windsor	3.3	10.0	6.3409	3.65913
Kitchener	4.4	8.8	8.5302	0.26979
Abbotsford	2.1	1.2	3.9525	-2.75250

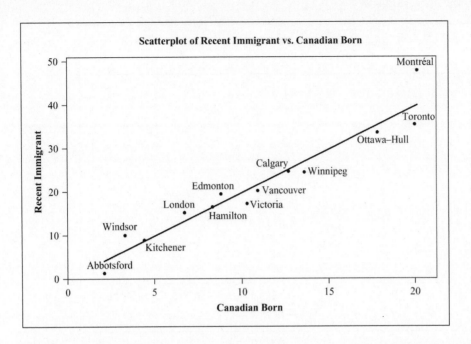

e) The plot of residuals vs. the Canadian born % is given below. It doesn't look quite random. For example, most of the points for Canadian born are negative. The model overestimates the immigrant percentage for cities with a Canadian born % of 10% or more. The plot of residuals vs. Canadian born % without Montreal is also given below. There is no clear pattern now.

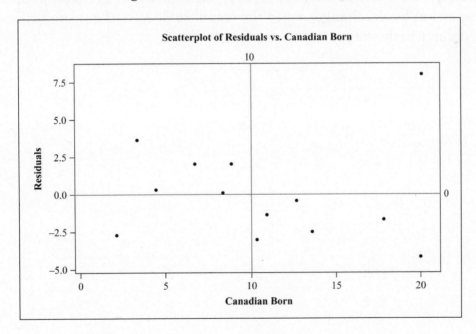

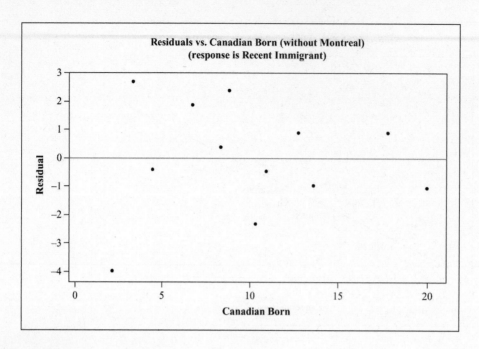

53. Regressing

a) 70 + .6(85 − 70) = 79

b) 70 + .6(79 − 70) = 75.4

c) For a test score of 85, you predict an exam score of 79, but if the exam score is 79, you predict a test score of 75.4, not 85. Predictions are not reversible in direction. The regression model changes when the response does, and our prediction is always nearer to the mean.

Chapter 8: Regression Wisdom

1. **Marriage age 2010**
   a) The trend in age at first marriage for American women is very strong over the entire time period recorded on the graph, but the direction and form are different for different time periods. The trend appears to be somewhat linear, and consistent at around 22 years, up until about 1940, when the age seemed to drop dramatically, to under 21. From 1940 to about 1970, the trend appears non-linear and slightly positive. From 1975 to the present, the trend again appears linear and positive. The marriage age rose rapidly during this time period.
   b) The association between age at first marriage for American women and year is strong over the entire time period recorded on the graph, but some time periods have stronger trends than others.
   c) The correlation, or the measure of the degree of linear association, is not high for this trend. The graph, as a whole, is non-linear. However, certain time periods, like 1975 to present, have a high correlation.
   d) Overall, the linear model is not appropriate. The scatterplot is not Straight Enough to satisfy the condition. You could fit a linear model to the time period from 1975 to 1995, but this seems unnecessary. The ages for each year are reported, and, given the fluctuations in the past, extrapolation seems risky.

3. **Gas mileage**
   a) The association between weight and gas mileage of cars is fairly linear, strong, and negative. Heavier cars tend to have lower gas mileage.
   b) For each additional thousand pounds of weight, the linear model predicts a decrease of 7.652 miles per gallon in gas mileage.
   c) The linear model is not appropriate. There is a curved pattern in the residuals plot. The model tends to underestimate gas mileage for cars with relatively low and high gas mileages, and overestimates the gas mileage of cars with average gas mileage.

5. **Good model?**
   a) The student's reasoning is not correct. A scattered residuals plot, not high R^2, is the indicator of an appropriate model. Once the model is deemed appropriate, R^2 is used as a measure of the strength of the model.
   b) The model may not allow the student to make accurate predictions. The data may be curved, in which case the linear model would not fit well.

7. **Movie dramas**
   a) The units for the slopes of these lines are millions of dollars per minute of running time.
   b) The slopes of the regression lines are the same. Dramas and movies from other genres have costs for longer movies that increase at the same rate.

c) The regression line for dramas has a lower *y*-intercept. Regardless of running time, dramas cost about 20 million dollars less than other genres of movies of the same running time.

9. Oakland passengers

a) According to the linear model, the use of the Oakland airport has been increasing by about 59 700 passengers per year, starting at about 282 000 passengers in 1990.

b) About 71% of the variability in the number of passengers can be accounted for by the model.

c) Errors in prediction based on the model have a standard deviation of 104 330 passengers.

d) No, the model would not be useful in predicting the number of passengers in 2010. This year would be an extrapolation too far from the years we have observed.

e) The negative residual is September of 2001. Air traffic was artificially low following the attacks on 9/11/2001.

11. Unusual points

a) 1) The point has high leverage and a small residual.

2) The point is not influential. It has the *potential* to be influential, because its position far from the mean of the explanatory variable gives it high leverage. However, the point is not *exerting* much influence, because it reinforces the association.

3) If the point were removed, the correlation would become weaker. The point heavily reinforces the positive association. Removing it would weaken the association.

4) The slope would remain roughly the same, since the point is not influential.

b) 1) The point has high leverage and probably has a small residual.

2) The point is influential. The point alone gives the scatterplot the appearance of an overall negative direction, when the points are actually fairly scattered.

3) If the point were removed, the correlation would become weaker. Without the point, there would be very little evidence of linear association.

4) The slope would increase, from a negative slope to a slope near 0. Without the point, the slope of the regression line would be nearly flat.

c) 1) The point has moderate leverage and a large residual.

2) The point is somewhat influential. It is well away from the mean of the explanatory variable, and has enough leverage to change the slope of the regression line, but only slightly.

3) If the point were removed, the correlation would become stronger. Without the point, the positive association would be reinforced.

4) The slope would increase slightly, becoming steeper after the removal of the point. The regression line would follow the general cloud of points more closely.

d) 1) The point has little leverage and a large residual.

2) The point is not influential. It is very close to the mean of the explanatory variable, and the regression line is anchored at the point $(\bar{x}, \bar{y})$, and would only pivot if it were possible to minimize the sum of the squared residuals. No amount of pivoting will reduce the residual for the stray point, so the slope would not change.

3) If the point were removed, the correlation would become slightly stronger, decreasing to become more negative. The point detracts from the overall pattern, and its removal would reinforce the association.

4) The slope would remain roughly the same. Since the point is not influential, its removal would not affect the slope.

13. The extra point
1) Point e is very influential. Its addition will give the appearance of a strong, negative correlation like $r = -0.90$.
2) Point d is influential (but not as influential as point e). Its addition will give the appearance of a weaker, negative correlation like $r = -0.40$.
3) Point c is directly below the middle of the group of points. Its position is directly below the mean of the explanatory variable. It has no influence. Its addition will leave the correlation the same, $r = 0.00$.
4) Point b is almost in the center of the group of points, but not quite. Its addition will give the appearance of a very slight positive correlation, like $r = 0.05$.
5) Point a is very influential. Its addition will give the appearance of a strong, positive correlation, like $r = 0.75$.

15. What's the cause?
1) High blood pressure may cause high body fat.
2) High body fat may cause high blood pressure.
3) Both high blood pressure and high body fat may be caused by a lurking variable, such as a genetic or lifestyle trait.

17. Reading
a) The principal's description of a strong, positive trend is misleading. First of all, "trend" implies a change over time. These data were gathered during one year, at different grade levels. To observe a trend, one class's reading scores would have to be followed through several years. Second, the strong, positive relationship only indicates the yearly improvement that would be expected, as children get older. For example, the Grade 4 students are reading at approximately a Grade 4 level, on average. This means that the school's students are progressing adequately in their reading, not extraordinarily.

Finally, the use of average reading scores instead of individual scores increases the strength of the association.

b) The plot appears very straight. The correlation between grade and reading level is very high, probably between 0.9 and 1.0.

c) If the principal had made a scatterplot of all students' scores, the correlation would have likely been lower. Averaging reduced the scatter, since each grade level has only one point instead of many, which inflates the correlation.

d) If a student is reading at grade level, then that student's reading score should equal his or her grade level. The slope of that relationship is about 1. That would be "acceptable," according to the measurement scale of reading level. Any slope greater than 1 would indicate above grade level reading scores, which would certainly be acceptable as well. A slope less than 1 would indicate below grade level average scores, which would be unacceptable.

19. Heating
a) The model predicts a decrease in $2.13 in heating cost for an increase in temperature of 1°F. Generally, warmer months are associated with lower heating costs.

b) When the temperature is 0°F, the model predicts a monthly heating cost of $133.

c) When the temperature is around 32°F, the predictions are generally too high. The residuals are negative, indicating that the actual values are lower than the predicted values.

d)

$$\hat{C} = 133 - 2.12(Temp)$$

$$\hat{C} = 133 - 2.13(10)$$

$$\hat{C} = \$111.70$$

According to the model, the heating cost in a month with average daily temperature 10°F is expected to be $111.70.

e) The residual for a 10° day is approximately –$6, meaning that the actual cost was $6 less than predicted, or $111.70 – $6 = $105.70.

f) The model is not appropriate. The residuals plot shows a definite curved pattern. The association between monthly heating cost and average daily temperature is not linear.

g) A change of scale from Fahrenheit to Celsius would not affect the relationship. Associations between quantitative variables are the same, no matter what the units.

21. Interest rates
a) $r = \sqrt{R^2} = \sqrt{0.774} = 0.88$. The correlation between rate and year is +0.88 since the scatterplot shows a positive association.

b) According to the model, interest rates during this period increased at about 0.25% per year, starting from an interest rate of about 0.64% in 1950.

c) The linear regression equation predicting interest rate from year is

$$\widehat{Rate} = 0.640282 + 0.247637(Year - 1950)$$

$$\widehat{Rate} = 0.640282 + 0.247637(50)$$

$$\widehat{Rate} = 13.022$$

According to the model, the interest rate is predicted to be about 13% in the year 2000.

d) This prediction is not likely to have been a good one. Extrapolating 20 years beyond the final year in the data would be risky, and unlikely to be accurate.

23. Interest rates revisited

a) The values of R^2 are approximately the same, so the models fit comparably well, but they have very different slopes.

b) The model that predicts the interest rate on 3-month Treasury bills from the number of years since 1950 is $\widehat{Rate} = 21.0688 - 0356578(Year - 1950)$. This model predicts the interest rate to be 3.24%, a rate much lower than the prediction from the other model.

c) We can trust the newer prediction, since it is in the middle of the data used to generate the model. Additionally, the model accounts for 74.5% of the variability in interest rate.

d) Since 2020 is at least 15 years after the last year included in the newer model, it would be extremely risky to use this, or any, model to make a prediction that far into the future.

25. Gestation

a) The association would be stronger if humans were removed. The point on the scatterplot representing human gestation and life expectancy is an outlier from the overall pattern and detracts from the association. Humans also represent an influential point. Removing the humans would cause the slope of the linear regression model to increase, following the pattern of the nonhuman animals much more closely.

b) The study could be restricted to nonhuman animals. This appears justifiable, since one could point to a number of environmental factors that could influence human life expectancy and gestation period, making them incomparable to those of animals.

c) The correlation is moderately strong. The model explains 72.2% of the variability in gestation period of nonhuman animals.

d) For every year increase in life expectancy, the model predicts an increase of approximately 15.5 days in gestation period.

e) $\widehat{Gest} = -39.5172 + 15.4980(LifEx)$

$\widehat{Gest} = -39.5172 + 15.4980\mathbf{(20)}$

$\widehat{Gest} \approx 270.4428$

According to the linear model, monkeys with a life expectancy of 20 years are expected to have gestation periods of about 270.4 days. Care should be taken when assessing the accuracy of this prediction. First of all, the residuals plot has not been examined, so the appropriateness of the model is questionable. Second, it is unknown whether or not monkeys were included in the original 17 nonhuman species studied. Since monkeys and humans are both primates, the monkeys may depart from the overall pattern as well.

27. Elephants and hippos

a) Hippos are more of a departure from the pattern. Removing that point would make the association appear to be stronger.

b) The slope of the regression line would increase, pivoting away from the hippos point.

c) Anytime data points are removed, there must be a justifiable reason for doing so, and saying, "I removed the point because the correlation was higher without it" is not a justifiable reason.

d) Elephants are an influential point. With the elephants included, the slope of the linear model is 15.4980 days gestation per year of life expectancy. When they are removed, the slope is <u>11.6</u> days per year. The decrease is significant.

29. Marriage age 2010 revisited

a) Modelling decisions may vary, but the important idea is using a subset of the data that allows us to make an accurate prediction for the year in which we are interested. We might model a subset to predict the marriage age in 2015, and model another subset to predict the marriage age in 1911. To predict the average marriage age of American women in 2015, use the data points from the most recent trend only. The data points from 1955–2010 look straight enough to apply the linear regression model. Even though there is still curvature, it is nowhere near as bad as the curvature when all the data points are used. Regression output from a computer program is given below, as well as a residuals plot.

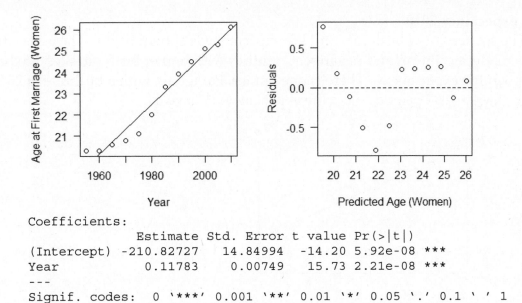

```
Coefficients:
              Estimate Std. Error t value Pr(>|t|)
(Intercept) -210.82727   14.84994  -14.20 5.92e-08 ***
Year           0.11783    0.00749   15.73 2.21e-08 ***
---
Signif. codes:  0 '***' 0.001 '**' 0.01 '*' 0.05 '.' 0.1 ' ' 1

Residual standard error: 0.4479 on 10 degrees of freedom
Multiple R-squared: 0.9612,    Adjusted R-squared: 0.9573
F-statistic: 247.5 on 1 and 10 DF,  p-value: 2.211e-08
```

The linear model used to predict average female marriage age from year is:
$\widehat{Age} = -210.827 + 0.127832(Year)$. The residuals plot shows a pattern, but the residuals are small, and the value of R^2 is high. 96.1% of the variability in average female age at first marriage is accounted for by variability in the year. The model predicts that each year that passes is associated with an increase of 0.118 years in the average female age at first marriage.

$\widehat{Age} = -210.827 + 0.127832(Year)$

$\widehat{Age} = -210.827 + 0.127832(2020)$

$\widehat{Age} = 27.19$

According to the model, the average age at first marriage for women in 2020 will be 27.19 years old. Care should be taken with this prediction, however. It represents an extrapolation of 10 years beyond the highest year, and the residuals plot shows a pattern.

b) This prediction is for a year that is 10 years higher than the highest year for which we have an average female marriage age. Don't place too much faith in this extrapolation.

c) An extrapolation of more than 50 years into the future would be absurd. There is no reason to believe the trend would continue. In fact, given the situation, it is very unlikely that the pattern would continue in this fashion. The model given in part a) predicts that the average marriage age will be 32.5 years in 2065. Realistically, that seems quite high.

31. Life expectancy 2010
   a) The scatterplot of birth rate and life expectancy is shown below. The association is moderate, linear, and negative. Countries with higher birth rates tend to have lower life expectancies. There is one outlier, Paraguay, with a birthrate of 28 births per 1000 people, and a life expectancy of 76 years.

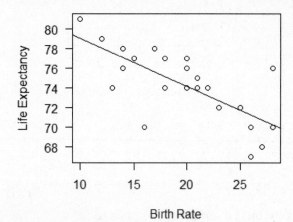

   b) The computer regression output is shown below:
```
Coefficients:
              Estimate Std. Error t value Pr(>|t|)
(Intercept) 83.95663    1.95686   42.904  < 2e-16 ***
birth.rate  -0.48704    0.09716   -5.013 5.11e-05 ***
---
Signif. codes:  0 '***' 0.001 '**' 0.01 '*' 0.05 '.' 0.1 ' ' 1

Residual standard error: 2.548 on 22 degrees of freedom
Multiple R-squared: 0.5332,      Adjusted R-squared: 0.512
F-statistic: 25.13 on 1 and 22 DF,p-value: 5.107e-05
```
   The linear regression equation that predicts life expectancy from birth rate is:

   $$\widehat{LifeExpect} = 83.9566 - 0.487037\,Birthrate$$

   c) $R^2 = 53.3\%$, so $r = \sqrt{R^2} = \sqrt{0.533} = -0.730$. 53.3% of the variability in life expectancy is explained by variability in the birthrate.
   d) The residuals plot shown below is reasonably scattered. Paraguay has a large residual. Its higher than average life expectancy continues to stand out.

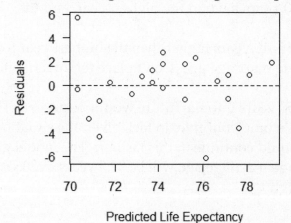

e) The linear model is appropriate.

f) The data point for Paraguay is not extraordinarily unusual. You may have chosen not to set it aside. If you did set it aside, the recomputed regression is $\overline{Life\ Expect} = 85.2955 - 0.57031 Birthrate$. The computer regression output is:

```
Coefficients:
            Estimate Std. Error t value Pr(>|t|)
(Intercept) 85.29547    1.78420  47.806  < 2e-16 ***
birth.rate  -0.57031    0.09044  -6.306 2.97e-06 ***
---
Signif. codes:  0 '***' 0.001 '**' 0.01 '*' 0.05 '.' 0.1 ' ' 1

Residual standard error: 2.235 on 21 degrees of freedom
Multiple R-squared: 0.6544,    Adjusted R-squared: 0.638
F-statistic: 39.77 on 1 and 21 DF,  p-value: 2.972e-06
```

$R^2 = 65.4\%$, so this new model accounts for 65.4% of the variability in life expectancy. The relationship is a little stronger, but that will always happen when setting outlying points aside from the regression. Whether or not to set Paraguay aside is a judgment call.

g) The government leaders should not suggest that women have fewer children in order to raise the life expectancy. Although there is evidence of an association between the birth rate and life expectancy, this does not mean that one causes the other. There may be lurking variables involved, such as economic conditions, social factors, or level of health care.

33. Rich getting richer 2011

a) The scatterplot of the percent earning over $150 000 versus year is given below. The percent earning over $150 000 increases over time, but not quite linearly. In the period 1986–1992, the relationship has a different form than that in the period 1993–2011. The relationship in the period 1993–2011 is strong and roughly linear.

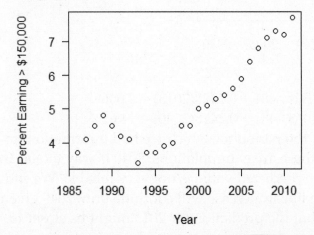

b) The regression equation is: $\overline{Percent > \$150000} = -288 + 0.14667 Year$.
76.03 percent of the variation in the percentage earning over $150 000 is explained by the straight line relationship (this is R^2).

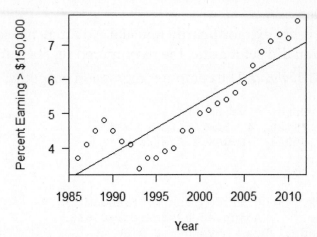

c) The predicted percent for 1997 is–288 + 0.14667(1997) = 4.87%. The actual value (in the data set) is 4.0 percent. The predicted value is a bit too high—the residual is –0.87%.

d) Dropping the first seven years (from 1986 to 1992), we obtain the following model: $\overline{Percent > \$150000} = -481 + 0.2428Year$. 98.4% of the variation in the percentage earning over $150 000 is explained by the linear relationship (this is R^2). This model fits the later data better, and is more appropriate because the trend after 1992 appears to be close to linear.

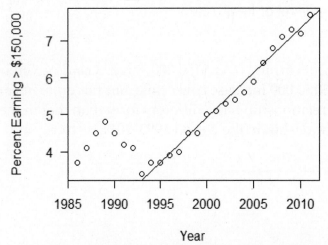

The predicted percent for 2013 is –481 + 0.2428(2013) = 8.066%
The predicted percent for 2040 is –481 + 0.2428(2040) = 14.62%
Both 2013 and 2040 are out of the period for which we had information (to estimate the model), so both these are extrapolations. 2040 is way too far from this period (1986–2001), so the predicted value is not very reliable. We should not be very confident that the linear increase will continue until 2040 given the earlier data from 1986–1992, but the prediction for 2013 might be closer to accurate.

35. Gender gap 2011

a) The scatterplot of the difference in median earnings versus year is shown below. The difference decreases with year, but not quite linearly. The relationship is strong and there are no outliers.

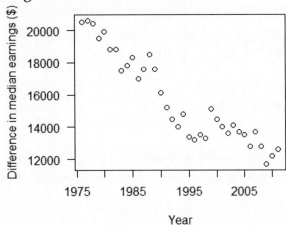

b) The linear regression model (the straight line fit) is:
Difference in Median earnings = 497 806 – 241.84(Year)
The R^2 for the model is 87.9%, i.e., 87.9% of the variation in the differences in medians is explained by this model. The plot of residuals versus year is shown below. The curving pattern on this plot indicates that a higher order model (e.g., a quadratic model) or some other curving pattern is needed to describe the relationship.

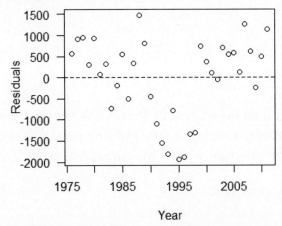

c) The predicted difference for 2070 (assuming that the form of the relationship will remain the same after 2011, and that the linear model is appropriate) is 497 806 – 241.84 × 2070 = –2804. The difference is likely to be higher than this as we can see that the decrease in the difference reduces with time.

d) It looks like a quadratic curve fits the data better.

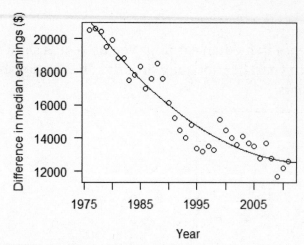

The plot of residuals for the quadratic model is much more random (free of patterns) than that for the linear model and so this is a more appropriate model than the linear model.

37. **Math and gender 2009**
    a) The plot of male mean score versus female mean score with the regression is shown below.

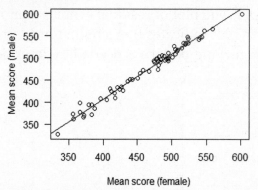

b) The plot of residuals versus the mean scores for females is shown below. The plot looks random (free of patterns) and so a straight line model seems appropriate. There is one large residual whose magnitude is greater than 20 points (Colombia).

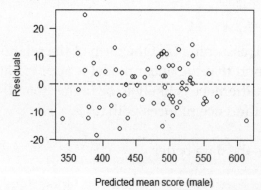

c) The histogram and the normal probability plot of residuals are shown below. The histogram looks roughly bell shaped and the normal probability plot is close to a straight, both suggesting that the residuals are approximately normally distributed. Colombia is noticeable in the right tail of the distribution.

Normal Q-Q Plot

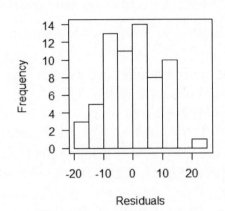 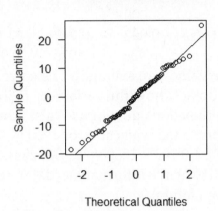

d) The plot of residuals versus row number is shown below. It seems that the OECD countries are less variable than the non-OECD countries.

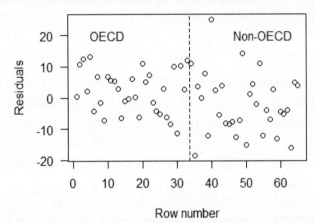

e) Shanghai-China and Kyrgyzstan have the greatest leverage on the fit. Their x-values (i.e., the means for females) are far from the average of all x-values. The average female score in Shanghai-China is 601 and the average female score in Kyrgyzstan is 334.

f) Slope close to 1 implies that when a country's average for females increases by a certain value, the country average for males also increases by approximately the same value.

When the slope is 1, the estimate of the Y-intercept is given by $\overline{Y} - \overline{X}$ = Average for males – Average for females = 471.98 – 463.18 = 8.80, so the estimated prediction equation is: Mean score (Male) = 8.80 + Mean score (Female).

g) The correlation will decrease, as we can see from the scatter plot that the points with both male and female averages greater than 500 look less linear. The other points following the same pattern on the scatterplot make it look more linear and so add strength to the linear relationship.

39. Bigger and bigger

a) All possible correlations are given below:

Correlations: Year, Bust size (in), Waist size (in), Hips size (in)

	Year	Bust size	Waist size
Bust size	1.000		
Waist size	0.992	0.992	
Hips size	0.936	0.936	0.929

b) These values are averages. The correlations with averages are usually high since they do not show the variability in individual observations.

c) The prediction equation is Waist size (in) = –182 + 0.105(Year).
The predicted waist size for 1950 is –182 + 0.105 × 1950 = 22.75 and that for 2080 is –182 + 0.105 × 2080 = 36.40 inches. The year 2080 is way outside the period for which we have information (data), so the prediction for 2080 is not very reliable (extrapolation).

d) The prediction equation is Waist size (in) = –47.6 + 2.10(Bust size (in))
The predicted waist size for UK women if their average bust size gets up to 44 is –47.6 + 2.10 × 44 = 44.8. The average bust size is outside the range of data on bust sizes, so this prediction is not very reliable.

41. Census at school

a) *Armspan* has the highest correlation (0.565) with *middlefinger* and so that is the best predictor of *middlefinger*. 32% (0.565^2) of the variation in *middlefinger* measurements can be explained by a linear association with *armspan*.

b) *Height* has the highest correlation (0.781) with *armspan* and so that is the best predictor of *armspan*. 61% (0.781^2) of the variation in *armspan* measurements can be explained by a linear association with *height*.

c) The prediction equation is *armspan* = 16.8 + 0.897 height. The slope of this regression equation is 0.987. That means for every unit in increase in height, the mean armspan increases by 0.987 units.

d) The histogram is close to a bell-shape and the normal scores plot is close to straight, both indicating that there are no serious violations of the assumption of normality of residuals.

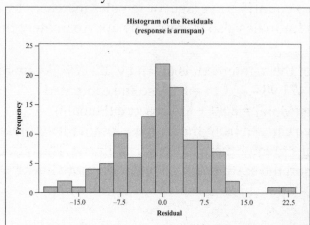

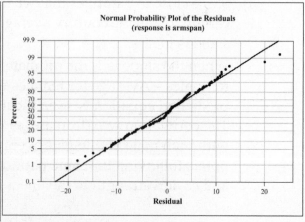

e) The plot of residuals against height and the plot of residuals against predicted values are shown below. They look pretty random, indicating a straight-line model should be adequate.

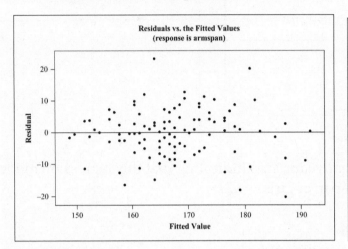

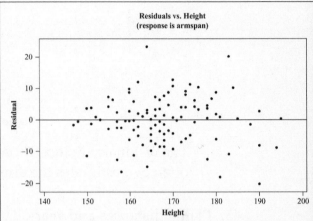

f) **Regression Analysis: armspan versus height**
The regression equation is
armspan = 16.8 + 0.897 height

Predictor	Coef	SE Coef	T	P
Constant	16.83	11.56	1.46	0.148
Height	0.89701	0.06866	13.06	0.000

S = 7.39858   R-Sq = 61.0%   R-Sq(adj) = 60.7%

Analysis of Variance

Source	DF	SS	MS	F	P
Regression	1	9342.0	9342.0	170.66	0.000
Residual Error	109	5966.6	54.7		
Total	110	15308.6			

Unusual Observations

Obs	height	armspan	Fit	SE Fit	Residual	St Resid
55	164	149.000	163.943	0.756	–14.943	–2.03R
58	181	161.000	179.192	1.131	–18.192	–2.49R
66	190	167.000	187.265	1.661	–20.265	–2.81R
69	183	201.000	180.986	1.242	20.014	2.74R
75	164	187.000	163.943	0.756	23.057	3.13R
89	194	182.000	190.853	1.913	–8.853	–1.24 X
92	195	192.000	191.750	1.977	0.250	0.04 X
98	158	142.000	158.561	0.986	–16.561	–2.26R

R denotes an observation with a large standardized residual.
X denotes an observation whose X value gives it large influence.

The individual with height 164 and armspan 187 (i.e., observation 75) has the most unusual armspan (3.13 units longer than the average armspan for individuals with height = 164).

g) The plot of residuals versus gender (with genderCode=0 for girls and 1 for boys) is shown below.

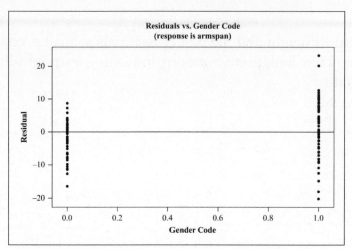

The residuals are mostly negative for girls indicating that this regression model often overestimates the armspan of girls.

43. Planet distances and years 2006

a) The association between distance from the sun and planet year is strong, positive, and curved concave upward. Generally, planets farther from the sun have longer years than closer planets.

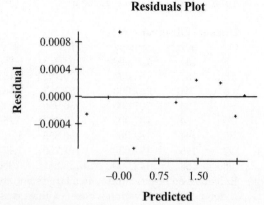

b) The rate of change in length of year per unit distance appears to be increasing, but not exponentially. Re-expressing with the logarithm of each variable may straighten a plot such as this. The scatterplot and residuals plot for the linear model relating log(Distance) and log(Length of Year) are shown below.

The regression model for the log-log re-expression is:

$$\widehat{\log(Length)} = -2.95 + 1.5\big(\log(Distance)\big).$$

c) $R^2 = 100\%$, so the model explains 100% of the variability in the log of the length of the planetary year, at least according to the accuracy of the statistical software. The residuals plot is scattered, and the residuals are all extremely small. This is a very accurate model.

45. Placental mammals

a) The plot of the neonatal brain weight versus gestation with the regression line is shown below. The value of R^2 for this straight line fit is 60.6%. That is, 60.6% of the variation in the neonatal brain weight is explained by the linear regression model. The plot of residuals versus predicted values is also shown below. It has a curving pattern, indicating that a straight line model is not appropriate. This can also be seen on the scatterplot of the neonatal brain weight versus gestation, but the curvature is more prominent in the plot of residuals versus predicted values.

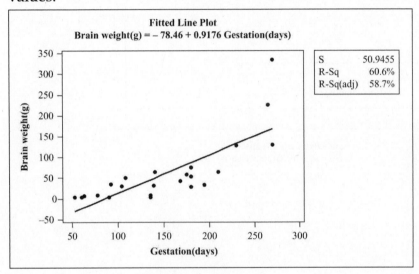

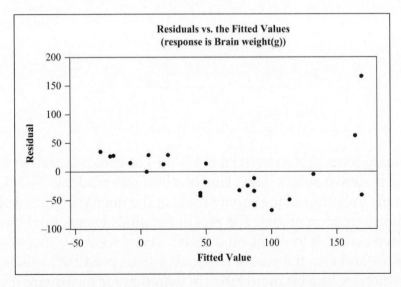

b) The plot of the logarithms of the neonatal brain weight versus gestation with the regression line is shown below. The value of R^2 for this model is 71.6%. That is, 71.6% of the variation in the logarithms of the neonatal brain weight is explained by this regression model. The plot of residuals versus predicted values is also shown below. It looks random (free of patterns), indicating that this model is appropriate for describing the relationship between the two variables.

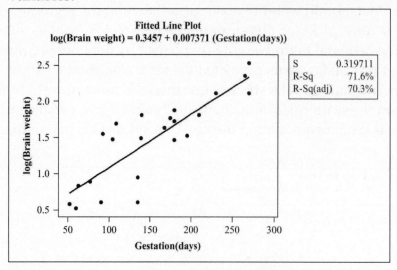

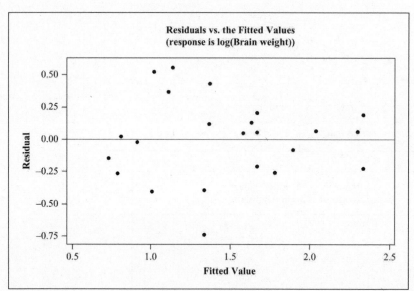

c) The plot of the square roots of the neonatal brain weight versus gestation with the regression line is shown below. The value of R^2 for this model is 74.09%. That is, 74.09% of the variation in the square roots of the neonatal brain weight is explained by this regression model. The plot of residuals versus predicted values is also shown below. It looks like the variability of residuals increases with the fitted value (and also the plot of residuals versus predicted values for the log transformation looks a bit more random than that for the square root model), and so the log transformation is a bit better (even though the R^2 is a bit higher for the square root transformation).

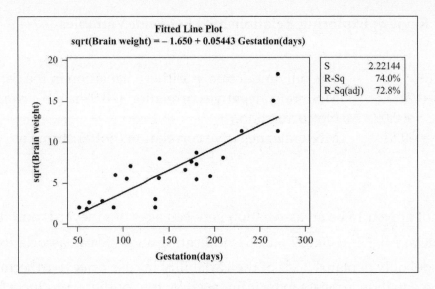

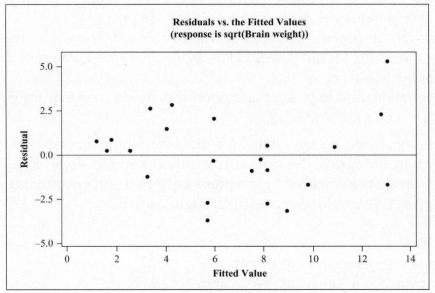

d) The predicted log brain weight for a 200-day gestation period is 0.346 + 0.00737 × 200 = 1.82, so the predicted brain weight is $10^{1.82}$ = 66.1g.

e) The predicted square root brain weight for a 200-day gestation period is –1.65 + 0.0544 × 200 = 9.23, so the predicted brain weight is 9.23^2 = 85.19g.

Part II Review: Exploring Relationships Between Variables

1. **College**

    % over 50: $r = 0.69$ The only moderate, positive correlation in the list.

    % under 20: $r = -0.71$ Moderate, negative correlation (-0.98 is too strong).

    % Full-time Fac.: $r = 0.09$ No correlation.

    % Gr. on time: $r = -0.51$ Moderate, negative correlation (not as strong as %under 20).

3. **Vineyards**

    a) There does not appear to be an association between ages of vineyards and the cost of products. $r = \sqrt{R^2} = \sqrt{0.027} = 0.164$, indicating a very weak association, at best. The model only explains 2.7% of the variability in case price. Furthermore, the regression equation appears to be influenced by two outliers, products from vineyards over 30 years old, with relatively high case prices.

    b) This analysis tells us nothing about vineyards worldwide. There is no reason to believe that the results for the Finger Lakes region are representative of the vineyards of the world.

    c) The linear equation used to predict case price from age of the vineyard is:
    $$\widehat{CasePrice} = 92.765 + 0.567284(\text{Years})$$

    d) This model is not useful because only 2.7% of the variability in case price is accounted for by the ages of the vineyards. Furthermore, the slope of the regression line seems influenced by the presence of two outliers, products from vineyards over 30 years old, with relatively high case prices.

5. **Lurking variables**

    a) (i) The correlation is negative, approximately $r = -0.7$.

    (ii) The correlation is negative, approximately $r = -0.85$.

    (iii) The correlation is essentially zero.

    (iv) The correlation is negative, approximately $r = -0.75$.

    b) (i) The correlation is positive, approximately $r = -1$.

    (ii) The correlation is essentially zero.

    (iii) The correlation is positive, approximately $r = -0.8$.

    (iv) The correlation is essentially zero.

7. **Acid rain**

    a) $r = \sqrt{R^2} = \sqrt{0.27} = 0.5196$. The association between pH and BCI appears negative in the scatterplot, so use the negative value of the square root.

    b) The association between pH and BCI is negative, moderate, and linear. Generally, higher pH is associated with lower BCI. Additionally, BCI appears more variable for higher values of pH.

    c) In a stream with average pH, the BCI would be expected to be average, as well.

d) In a stream where the pH is 3 standard deviations above average, the BCI is expected to be 1.56 standard deviations below the mean level of BCI.
$(r(3) = -0.5196(3) = -1.56)$

9. **A manatee model 2010**
   a) The association between the number of powerboat registrations and the number of manatees killed is straight enough to try a linear model.
   $\widehat{Kills} = 50.016 + 0.139245 ThousandBoats$ is the best fitting model. The residuals plot is scattered, so the linear model is appropriate.

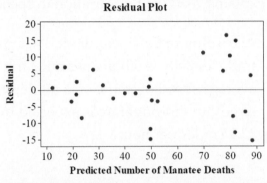

Residual Plot

   b) For every additional 10 000 powerboats registered, the model predicts that an additional 1.392 manatees will be killed.
   c) The model predicts that if no powerboats were registered, the number of manatee deaths would be approximately –50.016. This is an extrapolation beyond the scope of the data, and doesn't have much contextual meaning.
   d) $\widehat{Kills} = 50.016 + 0.139245 ThousandBoats$
   $\widehat{Kills} = -50.016 + 0.139245(914)$

   $\widehat{Kills} = 77.45$

   The model predicted 77.25 manatee kills in 2010, when the number of powerboat registrations was 914 000. The actual number of kills was 83. The model underpredicted the number of kills by 5.55.
   e) Negative residuals are better for the manatees. A negative residual suggests that the actual number of kills was below the number of kills predicted by the model.
   f) Over time, the number of powerboat registrations has increased and the number of manatees killed has increased. The trend may continue, resulting in a greater number of manatee deaths in the future. Extrapolation is risky, however. The very trend we are seeing may result in political or societal action to attempt to decrease the number of manatee deaths.

11. Traffic

a)

$$b_1 = \frac{rs_y}{s_x}$$

$$-0.352 = \frac{r(9.68)}{27.07}$$

$$r = -0.984$$

The correlation between traffic density and speed is $r = -0.984$

b) $R^2 = (-0.984)^2 = 0.969$.

The variation in the traffic density explains 96.9% of the variation in speed.

c)

$$\widehat{Speed} = 50.55 - 0.352cars$$

$$\widehat{Speed} = 50.55 - 0.352(20)$$

$$\widehat{Speed} = 32.95$$

According to the linear model, when traffic density is 50 cars per mile, the average speed of traffic on a moderately large city thoroughfare is expected to be 32.95 miles per hour.

d)

$$\widehat{Speed} = 50.55 - 0.352cars$$

$$\widehat{Speed} = 50.55 - 0.352(56)$$

$$\widehat{Speed} = 30.84$$

According to the linear model, when traffic density is 56 cars per mile, the average speed of traffic on a moderately large city thoroughfare is expected to be 30.84 miles per hour. If traffic is actually moving at 32.5 mph, the residual is $32.5 - 30.84 = 1.66$ miles per hour.

e)

$$\widehat{Speed} = 50.55 - 0.352cars$$

$$\widehat{Speed} = 50.55 - 0.352(125)$$

$$\widehat{Speed} = 6.55$$

According to the linear model, when traffic density is 125 cars per mile, the average speed of traffic on a moderately large city thoroughfare is expected to be 6.55 miles per hour. The point with traffic density 125 cars per minute and average speed 55 miles per hour is considerably higher than the model would predict. If this point were included in the analysis, the slope would increase.

f) The correlation between traffic density and average speed would become weaker. The influential point (125, 55) is a departure from the pattern established by the other data points.

g) The correlation would not change if kilometrer were used instead of miles in the calculations. Correlation is a "unitless" measure of the degree of linear association based on z-scores, and is not affected by changes in scale. The correlation would remain the same, $r = -0.984$.

13. Aboriginals rising 2011

a) A scatterplot of the 2011 versus 2001 census counts is shown below. The correlation appears to be very close to 1. In fact $r = 0.994$.

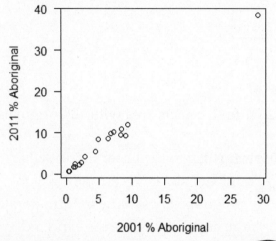

b) The fitted least squares line is $\widehat{2011Percent} = 0.19439 + 1.29105(2001Percent)$.

c) The slope tells us that for each increase of 1% in the percentage of Aboriginals in 2001 there is a 1.29% increase in the percentage of Aboriginals in 2011.

d) Prince Albert has the highest percentage of Aboriginals by a large margin. It also has the highest leverage of all census areas, and is influential to the fit. However, if you set Prince Albert aside, there appear to be two large outliers (or perhaps even a pattern in the residuals) so that the linear model no longer seems appropriate.

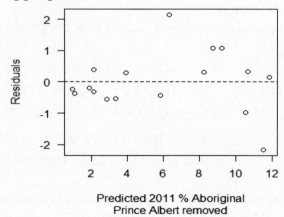

e) Taking logs seems to have resolved the issue. Prince Albert is less influential, and the residual plot appears patternless with no obvious outliers. The new model is $\log_{10}(\widehat{2011Percent}) = 0.21015 + 0.89062\log_{10}(2001Percent)$.

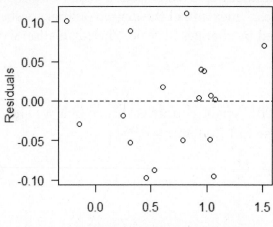

f) The predicted log₁₀ percentage in 2011 for a census area with 10% originals in 2001 is

$\log_{10}(\widehat{2011Percent}) = 0.21015 + 0.89062\log_{10}(10)$

$\log_{10}(\widehat{2011Percent}) = 0.21015 + 0.89062$

$\log_{10}(\widehat{2011Percent}) = 1.100772$

$\widehat{2011Percent}) = 10^{1.100772} = 12.61\%$

15. Cars with horses

a) The linear model that predicts the horsepower of an engine from the weight of the car is: *Horsepower* $= 3.49834 + 34.3144(Weight)$.

b) The weight is measured in thousands of pounds. The slope of the model predicts an increase of about 34.3 horsepower for each additional unit of weight. 34.3 horsepower for each additional thousand pounds makes more sense than 34.3 horsepower for each additional pound.

c) Since the residuals plot shows no pattern, the linear model is appropriate for predicting horsepower from weight.

d) *Horsepower* $= 3.49843 + 34.3144(Weight)$

Horsepower $= 3.49843 + 34.3144(2.595)$

Horsepower ≈ 92.544

According to the model, a car weighing 2595 pounds is expected to have 92.543 horsepower. The actual horsepower of the car is: $92.544 + 22.5 \approx 115.0$ horsepower.

17. Old Faithful

a) The association between the duration of eruption and the interval between eruptions of Old Faithful is fairly strong, linear, and positive. Long eruptions are generally associated with long intervals between eruptions. There are also two distinct clusters of data, one with many short eruptions followed by short intervals, the other with many long eruptions followed by long intervals, with only a few medium eruptions and intervals in between.

b) The linear model used to predict the interval between eruptions is:

$$\widehat{Interval} = 33.9668 + 10.3582(Duration).$$

c) As the duration of the previous eruption increases by one minute, the model predicts an increase of about 10.4 minutes in the interval between eruptions.

d) $R^2 = 77.0\%$, so the model accounts for 77% of the variability in the interval between eruptions. The predictions should be fairly accurate, but not precise. Also, the association appears linear, but we should look at the residuals plot to be sure that the model is appropriate before placing too much faith in any prediction.

e)

$$\widehat{Interval} = 33.9668 + 10.3582(Duration)$$
$$\widehat{Interval} = 33.9668 + 10.3582(4)$$
$$\widehat{Interval} \approx 75.4$$

According to the model, if an eruption lasts 4 minutes, the next eruption is expected to occur in approximately 75.4 minutes.

f) The actual eruption at 79 minutes is 3.6 minutes later than predicted by the model. The residual is 79 – 75.4 = 3.6 minutes. In other words, the model under-predicted the interval.

19. How old is that tree?

a) The correlation between tree diameter and tree age is $r = 0.888$. Although the correlation is moderately high, this does not suggest that the linear model is appropriate. We must look at a scatterplot to verify that the relationship is straight enough to try the linear model. After finding the linear model, the residuals plot must be checked. If the residuals plot shows no pattern, the linear model can be deemed appropriate.

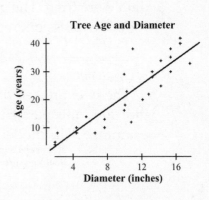

b) The association between diameter and age of these trees is fairly strong, somewhat linear, and positive. Trees with larger diameters are generally older.

c) The linear model that predicts age from diameter of trees is:
$\widehat{Age} = -0.974424 + 2.20552(Diameter)$. This model explains 78.9% of the variability in age of the trees.

d) The residuals plot shows a curved pattern, so the linear model is not appropriate. Additionally, there are several trees with large residuals.

e) The largest trees are generally above the regression line, indicating a positive residual. The model is likely to underestimate these values.

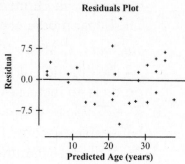

21. Big screen

TV screen sizes might vary from 19 to 70 inches. A TV with a screen that was 10 inches larger would be predicted to cost 10(0.03) = +0.3, 10(0.3) = +3, 10(3) = +30, or 10(30) = +300. Notice that the TV costs are measured in hundreds of dollars, so the potential price changes for getting a TV 10 inches larger are $30, $300, $3000, and $30,000. Only $300 seems reasonable, so the slope must be 0.3.

23. Down the drain The association between diameter of the drain plug and drain time of this water tank is strong, curved, and negative. Tanks with larger drain plugs have lower drain times. The linear model is not appropriate for the curved association, so several re-expressions of the data were tried. The best one was the reciprocal square root re-expression, resulting in the equation

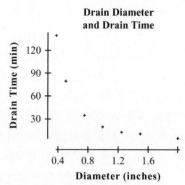

$$\frac{1}{\sqrt{DrainTime}} = 0.00243 + 0.219(Diameter).$$

The re-expressed data is nearly linear, and although the residuals plot might still indicate some pattern and has one large residual, this is the best of the models examined. The model explains 99.7% of the variability in drain time.

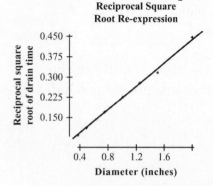

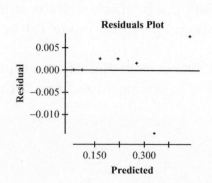

25. U.S. cities

There is a strong, roughly linear, negative association between mean January temperature and latitude. U.S. cities with higher latitudes generally have lower mean January temperatures. There are two outliers, cities with higher mean January temperatures than the pattern would suggest.

27. Winter in the city

a) $R^2 = (-0.848)^2 \approx 0.719$. The variation in latitude explains 71.9% of the variability in average January temperature.

b) The negative correlation indicates that as latitude increases, the average January temperature generally decreases.

c)

$$b_1 = \frac{rs_y}{s_x}$$

$$b_1 = \frac{(-0.848)(13.49)}{5.42}$$

$$b_1 = -2.1106125$$

$$\hat{y} = b_0 + b_1 x$$

$$\bar{y} = b_0 + b_1 \bar{x}$$

$$26.44 = b_0 - 2.1106125(39.02)$$

$$b_0 = 108.79610$$

The equation of the linear model for predicting January temperature from latitude is: $JanTemp = 108.796 - 2.111(Latitude)$

d) For each additional degree of latitude, the model predicts a decrease of approximately 2.1 °F in average January temperature.

e) The model predicts that the mean January temperature will be approximately 108.8 °F when the latitude is 0°. This is an extrapolation, and may not be meaningful.

f)

$$JanTemp = 108.796 - 2.111(Latitude)$$

$$JanTemp = 108.796 - 2.111(40)$$

$$JanTemp \approx 24.4$$

According to the model, the mean January temperature in Denver is expected to be 24.4 °F.

g) In this context, a positive residual means that the actual average temperature in the city was higher than the temperature predicted by the model. In other words, the model underestimated the average January temperature.

h) Temperature in degrees Celsius are related to temperature in degrees Fahrenheit as

$$C = \frac{5}{9}(F - 32)$$

(i) Correlation is unchanged by linear transformations in either or both variables.

(ii) The equation of the line in Celsius is:

$$C = \frac{5}{9}(F - 32)$$

$$\approx \frac{5}{9}(108.796 - 2.111 \, Latitude - 32)$$

$$\approx 42.664 - 1.173 \, Latitude$$

(iii) The predicted temperature at Denver is $42.664 - 1.173(40) = -4.256 \, °C$.

Note that this is the same as the previous prediction, once it is converted from °F to °C, i.e., $\frac{5}{9}(24.4 - 32) = -4.22$ apart from round-off error.

29. Jumps 2008

a) The association between Olympic long jump distances and high jump heights is strong, linear, and positive. Years with longer long jumps tended to have higher high jumps. There is one departure from the pattern. The year in which the Olympic gold medal long jump was the longest had a shorter gold medal high jump than we might have predicted.

b) There is an association between long jump and high jump performance, but it is likely that training and technique have improved over time and affected both jump performances.

c) The correlation would be the same, 0.920. Correlation is a measure of the degree of linear association between two quantitative variables and is unaffected by changes in units.

d) In a year when the high jumper jumped one standard deviation better than the average jump, the long jumper would be predicted to jump $r = 0.920$ standard deviations above the average long jump.

31. French

a) Most of the students would have similar weights. Regardless of their individual French vocabularies, the correlation would be near 0.

b) There are two possibilities. If the school offers French at all grade levels, then the correlation would be positive and strong. Older students, who typically weigh more, would have higher scores on the test, since they would have learned more French vocabulary. If French is not offered, the correlation between weight and test score would be near 0. Regardless of weight, most students would have horrible scores.

c) The correlation would be near 0. Most of the students would have similar weights and vocabulary test scores. Weight would not be a predictor of score.

d) The correlation would be positive and strong. Older students, who typically weigh more, would have higher test scores, since they would have learned more French vocabulary.

33. Lunchtime
The association between time spent at the table and number of calories consumed by toddlers is moderate, roughly linear, and negative. Generally,

toddlers who spent a longer time at the table consumed fewer calories than toddlers who left the table quickly. The scatterplot between time at the table and calories consumed is straight enough to justify the use of the linear model. The linear model that predicts the number of calories consumed by a toddler from the time spent at the table is $\widehat{Calories} = 560.7 - 3.08(Time)$. For each additional minute spent at the table, the model predicts that the number of calories consumed will be approximately 3.08 fewer. Only 42.1% of the variability in the number of calories consumed can be explained by the variability in time spent at the table. The residuals plot shows no pattern, so the linear model is appropriate, if not terribly useful, for prediction.

35. **Tobacco and alcohol** The first concern about these data is that they consist of averages for regions in Great Britain, not individual households. Any conclusions reached can only be about the regions, not the individual households living there. The second concern is the data point for Northern Ireland. This point has high leverage, since it has the highest household tobacco spending and the lowest household alcohol spending. With this point included, there appears to be only a weak positive association between tobacco and alcohol spending. Without the point, the association is much stronger. In Great Britain, with the exception of Northern Ireland, higher levels of household spending on tobacco are associated with higher levels of household spending on tobacco. It is not necessary to make the linear model, since we have the household averages for the regions in Great Britain, and the model wouldn't be useful for predicting in other countries or for individual households in Great Britain.

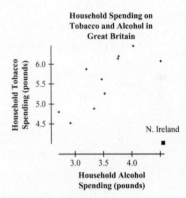

37. **Vehicle weights**
   a)
   $$\widehat{Wt} = 10.85 + 0.64 scale$$
   $$\widehat{Wt} = 10.85 + 0.64(31.2)$$
   $$\widehat{Wt} = 30.818$$

   According to the model, a truck with a scale weight of 31 200 pounds is expected to weigh 30 818 pounds.

   b) If the actual weight of the truck is 32,120 pounds, the residual is 32 120 – 30 818 = 1302 pounds. The model underestimated the weight.

c)

$$\widehat{Wt} = 10.85 + 0.64 scale$$

$$\widehat{Wt} = 10.85 + 0.64(35.590)$$

$$\widehat{Wt} = 33.6276 \; thousand \; pounds$$

The predicted weight of the truck is 33,627.6 pounds. If the residual is –2440 pounds, the actual weight of the truck is 33 627.6 – 2440 = 31 187.6 pounds.

d) $R^2 = 93\%$, so the model explains 93% of the variability in weight, but some of the residuals are 1000 pounds or more. If we need to be more accurate than that, then this model will not work well.

e) Negative residuals will be more of a problem. Police would be issuing tickets to trucks whose weights had been overestimated by the model. The U.S. justice system is based upon the principle of innocence until guilt is proven. These truckers would be unfairly ticketed, and that is worse than allowing overweight trucks to pass.

Chapter 9: Understanding Randomness

1. **Coin toss** A coin flip is random because the outcome cannot be predicted beforehand.

3. **Random outcomes**
   a) Yes, who takes out the trash cannot be predicted before the flip of a coin.
   b) No, it is not random, since you will probably name a favourite team.
   c) Yes, your new roommate cannot be predicted before names are drawn.

5. **Components**
   Rolling the pair of dice is the component.

7. **Response variable**
   To simulate, you could roll dice and note whether or not "doubles" came up. A trial would be completed once "doubles" came up. You would count up the number of rolls until "doubles" for the response variable. Alternatively, you could **use** the digits 1, 2,3, 4, 5, 6 on random digits table and disregard digits 7, 8, **9**, and 0. Using the table, note the first legal digit for the first die, and then note the next legal digit for the second die. A double would indicate rolling double.

9. **The lottery** In provincial lotteries, a machine pops up numbered balls. If the lottery were truly random, the outcome could not be predicted and the outcomes would be equally likely. It is random only if the balls generate numbers in equal probabilities.

11. **Birth defects** Answers may vary. Generate two-digit random numbers, 00–99. Let 00–02 represent a defect. Let 03–99 represent no defect.

13. **Geography**
    a) Looking at pairs of digits, the first province number is 88. Ignore this since there is no 88th province. The next set is 06 and so select Northwest Territories for the first student. Ignore the pairs 35, 65, as there are no provinces with these names. The next province number is 13, Yukon.

    b) Continuing along, ignore the pair 31. The next pair is 05, Newfoundland and Labrador. Then ignore 63, 21, 05. We ignore 05 because that province is already selected. The next pair is 08, Nunavut. Continue this way until you find the next number within the range 01–13.

15. **Play the lottery** If the lottery is random, it doesn't matter if you play the same favourite "lucky" numbers or if you play different numbers each time. All numbers are equally likely (or, rather, UNLIKELY) to win.

17. Bad simulations

a) The outcomes are not equally likely. For example, the probability of getting 5 heads in 9 tosses is not the same as the probability of getting 0 head, but the simulation assumes they are equally likely.

b) The even-odd assignment assumes that the player is equally likely to score or miss the shot. In reality, the likelihood of making the shot depends on the player's skill.

c) Suppose a hand has four aces. This might be represented by 1,1,1,1, and any other number. The likelihood of getting the first ace in the hand again is not the same for the second or third or fourth ace. But with this simulation, the likelihood is the same for each.

19. Wrong conclusion The conclusion should indicate that the simulation suggests that the average length of the line would be 3.2 people. Future results might not match the simulated results exactly.

21. Election

a) Answers will vary. A component is one voter voting. An outcome is a vote for your candidate. Using two random digits, 00–99, let 01–55 represent a vote for your candidate, and let 55–99 and 00 represent a vote for the underdog.

b) A trial is 100 votes. Examine 100 two-digit random numbers and count how many simulated votes are cast for each candidate. Whoever gets the majority of the votes wins the trial.

c) The response variable is whether the underdog wins or not. To calculate the experimental probability, divide the number of trials in which the simulated underdog wins by the total number of trials.

23. Cereal Answers will vary. A component is the simulation of the picture in one box of cereal. One possible way to model this component is to generate random digits 0–9. Let 0 and 1 represent Sidney Crosby, 2–4 represent Christine Sinclair, and 5–9 represent Clara Hughes. Each trial will consist of 5 random digits, and the response variable will be whether or not a complete set of pictures is simulated. Trials in which at least one of each picture is simulated will be a success. The total number of successes divided by the total number of trials will be the simulated probability of ending up with a complete set of pictures. According to the simulation, the probability of getting a complete set of pictures is expected to be about 51.5%.

25. Multiple choice Answers will vary. A component is one multiple-choice question. One possible way to model this component is to generate random digits 0–9. Let digits 0–7 represent a correct answer, and let digits 8 and 9 represent an incorrect answer. Each trial will consist of 6 random digits. The response variable is whether or not all 6 simulated questions are answered correctly (all 6 digits are 0–7). The total number of successes divided by the total number of trials will be the simulated probability of getting all 6 questions right. According to the simulation, the probability of getting all 6 multiple-choice questions correct is expected to be about 26%.

27. **Beat the lottery**
    a) Answers based on your simulation will vary, but you should win about 10% of the time.
    b) You should win at the same rate with any number.

29. **It evens out in the end** Answers based on your simulation will vary, but you should win about 10% of the time. Playing lottery numbers that have turned up the least in recent lottery drawers offers no advantage. Each new drawing is independent of recent drawings.

31. **Driving test** Answers will vary. A component is one driver's test, but this component will be modelled differently, depending on whether or not it is the first test taken. One possible way to model this component is to generate pairs of random digits 00–99. Let 01–34 represent passing the first test and let 35–99 and 00 represent failing the first test. Let 01–72 represent passing a retest, and let 73–99 and 00 represent failing a retest. To simulate one trial, generate random numbers until one is generated that represents passing a test. Begin each trial using the "first test" representation, and switch to the "retest" representation if failure is indicated on the first simulated test. The response variable is the number of simulated tests required to achieve the first passing test. The total number of simulated tests taken divided by the total number of trials is the simulated average number of tests required to pass. According to the simulation, the number of driving tests required to pass is expected to be about 1.9.

33. **Basketball strategy** Answers will vary. A component is one foul shot. One way to model this component would be to generate pairs of random digits 00–99. Let 01–72 represent a made shot, and let 73–99 and 00 represent a missed shot. The response variable is the number of shots made in a "one and one" situation. If the first shot simulated represents a made shot, simulate a second shot. If the first shot simulated represents a miss, the trial is over. The simulated average number of points is the total number of simulated points divided by the number of trials. According to the simulation, the player is expected to score about 1.24 points.

35. **Free groceries** Answers will vary. A component is the selection of one card with the prize indicated. One possible way to model the prize is to generate pairs of random digits 00–99. Let 01–10 represent $200, let 11–20 represent $100, let 21–40 represent $50, and let 41–99 and 00 represent $20. Repeated pairs of digits must be ignored. (For this reason, a simulation in which random digits 0–9 are generated with 0 representing $200, 1 representing $100, etc., is NOT acceptable. Each card must be

individually represented.) A trial continues until the total simulated prize is greater than $500. The response variable is the number of simulated customers until the payoff is greater than $500. The simulated average number of customers is the total number of simulated customers divided by the number of trials. According to the simulation, about 10.2 winners are expected each week.

37. **The family** Answers will vary. Each child is a component. One way to model the component is to generate random digits 0–9. Let 0–4 represent a boy and let 5–9 represent a girl. A trial consists of generating random digits until a child of each gender is simulated. The response variable is the number of children simulated until this happens. The simulated average family size is the number of digits generated in each trial divided by the total number of trials. According to the simulation, the expected number of children in the family is about 3.

39. **Dice game** Answers will vary. Each roll of the die is a component. One way of modelling this component is to generate random digits 0–9. The digits 1–6 correspond to the numbers on the faces of the die, and digits 7–9 and 0 are ignored. A trial consists of generating random numbers until the sum of the numbers is exactly 10. If the sum exceeds 10, the last roll must be ignored and simulated again, but still counted as a roll. The response variable is the number of rolls until the sum is exactly 10. The simulated average number of rolls until this happens is the total number of rolls simulated divided by the number of trials. According to the simulation, expect to roll the die about 7.5 times.

41. **The Stanley Cup final** Answers may vary. Each game is a component. One way of modelling this component is to generate pairs of random digits 00–99. Let 01–55 represent a win by the favoured team, and let 56–99 and 00 represent a win by the underdog. A trial consists of generating numbers until one team has 4 simulated wins. The response variable is whether or not the underdog wins. The simulated percentage of Stanley Cup wins is the total number of successes divided by the total number of trials. According to simulation, the underdog is expected to win the Stanley Cup about 39% of the time.

43. **Second team** Answers will vary. Each player chosen is a component. One way to model this component is to generate random numbers 0–9. Let digits 1–4 represent the four players who are to be chosen. Ignore digits 5–9 and 0. A trial consists of generating a sequence of random numbers that represents the order in which the cards are chosen. Since each number represents a person, and people cannot be chosen more than once, ignore repeated numbers. The response variable is whether or not any digit in the generated sequence matches the corresponding digit in the sequence 1 2 3 4 (or any other sequence of the four numbers, as long as it is determined ahead of time). The simulated percentage of the time this is expected to happen is the total number of successes (times that the sequences have no matching

corresponding digits) divided by the number of trials. According to the simulation, all players are expected to be paired with someone other than the person with whom he or she came to the party about 37.5% of the time.

45. **Smartphones** Answers will vary. Each driver is a component. One way to model this component is to generate random digits 00–99. Let 01–12 represent a driver who is either texting or talking (i.e., using) on his or her (hand-held) cell phone, and let 13–99 and 00 represent a driver who is neither texting nor talking on his or her cell phone. A trial consists of 10 numbers. The response variable is whether or not at least 4 of the simulated drivers are using their cell phones. The simulated percentage of the time that 4 or more drivers out of ten are using their cell phones (if the true rate of usage is 12%) is the number of successes divided by the total number of trials. You should expect to find 4 or more drivers texting or talking among 10 drivers only about 2% of the time. Based on what you saw at the bus stop, you'd suspect that the legislator's claim of 12% usage is probably too low.

47. **Tires** Answers may vary. Each tire is a component. According to the Normal model, the probability that one tire will last at least 30 000 km is about 78.8%. To model this component, generate triples of random digits 000–999. Let 001–788 represent a tire that lasts at least 30 000 km, and let 789–999 and 000 represent a tire that does not last at least 30 000 km. (The simulation can be adjusted with fewer or more digits, depending on the accuracy desired and the rounding of the probability of a tire lasting 30 000 km.) A trail consists of four simulated tires. The response variable is whether or not all four simulated tires lasted at least 30 000 km. The percentage of the time all four tires are expected to last more than 30 000 km is the total number of successes divided by the number of trials. According to the simulation, all four tires are expected to last this long about 38.5% of the time.

Chapter 10: Sample Surveys

1. **Roper**
   a) Roper is not using a simple random sample. The samples are designed to get 500 males and 500 females. This would be very unlikely to happen in a simple random sample.
   b) They are using stratified sample, with two strata, males and females.

3. **Emoticons**
   a) This is a voluntary response sample.
   b) We have absolutely no confidence in estimates made from voluntary response samples.

5. **Gallup**
   a) The population of interest is all adults in the United States aged 18 and older.
   b) The sampling frame is U.S. adults with land-line telephones.
   c) Some members of the population (e.g., many college students) don't have land-line telephones, so they can never be chosen in the sample. This may create a bias.

7. *Population* – human resources directors of Fortune 500 companies
   Parameter – proportion of directors who don't feel that surveys intruded on their workday
   Sampling frame – list of HR directors at Fortune 500 companies
   Sample – the 23% of the HR directors who responded
   Method – questionnaire mailed to all (non-random)
   Bias – nonresponse. It is impossible to generalize about the opinions all HR directors because the nonrespondents would likely have a different opinion about the question. HR directors who feel that surveys intrude on their workday would be less likely to respond.

9. *Population* – all U.S. adults
   Parameter – proportion who have used and benefited from alternative medical treatments.
   Sampling frame – all Consumers Union subscribers
   Sample – those subscribers who responded
   Method – not specified, but probably a questionnaire mailed to all subscribers
   Bias – nonresponse bias, specifically voluntary response bias. Those who respond may have strong feelings one way or another.

11. *Population* – adults
    Parameter – proportion who think drinking and driving is a serious problem
    Sampling frame – bar patrons

Sample – every 10th person leaving the bar
Method – systematic sampling
Bias – undercoverage. Those interviewed had just left a bar, and may have opinions about drinking and driving that differ from the opinions of the population in general.

13. *Population* – soil around a former waste dump
Parameter – proportion with elevated levels of harmful substances
Sampling frame – accessible soil around the dump
Sample – 16 soil samples
Method – not clear. There is no indication that the samples were selected randomly.
Bias – possible convenience sample. Since there is no indication of randomization, the samples may have been taken from easily accessible areas. Soil in these areas may be more or less polluted than the soil in general.

15. *Population* – snack food bags
Parameter – proportion passing inspection
Sampling Frame – all bags produced each day
Sample – 10 bags, one from each of 10 randomly selected cases
Method – multistage sampling. Presumably, they take a simple random sample of 10 cases, followed by a simple random sample of one bag from each case.
Bias – no indication of bias

17. **Mistaken poll** The station's faulty prediction is more likely to be the result of bias. Only people watching the news were able to respond, and their opinions were likely to be different from those of other voters. The sampling method may have systematically produced samples that did not represent the population of interest.

19. **Parent opinion, part 1**
    a) This is a voluntary response sample. Only those who see the ad, feel strongly about the issue, and have Web access will respond.
    b) This is cluster sampling, but probably not a good idea. The opinions of parents in one school may not be typical of the opinions of all parents.
    c) This is an attempt at a census, and will probably suffer from nonresponse bias.
    d) This is stratified sampling. If the follow-up is carried out carefully, the sample should be unbiased.

21. **Roller coasters**
    a) This is a systematic sample.
    b) This sample is likely to be representative of those waiting in line for the roller coaster, especially if those people at the front of the line (after their long wait) respond differently from those at the end of the line.

c) The sampling frame is patrons willing to wait in line for the roller coaster. The sample should be representative of the people in line, but not of all the people at Canada's Wonderland.

23. Social life
a) This is a voluntary response sample.
b) We have absolutely no confidence in estimates made from voluntary response samples. Voluntary response bias exists. Additionally, undercoverage limits the scope of any conclusions to the population of young adults, since only those that visited gamefaqs.com could have been chosen to be in the sample.

25. Wording the survey
a) Responses to these questions will differ. Question 1 will probably get "no" answers, and Question 2 will probably get "yes" answers. This is response bias, based on the wording of the questions.
b) A question with neutral wording might be: "Do you think standardized tests are appropriate for deciding whether a student should be promoted to the next grade?"

27. Another ride
Only those who think it is worth the wait are likely to be in line. Those who don't like roller coasters are unlikely to be in the sampling frame, so the poll won't get a fair picture of whether park patrons overall would still favour more roller coasters.

29. Survey questions
a) The question is biased toward "yes" answers because of the word "pollute." A better question might be: "Should companies be responsible for any costs of environmental clean up?"
b) The question is biased toward "no" because of the preamble "18-year-olds are old enough to serve in the military." A better question might be: "Do you think the drinking age should be lowered from 21?"

31. Phone surveys
a) A simple random sample is difficult in this case because there is a problem with undercoverage. People with unlisted phone numbers and those without phones are not in the sampling frame. People who are at work, or otherwise away from home, are not included in the sampling frame. These people could never be in the sample itself.
b) One possibility is to generate random phone numbers and call at random times, although obviously not in the middle of the night! This would take care of the undercoverage of people at work during the day, as well as people with unlisted phone numbers, although there is still a problem avoiding undercoverage of people without phones.

c) Under the original plan, those families in which one person stays home are more likely to be included. Under the second plan, many more are included. People without phones are still excluded.

d) Follow-up of this type greatly improves the chance that a selected household is included, increasing the reliability of the survey.

e) Random dialers allow people with unlisted phone numbers to be selected, although they may not be the most willing participants. There is a reason that the phone number is unlisted. Time of day will still be an issue, as will people without phones.

33. Arm length
a) Answers will vary. My arm length is 3 hand widths and 2 finger widths.
b) The parameter estimated by 10 measurements is the true length of your arm. The population is all possible measurements of your arm length.
c) The population is now the arm lengths of your friends. The average now estimates the mean of the arm lengths of your friends.
d) These 10 arm lengths are unlikely to be representative of the community, or the country. Your friends are likely to be of the same age, and not very diverse.

35. Accounting
a) Assign numbers 001–120 to each order. Generate 10 random numbers 001–120, and select those orders to recheck.
b) The supervisor should perform a stratified sample, randomly checking a certain percentage of each type of sales, retail and wholesale.

37. Quality control
a) Select three cases at random, then select one jar randomly from each case.
b) Generate three random numbers between 61–80, with no repeats, to select three cases. Then assign each of the jars in the case a number 01–12, and generate one random number for each case to select the three jars, one from each case.
c) This is not a simple random sample, since there are groups of three jars that cannot be the sample. For example, it is impossible for three jars in the same case to be the sample. This would be possible if the sample were a simple random sample.

39. Sampling methods
a) This method would probably result in undercoverage of those doctors that are not listed in the Yellow Pages. Using the "line listings" seems fair, as long as all doctors are listed, but using the advertisements would not be a typical list of doctors.
b) This method is not appropriate. This cluster sample will probably contain listings for only one or two types of businesses, not a representative cross-section of businesses.

1. **Tips**   Each of the 40 deliveries is an experimental unit. He has randomized the experiment by flipping a coin to decide whether or not to phone.

3. **Tips II**   The factor is calling, and the levels are whether or not he calls the customer. The response variable is the tip percentage for each delivery.

5. **Tips again**   By calling some customers but not others during the same run, the driver has controlled many variables, such as the day of the week, season, and weather. The experiment was randomized because he flipped a coin to determine whether or not to phone and it was replicated because he did this for 40 deliveries.

7. **More tips**   Because customers don't know about the experiment, those that are called don't know that others are not, and vice versa. Thus, the customers are blind. That would make this a single-blind study. It can't be double-blind because the delivery driver must know whether or not he phones.

9. **Block that tip**   Yes. Driver is now a block. The experiment is randomized within each block. This is a good idea because some drivers might generally get higher tips than others, but the goal of the experiment is to study the effect of phone calls. Blocking on driver eliminates the variability in tips inherent to the driver.

11. **Confounded tips**   Answers may vary. The cost or size of a delivery may confound his results. Larger orders may generally tip a higher or lower percentage of the bill.

13. **Facebook usage and grades**
    a) No, this is not an experiment. There are no imposed treatments. This is an observational study.
    b) We cannot conclude that the differences in average graduating grades are caused by differences in average amount of time spent on Facebook each day. Since this is not an experiment, there can be other factors related to Facebook usage that are common among Facebook users and some of those factors might be the reason for the differences in average graduating grades.

15. **MS and vitamin D**
    a) This is a retrospective observational study.
    b) This is an appropriate choice, since MS is a relatively rare disease.
    c) The subjects were U.S. military personnel, some of whom had developed MS.
    d) The variables were the vitamin D blood levels and whether or not the subject developed MS.

17. **Menopause**
    a) This was a randomized, comparative, placebo-controlled experiment.
    b) Yes, such an experiment is the right way to determine whether black cohosh is an effective treatment for hot flashes.
    c) The subjects were 351 women, aged 45 to 55, who reported at least two hot flashes a day.
    d) The treatments were black cohosh, a multi-herb supplement, plus advice to consume more soy foods, estrogen, and a placebo. The response was the women's self-reported symptoms, presumably the frequency of hot flashes.

19. a) This is an experiment, since treatments were imposed.
    b) The subjects studied were 30 patients with bipolar disorder.
    c) The experiment has 1 factor (omega-3 fats from fish oil) at 2 levels (high dose of omega-3 fats from fish oil and no omega-3 fats from fish oil).
    d) One factor at 2 levels gives a total of 2 treatments.
    e) The response variable is "improvement," but there is no indication of how the response variable was measured.
    f) There is no information about the design of the experiment.
    g) The experiment is blinded, since the use of a placebo keeps the patients from knowing whether or not they received the omega-3 fats from fish oils. It is not stated whether or not the evaluators of the "improvement" were blind to the treatment, which would make the experiment double-blind.
    h) Although it needs to be replicated, the experiment can determine whether or not omega-3 fats from fish oils cause improvements in patients with bipolar disorder, at least over the short term. The experiment design would be stronger if it were double-blind.

21. a) This is an observational study. The researchers are simply studying traits that already exist in the subjects, not imposing new treatments.
    b) This is a prospective study. The subjects were identified first, then traits were observed.
    c) The subjects are roughly 200 men and women with moderately high blood pressure and normal blood pressure. There is no information about the selection method.
    d) The parameters of interest are difference in memory and reaction time scores between those with normal blood pressure and moderately high blood pressure.
    e) An observational study has no random assignment, so there is no way to know whether high blood pressure caused subjects to do worse on memory and reaction time tests. A lurking variable, such as age or overall health, might have been the cause. The most we can say is that there was an association between blood pressure and scores on memory and reaction time tests in this group, and recommend a controlled experiment to attempt to determine whether or not there is a cause-and-effect relationship.

23. a) This is an observational study. The researchers are simply studying traits that already exist in the subjects, not imposing new treatments.
    b) This is a retrospective study. Researchers studied medical records that already existed.
    c) The subjects were 360 000 Swedish men. The selection process is not stated.
    d) The parameter of interest is the difference in percentage of men with kidney cancer among different groups.
    e) There is no random assignment, so a cause-and-effect relationship between weight or high blood pressure and kidney cancer cannot be established.

25. a) This is an experiment, since treatments were imposed on randomly assigned groups.
    b) Twenty-four post-menopausal women were the subjects in this experiment.
    c) There is 1 factor (type of drink) at 2 levels (alcoholic and non-alcoholic). (Supplemental estrogen is not a factor in the experiment, but rather a blocking variable. The subjects were not given estrogen supplements as part of the experiment.)
    d) One factor, with 2 levels, is 2 treatments.
    e) The response variable is an increase in estrogen level.
    f) This experiment uses a blocked design. The subjects were blocked by whether or not they used supplemental estrogen. This design reduces variability in the response variable of estrogen level that may be associated with the use of supplemental estrogen.
    g) This experiment does not use blinding.
    h) This experiment indicates that drinking alcohol leads to increased estrogen level among those taking estrogen supplements.

27. a) This is an experiment, since treatments were imposed on randomly assigned groups.
    b) The experimental units were 10 randomly selected garden locations.
    c) There is 1 factor (type of trap) at 2 levels (the different types of trap).
    d) One factor, at 2 levels, results in 2 treatments.
    e) The response variable is the number of bugs in each trap.
    f) This experiment incorporates a blocked design. Placing one of each type of trap at each location reduces variability due to location in the garden.
    g) There is no mention of blinding.
    h) This experiment can determine which trap is more effective at catching bugs.

29. a) This is an observational study.
    b) The study is retrospective. Results were obtained from pre-existing church records.

c) The subjects of the study are women in Finland. The data were collected from church records dating 1640 to 1870, but the selection process is unknown.

d) The parameter of interest is difference in average lifespan between mothers of sons and daughters.

e) For this group, having sons was associated with a decrease in lifespan of an average of 34 weeks per son, while having daughters was associated with an unspecified increase in lifespan. As there is no random assignment, there is no way to know that having sons caused a decrease in lifespan.

31. a) This is an observational study. (Although some might say that the sad movie was "imposed" on the subjects, this was merely a stimulus used to trigger a reaction, not a treatment designed to attempt to influence some response variable. Researchers merely wanted to observe the behaviour of two different groups when each was presented with the stimulus.)

 b) The study is prospective. Researchers identified subjects, and then observed them after the sad movie.

 c) The subjects in this study were the 71 people with and the 33 without depression. The selection process is not stated.

 d) The parameter of interest is the difference in crying response between depressed and non-depressed people exposed to sad situations.

 e) There is no apparent difference in crying response to sad movies for the depressed and non-depressed groups.

33. a) This is an experiment.
 b) The subjects were rats used in this experiment.
 c) There is 1 factor (sleep deprivation) at 4 levels (no deprivation, 6, 12, and 24 hours).
 d) One factor, at 4 levels, results in 4 treatments.
 e) The response variable is glycogen level in the brain.
 f) There is no mention of randomness in the design. Hopefully, the rats were randomly assigned to treatment groups.
 g) No blinding is indicated, and seems unlikely. The rats know that they have been kept awake.
 h) As long as proper randomization was employed, the conclusion could be that rats deprived of sleep have significantly lower glycogen levels and may need sleep to restore that brain energy fuel. Extrapolating to humans would be speculative.

35. a) This is an experiment. Subjects were randomly assigned to treatments.
 b) The subjects were the people in this experiment experiencing migraines.
 c) There are 2 factors (pain reliever and water temperature). The pain reliever factor has 2 levels (pain reliever or placebo), and the water temperature factor has 2 levels (ice water and regular water).

d) Two factors, at 2 levels each, results in 4 treatments.

e) The response variable is the level of pain relief.

f) The experiment is completely randomized.

g) The subjects are blinded to the pain reliever factor through the use of a placebo. The subjects are not blinded to the water factor. They will know whether they are drinking ice water or regular water.

h) The experiment may indicate whether pain reliever alone or in combination with ice water give pain relief, but patients are not blinded to ice water, so the placebo effect may also be the cause of any relief seen due to ice water.

37. a) This is an experiment. Athletes were randomly assigned to one of two exercise programs.

b) The subjects are athletes suffering hamstring injuries.

c) There is one factor (type of exercise) at 2 levels (static stretching, and agility and trunk stabilization).

d) One factor, at 2 levels, results in 2 treatments.

e) The response variable is the time before the athletes were able to return to sports.

f) The experiment is completely randomized.

g) The experiment employs no blinding. The subjects know what kind of exercise they do.

h) Assuming that the athletes actually followed the exercise program, this experiment can help determine which of the two exercise programs is more effective at rehabilitating hamstring injuries.

39. **Omega-3** The experimenters need to compare omega-3 results to something. Perhaps bipolarity is seasonal and would have improved during the experiment anyway.

41. **Omega-3 revisited**

a) Subjects' responses might be related to other factors, like diet, exercise, or genetics. Randomization should equalize the two groups with respect to unknown factors.

b) More subjects would minimize the impact of individual variability in the responses, but the experiment would become more costly and time-consuming.

43. **Omega-3, finis** The researchers believe that people who engage in regular exercise might respond differently to the omega-3. This additional variability could obscure the effectiveness of the treatment.

45. **Tomatoes, next season** Answers may vary. Number the tomatoes plants 1 to 24. Use a random number generator to randomly select 24 numbers from 1 to 24 without replication. Assign the tomato plants matching the first 8 numbers to the

first group, the second 8 numbers to the second group, and the third group of 8 numbers to the third group.

47. Shoes
a) First, the manufacturers are using athletes who have a vested interest in the success of the shoe by virtue of their sponsorship. They should try to find some volunteers that aren't employed by the company! Second, they should randomize the order of the runs, not run all the races with the new shoes second. They should blind the athletes by disguising the shoes, if possible, so they don't know which is which. The experiment could be double blinded, as well, by making sure that the timers don't know which shoes are being tested at any given time. Finally, they should replicate this rather small experiment several times since dash times will vary under both shoe conditions even for the same athlete.
b) First of all, the problems identified in part a) would have to be remedied before *any* conclusions can be reached. Even if this is the case, the results cannot be generalized to all runners. This experiment compares effects of the shoes on speed for Olympic class runners, not runners in general.

49. Hamstrings
a) Allowing the athletes to choose their own treatments could confound the results. Other issues such as severity of injury, diet, age, etc., could also affect time to heal, and randomization should equalize the two treatment groups with respect to any such variables.
b) A control group could have revealed whether either exercise program was better (or worse) than just letting the injury heal without exercise.
c) Although the athletes cannot be blinded, the doctors who approve their return to sports should not know which treatment the subject had engaged in.
d) It's difficult to say with any certainty, since we aren't sure if the distributions of return times are unimodal and roughly symmetric, and contain no outliers. Otherwise, the use of mean and standard deviation as measures of center and spread is questionable. Assuming mean and standard deviation are appropriate measures, the subjects who exercised with agility and trunk stabilization had a mean return time of 22.2 days compared to the static stretching group, with a mean return time of 37.4 days. The agility and trunk stabilization group also had a much more consistent distribution of return times, with a standard deviation of 8.3 days, compared to the standard deviation of 27.6 days for the static stretching group.

51. Mozart
a) The differences in spatial reasoning scores between the students listening to Mozart and the students sitting quietly were more than would have been expected from ordinary sampling variation.

b)

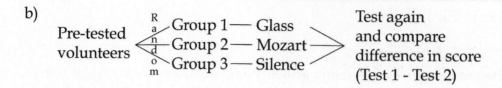

c) The Mozart group seems to have the smallest median difference in spatial reasoning test score and thus the *least* improvement, but there does not appear to be a significant difference.

d) No, the results do not prove that listening to Mozart is beneficial. If anything, there was generally less improvement. The difference does not seem significant compared with the usual variation one would expect between the three groups. Even if type of music has no effect on test score, we would expect some variation between the groups.

53. Frumpies

a) They should perform a survey. Randomly select a group of children, ages 10 to 13, and have them taste the cereal. Ask the children a question with as little bias as possible, such as, "Do you like the cereal?"

b) Answers may vary. Get volunteers age 10 to 13. Randomly select half of the volunteers to taste the new flavour, and half to taste the old flavour. Compare the percentage of favourable ratings for each cereal.

c) Answers may vary. From the volunteers, identify the children who watch Frump, and identify the children who do not watch Frump. Use a blocked design to reduce variation in cereal preference that may be associated with watching the Frump cartoon.

55. Wine

a) This is a prospective observational study. The researchers followed a group of individuals born at a Copenhagen hospital between 1959 and 1961.

b) The results of the Danish study report a link between high socioeconomic status, education, and wine drinking. Since people with high levels of education and higher socioeconomic status are also more likely to be healthy, the relation between health and wine consumption might be explained by the confounding variables of socioeconomic status and education.

c) Studies such as these prove none of these links. While the variables have a relation, there is no indication of a cause-and-effect relationship. The only way to determine causation is through a controlled, randomized, and replicated experiment.

57. Dowsing

a) Arrange the 20 containers in 20 separate locations. Number the containers 01–20, and use a random number generator to identify the 10 containers that should be filled with water.

b) We would expect the dowser to be correct about 50% of the time, just by guessing. A record of 60% (12 out of 20) does not appear to be significantly different than the 10 out of 20 expected.

c) Answers may vary. A high level of success would need to be observed. 90% to 100% success (18 to 20 correct identifications) would be convincing.

59. Reading

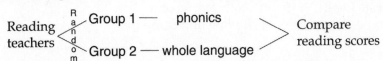

Answers may vary. This experiment has 1 factor (reading program), at 2 levels (phonics and whole language), resulting in 2 treatments. The response variable is reading score on an appropriate reading test after a year in the program. After randomly assigning students to teachers, randomly assign half the reading teachers in the district to use each method. There may be variation in reading score based on school within the district, as well as by grade. Blocking by both school and grade will reduce this variation.

61. Weekend deaths

a) The difference between death rate on the weekend and death rate during the week is greater than would be expected due to natural sampling variation.

b) This was a prospective observational study. The researchers identified hospitals in Ontario, Canada, and tracked admissions to the emergency rooms. This certainly cannot be an experiment. People can't be assigned to become injured on a specific day of the week!

c) Waiting until Monday, if you were ill on Saturday, would be foolish. There are likely to be confounding variables that account for the higher death rate on the weekends. For example, people might be more likely to engage in risky behaviour on the weekend.

d) Alcohol use might have something to do with the higher death rate on the weekends. Perhaps more people drink alcohol on weekends, which may lead to more traffic accidents, and higher rates of violence during these days of the week. Another explanation is that people have more time to engage in risky behaviour on the weekends; working hours (which for most people is Monday to Friday) tend to be much more regulated, so they are not as risky as 'free time.'

63. **Beetles** Answers may vary. This experiment has 1 factor (pesticide), at 3 levels (pesticide A, pesticide B, no pesticide), resulting in 3 treatments. The response variable is the number of beetle larvae found on each plant. Randomly select a third of the plots to be sprayed with pesticide A, a third with pesticide B, and a third to be sprayed with no pesticide (since the researcher also wants to know whether the pesticides even work at all). To control the experiment, the plots of land should be as similar as possible, with regard to amount of sunlight, water, proximity to other plants, etc. If not, plots with similar characteristics should be blocked together. If possible, use some inert substance as a placebo pesticide on the control group, and do not tell the counters of the beetle larvae which plants have been treated with pesticides. After a given period of time, count the number of beetle larvae on each plant and compare the results.

Plots of corn → Random → Group 1 — pesticide A / Group 2 — pesticide B / Group 3 — no pesticide → Count the number of beetle larvae on each plant and compare

65. **Safety switch** Answers may vary. This experiment has 1 factor (hand), at 2 levels (right, left), resulting in 2 treatments. The response variable is the difference in deactivation time between left and right hand. Find a group of volunteers. Using a matched design, require each volunteer to deactivate the machine with his or her left hand, as well as with his or her right hand. Randomly assign the left or right hand to be used first. Hopefully, this will equalize any variability in time that may result from experience gained after deactivating the machine the first time. Complete the first attempt for the whole group. Now repeat the experiment with the alternate hand. Check the differences in time for the left and right hands. Since the response variable is difference in times for each hand, workers should be blocked into groups based on their dominant hand. Another way to account for this difference would be to use the absolute value of the difference as the response variable. We are interested in whether or not the difference is significantly different from the zero difference we would expect if the machine were just as easy to operate with either hand.

67. **Skydiving, anyone?**
   a) There is 1 factor, jumping, with 2 levels, with and without a working parachute.
   b) You would need some (dim-witted) volunteer skydivers as the subjects.
   c) A parachute that looked real, but didn't open, would serve as the placebo.

d) One factor at 2 levels is 2 treatments, a good parachute and a placebo parachute.

e) The response variable is whether the skydiver survives the jump (or the extent of injuries).

f) All skydivers should jump from the same altitude, in similar weather conditions, and land on similar surfaces.

g) Make sure that you randomly assign the skydivers to the parachutes.

h) The skydivers (and the distributers of the parachutes) shouldn't know who got a working chute. Additionally, the people evaluating the subjects after the jumps should not be told who had a real chute, either.

Part III Review: Gathering Data

1. The researchers performed a prospective observational study, since the children were identified at birth and examined at ages 8 and 20. There were indications of behavioural differences between the group of "preemies" and the group of full-term babies. The "preemies" were less likely to engage in risky behaviours, like use of drugs and alcohol, teen pregnancy, and conviction of crimes. This may point to a link between premature birth and behaviour, but there may be lurking variables involved. Without a controlled, randomized, and replicated experiment, a cause-and-effect relationship cannot be determined.

3. The researchers at the Purina Pet Institute performed an experiment, matched by gender and weight. The experiment had one factor (diet), at two levels (allowing the dogs to eat as much as they want, or restricted diet), resulting in two treatments. One of each pair of similar puppies was randomly assigned to each treatment. The response variable was length of life. The researchers were able to conclude that, on average, dogs with a lower-calorie diet live longer.

5. This is a completely randomized experiment, with the treatment being receiving folic acid or not (one factor, two levels). Treatments were assigned randomly and the response variable is the number of precancerous growths, or simply the occurrence of additional precancerous growths. Neither blocking nor matching is mentioned, but in a study such as this one, it is likely that researchers and patients are blinded. Since treatments were randomized, it seems reasonable to generalize results to all people with precancerous polyps, though caution is warranted since these results contradict a previous study.

7. The fireworks manufacturers are sampling. No information is given about the sampling procedure, so hopefully the tested fireworks are selected randomly. It would probably be a good idea to test a few of each type of firework, so stratification by type seems likely. The population is all fireworks produced each day, and the parameter of interest is the proportion of duds. With a random sample, the manufacturers can make inferences about the proportion of duds in the entire day's production, and use this information to decide whether or not the day's production is suitable for sale.

9. This is an observational retrospective study. Researcher can conclude that for anyone's lunch, even when packed with ice, food temperatures are rising to unsafe levels.

11. This is an experiment, with a control group being the genetically engineered mice who received no antidepressant and the treatment group being the mic who received the drug. The response variable is the amount of plaque in their brains

after one dose and after four months. There is no mention of blinding or matching. Conclusions can be drawn to the general population of mice and we should assume treatments were randomized. To conclude the same for humans would be risky, but researchers might propose an experiment on humans based on this study.

13. The researchers performed an experiment. There is 1 factor (gene therapy), at 2 levels (gene therapy and no gene therapy), resulting in 2 treatments. The experiment is completely randomized. The response variable is heart muscle condition. The researchers can conclude that gene therapy is responsible for stabilizing heart muscle in laboratory rats.

15. The orange juice plant depends on sampling to ensure the oranges are suitable for juice. The population is all of the oranges on the truck, and the parameter of interest is the proportion of unsuitable oranges. The procedure used is a random sample, stratified by location in the truck. Using this well-chosen sample, the workers at the plant can estimate the proportion of unsuitable oranges on the truck, and decide whether or not to accept the load.

17. The researchers performed a prospective observational study, since the subjects were identified ahead of time. Physically fit men may have a lower risk of death from cancer than other men.

19. **Point spread** Answers may vary. Perform a simulation to determine the gambler's expected winnings. A component is one game. To model that component, generate random digits 0 to 9. Since the outcome after the point spread is a toss-up, let digits 0–4 represent a loss, and let digits 5–9 represent a win. A run consists of 5 games, so generate 5 random digits at a time. The response variable is the profit the gambler makes, after accounting for the $10 bet. If the outcome of the run is 0, 1, or 2 simulated wins, the profit is –$10. If the outcome is 3, 4, or 5 simulated wins, the profit is $0, $10, or $40, respectively. The total profit divided by the number of runs is the average weekly profit. According to the simulation (80 runs were performed), the gambler is expected to break even. His simulated losses equaled his simulated winnings. (In theory, the gambler is expected to lose about $2.19 per game.)

21. **Everyday randomness** Answers will vary. Most of the time, events described as "random" are anything but truly random.

23. **Tips**
   a) The waiters performed an experiment, since treatments were imposed on randomly assigned groups. This experiment has one factor (candy), at two levels (candy or no candy), resulting in two treatments. The response variable is the percentage of the bill given as a tip.

   b) If the decision whether to give candy or not was made before the people were served, the server may have subconsciously introduced bias by treating the customers better. If the decision was made just before the check was delivered, then it is reasonable to conclude that the candy was the cause of the increase in the percentage of the bill given as a tip.

   c) "Statistically significant" means that the difference in the percentage of tips between the candy and no candy groups was more than expected due to sampling variability.

25. Timing Everyone responding to the poll is 'online', and probably younger than the average voter. It may be the case that younger Canadians vote Liberal at a higher proportion than older Canadians. Older citizens also tend to show up to vote on election day in a higher proportion than young Canadians.

27. When to stop?

   a) Answers may vary. A component in this simulation is rolling 1 die. To simulate this component, generate a random digit 1 to 6. To simulate a run, simulate 4 rolls, stopping if a 6 is rolled. The response variable is the sum of the 4 rolls, or 0 if a 6 is rolled. The average number of points scored is the sum of all rolls divided by the total number of runs. According to the simulation, the average number of points scored will be about 5.8.

   b) Answers may vary. A component in this simulation is rolling 1 die. To simulate this component, generate a random digit 1 to 6. To simulate a run, generate random digits until the sum of the digits is at least 12, or until a 6 is rolled. The response variable is the sum of the digits, or 0 if a 6 is rolled. The average number of points scored is the sum of all rolls divided by the total number of runs. According to the simulation, the average number of points scored will be about 5.8, similar to the outcome of the method described in part a).

   c) Answers may vary. Be careful when making your decision about the effectiveness of your strategy. If you develop a strategy with a higher simulated average number of points than the other two methods, this is only an indication that you may win in the long run. If the game is played round by round, with the winner of a particular round being declared as the player with the highest roll made during that round, the game is much more variable. For example, if Player B rolls a 12 in a particular game, Player A will always lose that game, provided he or she sticks to the strategy. A better way to get a feel for your chances of winning this type of game might be to simulate several rounds, recording whether each player won or lost the round. Then estimate the percentage of the time that each player is expected to win, according to the simulation.

29. **Homecoming**
    a) Since telephone numbers were generated randomly, every number that could possibly occur in that community had an equal chance of being selected. This method is "better" than using the phone book, because unlisted numbers are also possible. Those community members who deliberately do not list their phone numbers might not consider this method "better"!
    b) Although this method results in a simple random sample of phone numbers, it does not result in a simple random sample of residences. Residences without a phone are excluded, and residences with more than one phone have a greater chance of being included.
    c) No, this is not a SRS of local voters. People who respond to the survey may be of the desired age, but not registered to vote. Additionally, some voters who are contacted may choose not to participate.
    d) This method does not guarantee an unbiased sample of households. Households in which someone answered the phone may be more likely to have someone at home when the phone call was generated. The attitude about homecoming of these households might not be the same as the attitudes of the community at large.

31. **Smoking and Alzheimer's**
    a) The studies do not prove that smoking offers any protection from Alzheimer's. The studies merely indicate an association. There may be other variables that can account for this association.
    b) Alzheimer's usually shows up late in life. Since smoking is known to be harmful, perhaps smokers have died of other causes before Alzheimer's can be seen.
    c) The only way to establish a cause-and-effect relationship between smoking and Alzheimer's is to perform a controlled, randomized, and replicated experiment. This is unlikely to ever happen, since the factor being studied, smoking, has already been proven harmful. It would be unethical to impose this treatment on people for the purposes of this experiment. A prospective observational study could be designed in which groups of smokers and nonsmokers are followed for many years and the incidence of Alzheimer's disease is tracked.

33. **Sex and violence**

Volunteers → (Random) Group 1 — Violent content / Group 2 — Sexual content / Group 3 — Neutral content → Compare number of brand names recalled

This experiment has one factor (program content), at three levels (violent, sexual, and neutral), resulting in three treatments. The response variable is the number of brand names recalled after watching the program. Numerous subjects will be randomly assigned to see shows with violent, sexual, or neutral content. They will see the same commercials. After the show, they will be interviewed for their recall of brand names in the commercials.

35. Age and party 2008.

a) The number of respondents is roughly the same for each age category. This may indicate a sample stratified by age category, although it may be a simple random sample.

b) 1530 Democrats were surveyed. $\frac{1530}{4002} \approx 38.2\%$ of the people surveyed were Democrats.

c) We don't know. If data were collected from voting precincts that are primarily Democratic or primarily Republican, that would bias the results. Because the survey was commissioned by NBC News, we can assume the data collected are probably reliable. Also, many who say their party preference is Democrat may not necessarily be 'registered' Democrats, so party preference and actual party registration are not exactly the same thing.

d) The pollsters were probably attempting to determine whether or not political party is associated with age.

37. Save the grapes This experiment has one factor (bird control device), at three levels (scarecrow, netting, and no device), resulting in three treatments. Randomly assign different plots in the vineyard to the different treatments, making sure to ensure adequate separation of plots so that the possible effect of the scarecrow will not be confounded with the other treatments. The response variable to be measured at the end of the season is the proportion of bird-damaged grapes in each plot.

39. Acupuncture

a) The "fake" acupuncture was the control group. In an experiment, all subjects must be treated as alike as possible. If there were no "fake" acupuncture, subjects would know that they had not received acupuncture, and might react differently. Of course, all volunteers for the experiment must be aware of the possibility of being randomly assigned to the control group.

b) Experiments always use volunteers. This is not a problem, since experiments are testing response to a treatment, not attempting to determine an unknown population parameter. The randomization in an experiment is random assignment to treatment groups, not random selection from a population. Voluntary response is a problem when sampling, but is not an issue in experimentation. In this case, it is probably reasonable to assume that the volunteers have similar characteristics to others in the population of people with chronic lower back pain.

c) There were differences in the amount of pain relief experienced by the two groups, and these differences were large enough that they could not be explained by natural variation alone. Researchers concluded that both proper and "fake" acupuncture reduced back pain.

41. Security

a) To ensure that passengers from first-class, as well as coach, get searched, select two passengers from first-class and 12 from coach. Using this stratified random sample, 10% of the first-class passengers are searched, as are 10% of the coach passengers.

b) Answers will vary. Number the passengers alphabetically, with 2-digit numbers. Bergman = 01, Bowman = 02, and so on, ending with Testut = 20. Read the random digits in pairs, ignoring pairs 21 to 99 and 00, and ignoring repeated pairs.

```
65|43|67|11|        27|04|          The passengers selected for
XX XX XX Fontana    XX Castillo      search from first-class are
                                     Fontana and Castillo.
```

c) Number the passengers alphabetically, with 3 digit numbers, 001 to 120. Use the random number table to generate 3-digit numbers, ignoring numbers 121 to 999 and 000, and ignoring repeated numbers. Search the passengers corresponding to the first 12 valid numbers generated.

43. Par 4 Answers may vary. A component in this simulation is a shot. Use pairs of random digits 00 to 99 to represent a shot. The way in which this component is simulated depends on the type of shot. For the first shot, let pairs of digits 01 to 70 represent hitting the fairway, and let pairs of digits 71 to 99, and 00, represent not hitting the fairway.

If the first simulated shot hits the fairway, let 01 to 80 represent landing on the green on the second shot, and let 81 to 99, and 00, represent not landing on the green on the second shot. If the first simulated shot does not hit the fairway, let 01 to 40 represent landing on the green on the second shot, and let 41 to 99, and 00, represent not landing on the green on the second shot.

If the second simulated shot does not land on the green, let 01 to 90 represent landing on the green, and 91 to 99, and 00, represent not landing on the green. Keep simulating shots until the shot lands on the green.

Once on the green, let 01 to 20 represent sinking the putt on the first putt, and let 21 to 99, and 00, represent not sinking the putt on the first putt. If second putts are required, continue simulating putts until a putt goes in, with 01 to 90 representing making the putt, and 91 to 99, and 00, representing not making the putt.

A run consists of following the guidelines above until the final putt is made. The response variable is the number of shots required until the final putt is made. The simulated average score on the hole is the total number of shots required divided by the total number of runs. According to 40 runs of this simulation, a pretty good golfer can be expected to average about 4.2 strokes per hole. Your simulation results may vary.

Chapter 12: From Randomness to Probability

1. **Sample spaces**
   a) S = {HH, HT, TH, TT} All of the outcomes are equally likely to occur.
   b) S = {0, 1, 2, 3} All outcomes are not equally likely. A family of three is more likely to have, for example, two boys than three boys. There are three equally likely outcomes that result in two boys (BBG, BGB, and GBB), and only one that results in three boys (BBB).
   c) S = {H, TH, TTH, TTT} All outcomes are not equally likely. For example the probability of getting heads on the first try is $\frac{1}{2}$. The probability of getting three tails is $\left(\frac{1}{2}\right)^3 = \frac{1}{8}$.
   d) S = {1, 2, 3, 4, 5, 6} All outcomes are not equally likely. Since you are recording only the larger number of two dice, 6 will be the larger when the other die reads 1, 2, 3, 4, or 5. The outcome 2 will only occur when the other die shows 1 or 2.

3. **Roulette** If a roulette wheel is to be considered truly random, then each outcome is equally likely to occur, and knowing one outcome will not affect the probability of the next. Additionally, there is an implication that the outcome is not determined through the use of an electronic random number generator.

5. **Winter** Although acknowledging that there is no law of averages, Knox attempts to use the law of averages to predict the severity of the winter. Some winters are harsh and some are mild over the long run, and knowledge of this can help us to develop a long-term probability of having a harsh winter. However, probability does not compensate for odd occurrences in the short term. Suppose that the probability of having a harsh winter is 30%. Even if there are several mild winters in a row, the probability of having a harsh winter is still 30%.

7. **Cold streak** There is no such thing as being "due for a hit." This statement is based on the so-called law of averages, which is a mistaken belief that probability will compensate in the short term for odd occurrences in the past. The batter's chance for a hit does not change based on recent successes or failures.

9. **Fire insurance**
   a) It would be foolish to insure your neighbour's house for $300. Although you would probably simply collect $300, there is a chance you could end up paying much more than $300. That risk probably is not worth the $300.
   b) The insurance company insures many people. The overwhelming majority of customers pay the insurance and never have a claim. The few customers who do have a claim are offset by the many who simply pay their premiums without a claim. The relative risk to the insurance company is low.

11. **Spinner**
    a) This is a legitimate probability assignment. Each outcome has probability between 0 and 1, inclusive, and the sum of the probabilities is 1.
    b) This is a legitimate probability assignment. Each outcome has probability between 0 and 1, inclusive, and the sum of the probabilities is 1.
    c) This is not a legitimate probability assignment. Although each outcome has probability between 0 and 1, inclusive, the sum of the probabilities is greater than 1.
    d) This is a legitimate probability assignment. Each outcome has probability between 0 and 1, inclusive, and the sum of the probabilities is 1. However, this game is not very exciting!
    e) This probability assignment is not legitimate. Although the sum of the probabilities is 1, there is one probability, –0.50, that is not between 0 and 1, inclusive.

13. **Car repairs** Since all of the events listed are disjoint, the addition rule can be used.
    a) P(no repairs) = 1 – P(some repairs) = 1 – (0.17 + 0.07 + 0.04) = 1 – (0.28) = 0.72
    b) P(no more than one repair) = P(no repairs *or* one repair) = 0.72 + 0.17 = 0.89
    c) P(some repairs) = P(one *or* two *or* three *or* more repairs) = 0.17 + 0.07 + 0.04 = 0.28

15. **Ethanol**
    a) 160/500 = 0.32
    b) (155 + 49)/500 = 0.408

17. **Census.** Denote the two events as T (on time) and L (online) respectively. We know that $P(T)$ = 0.66, $P(L)$ = 0.19, $P(T$ and $L)$ = 0.16.
    a) $P(T$ or $L)$ = $P(T)$ + $P(L)$ – $P(T$ and $L)$ = 0.66 +0.19 – 0.16 = 0.69 (69%)
    b) $P(L)$ – $P(L$ and $T)$ = 0.19 – 0.16 = 0.03 (3%)
    c) $P(T)$ – $P(L$ and $T)$ = 0.66 – 0.16 = 0.50 (50%)
    d) $P(T^c$ and $L^c)$ = $P(T$ or $L)^c$ = $1 - P(T$ or $L)$ = $1 - 0.69 = 0.31$ (31%)

19. **Homes** Using obvious notations, we know that $P(G)$ = 0.64, $P(S)$ = 0.21, $P(G$ and $S)$ = 0.17.
    a) $P(S$ or $G)$ = $P(S)$ + $P(G)$ – $P(S$ and $G)$ = 0.21 + 0.64 – 0.17 = 0.68
    b) $P(S^c$ and $G^c)$ = $P(S$ or $G)^c$ = $1 - P(S$ or $G)$ = $1 - 0.68 = 0.32$
    c) $P(S$ and $G^c)$ = $P(S)$ – $P(S$ and $G)$ = $0.21 - 0.17 = 0.04$

21. **Emigration** Let M = randomly selected emigrant is male, O = randomly selected immigrant is at least 18 years old. We know that $P(M)$ = 0.51, $P(O)$ = 0.82, $P(M$ and $O)$ = 0.42.
    a) $P(M$ and $O^c)$ = $P(M)$ – $P(M$ and $O)$ = $0.51 - 0.42 = 0.09$
    b) $P(M$ or $O)$ = $P(M)$ + $P(O)$ – $P(M$ and $O)$ = 0.51 +0.82 - 0.42 =0.91

c) $P(M^c \text{ and } O^c) = P(M \text{ or } O)^c = 1 - P(M \text{ or } O) = 1 - 0.91 = 0.09$

23. **Movies**
    a) 51/240 = 0.2125
    b) (31 + 38)/240 = 0.2875
    c) (14 + 15)/240 = 0.1208
    d) 0.2875 + 0.2125 – 0.1208 = 0.3792

25. **Gambling** There are 36 possible equally likely outcomes of which only {(1,1), (1, 2), (2,1), (6,6)} give a sum of 2, 3, or 12, so the required probability is 4/36 = 1/9.

27. **Poker** If the cards are selected at random (simple random sampling) from the deck of 52 cards, all sets of five cards have the same chance to be the selected set.

29. **Simulations** Answers may vary. Author simulations estimate the probability to be about 90%.

Chapter 13: Probability Rules!

1. **Pet Ownership**

   $$P(\text{dog or cat}) = P(\text{dog}) + P(\text{cat}) - P(\text{dog and cat})$$
   $$= 0.25 + 0.29 - 0.12 = 0.42$$

3. **Sports again**

   $$P(\text{football} \mid \text{no basketball}) = \frac{P(\text{football and no basketball})}{P(\text{no basketball})} = \frac{\frac{38}{100}}{\frac{60}{100}} \approx 0.633s$$

   (Or, use the table. Of the 60 people who don't like to watch basketball, 38 people like to watch football. $38/60 \approx 0.633$)

5. **Titanic** The overall survival rate, $P(S)$, was 0.323, yet the survival rate for first class passengers, $P(S \mid FC)$, was 0.625. Since $P(S) \neq P(S \mid FC)$, survival and ticket class are not independent. Rather, survival rate depended on class.

7. **Ethanol**
   a) $(136/500) \times (135/499) \times (134/498) = 0.01980056499$
   b) There are 345 (500 – 155) adults who did not respond "Maintain current production level" and so none responded, "Maintain current production level" $(345/500) \times (344/499) \times (343/498) = 0.328$
   c) We assume that every group of three people in this group of 500 people has the same chance to be the group selected. (i.e., a simple random sample of size 3).
   d) We assume that random sampling here will ensure it

9. **M&M's**
   a) Proportion of brown candies = 1 – (0.20 + 0.20 + 3 × 0.10) = 0.30.
      $P(\text{all three are brown}) = 0.3 \times 0.3 \times 0.3 = 0.027$
   b) $P(R^c \text{ and } R^c \text{ and } R) = 0.8 \times 0.8 \times 0.2 = 0.128$
   c) $P(Y^c \text{ and } Y^c \text{ and } Y^c) = 0.8 \times 0.8 \times 0.8 = 0.512$
   d) $P(\text{at least one green}) = 1 - P(\text{none are green}) =$
      $1 - P(G^c \text{ and } G^c \text{ and } G^c) = 1 - 0.9 \times 0.9 \times 0.9 = 0.271$

11. **Disjoint or independent?**
    a) They are disjoint because two events cannot happen simultaneously.
    b) They are independent because the second being red does not depend on whether the first was red.
    c) No. If two events are disjoint, whenever one event occurs, the other one cannot occur and so they are not independent.

13. Dice
a) $(1/6)(1/6)(1/6) = 1/216 = .0046$
b) $(3/6)(3/6)(3/6) = 27/216 = 0.125$
c) $(4/6)(4/6)(4/6) = 64/216 = 0.2963$
d) $1 - (5/6)(5/6)(5/6) = 1 - 125/216 = 91/216 = 0.4213$
e) $1 - (1/6)(1/6)(1/6) = 1 - 1/216 = 215/216 = 0.9954$

15. Champion bowler
a) $0.3 \times 0.3 \times 0.3 = 0.027$
b) $0.3 \times 0.3 \times 0.7 = 0.063$
c) $P(\text{at least one strike}) = 1 - P(\text{no strikes}) = 1 - 0.027 = 0.973$
d) $0.7 \times 0.7 \times 0.7 \times 0.7 \times 0.7 \times 0.7 \times 0.7 \times 0.7 \times 0.7 \times 0.7 \times 0.7 \times 0.7 = 0.7^{12} = 0.013841287201$

17. Ottawa
a) $0.76 \times 0 76 \times 0.76 = 0.438976$
b) $0.9 \times 0.9 \times 0.9 = 0.729$
c) $P(\text{at least one person speaks a non-official language at home}) = 1 - P(\text{no one speaks a non-official language}) = 1 - (1 - 0.11) \times (1 - 0.11) \times (1 - 0.11) = 0.295031$

19. Tires
$P(\text{at least one defective}) = 1 - P(\text{none is defective}) = 1 - 0.98 \times 0.98 \times 0.98 \times 0.98 = 0.07763184$

21. Lottery tickets
a) $1 - 0.324 = 0.676$
b) $0.676^6 = 0.0954$
c) 0.324 since these events are independent.
d) $P(WWL) + P(WLW) + P(LWW) = 3 \times 0.324 \times 0.324 \times 0.676 = 0.213$

23. Cards
a) $P(\text{Heart|Red}) = \dfrac{P(\text{Heart and Red})}{P(\text{Red})} = \dfrac{13/52}{26/52} = \dfrac{1}{2}$

b) $P(\text{Red|Heart}) = \dfrac{P(\text{Heart and Red})}{P(\text{Heart})} = \dfrac{13/52}{13/52} = 1$

c) $P(\text{Ace|Red}) = \dfrac{P(\text{Ace and Red})}{P(\text{Red})} = \dfrac{2/52}{26/52} = \dfrac{2}{26} = \dfrac{1}{13}$

d) $P(\text{Queen|Face}) = \dfrac{P(\text{Queen and Face})}{P(\text{Face})} = \dfrac{4/52}{12/52} = \dfrac{1}{3}$ (Jack, King and Queen are face cards)

25. Men's health Construct a two-way table of the conditional probabilities, including the marginal probabilities.

Cholesterol	Blood Pressure		
	High	OK	Total
High	0.11	0.21	0.32
OK	0.16	0.52	0.68
Total	0.27	0.73	1.00

a) P(both conditions) = 0.11

b) P(high blood pressure) = 0.11 + 0.16 = 0.27

c) $P(\text{high chol.}|\text{high BP}) = \dfrac{P(\text{high chol. and high BP})}{P(\text{high BP})} = \dfrac{0.11}{0.27} \approx 0.407$

Consider only the High Blood Pressure column. Within this column, the probability of having high cholesterol is 0.11 out of a total of 0.27.

d) $P(\text{high BP}|\text{high chol.}) = \dfrac{P(\text{high BP and high chol.})}{P(\text{high chol.})} = \dfrac{0.11}{0.32} \approx 0.344$

This time, consider only the High Cholesterol row. Within this row, the probability of having high blood pressure is 0.11, out of a total of 0.32.

27. Movies

a) $15/38 = 0.3947$

b) $15/51 = 0.2941$

c) $(21 + 22 + 38)/(69 + 46 + 56) = 81/171 = 0.4737$

d)

$$P(\text{over 30}|\text{didn't select a comedy}) = \frac{P(\text{over 30 and didn't select a comedy})}{P(\text{didn't select a comedy})}$$

$$= \frac{(9+7+6+39+17+12)/240}{(51+85)/240} = \frac{90}{136} = \frac{45}{68} = 0.6618$$

29. Sick kids Having a fever and having a sore throat are not independent events, so: P(fever and sore throat) = P(Fever) × P(Sore Throat | Fever) = $(0.70)(0.30) = 0.21$. The probability that a kid with a fever has a sore throat is 0.21.

31. Cards again

a)

$$P(\text{first heart drawn is on the third card}) = P(\text{no heart})P(\text{no heart})P(\text{heart})$$

$$= \left(\frac{39}{52}\right)\left(\frac{38}{51}\right)\left(\frac{13}{50}\right) \approx 0.145$$

b)

$$P(\text{all three cards drawn are red}) = P(\text{red})P(\text{red})P(\text{red})$$

$$= \left(\frac{26}{52}\right)\left(\frac{25}{51}\right)\left(\frac{24}{50}\right) \approx 0.118$$

c)

$$P(\text{none of the cards are spades}) = P(\text{no spade})P(\text{no spade})P(\text{no spade})$$

$$= \left(\frac{39}{52}\right)\left(\frac{38}{51}\right)\left(\frac{37}{50}\right) \approx 0.414$$

d)

$$P(\text{at least one of the cards is an ace}) = 1 - P(\text{none of the cards are aces})$$

$$= 1 - \left[P(\text{no ace})P(\text{no ace})P(\text{no ace})\right]$$

$$= 1 - \left(\frac{48}{52}\right)\left(\frac{47}{51}\right)\left(\frac{46}{50}\right) \approx 0.217$$

33. **Batteries** We assume the batteries are not being replaced, use conditional probabilities throughout.

a)

$$P(\text{the first two batteries are good}) = P(\text{good})P(\text{good})$$

$$= \left(\frac{7}{12}\right)\left(\frac{6}{11}\right) \approx 0.318$$

b)

$$P(\text{at least one of the first three batteries works}) = 1 - P(\text{none of the first three batt. work})$$

$$= 1 - \left[P(\text{no good})P(\text{no good})P(\text{no good})\right]$$

$$= 1 - \left(\frac{5}{12}\right)\left(\frac{4}{11}\right)\left(\frac{3}{10}\right) \approx 0.955$$

c)

$$P(\text{the first four batteries are good}) = P(\text{good})P(\text{good})P(\text{good})P(\text{good})$$

$$= \left(\frac{7}{12}\right)\left(\frac{6}{11}\right)\left(\frac{5}{10}\right)\left(\frac{4}{9}\right) \approx 0.071$$

d)

$$P(\text{pick five to find one good}) = P(\text{no good})P(\text{no good})P(\text{no good})P(\text{no good})P(\text{good})$$

$$= \left(\frac{5}{12}\right)\left(\frac{4}{11}\right)\left(\frac{3}{10}\right)\left(\frac{2}{9}\right)\left(\frac{7}{8}\right) \approx 0.009$$

35. **Cards III** Yes, getting an ace is independent of the suit when drawing one card from a well shuffled deck. The overall probability of getting an ace is 4/52, or 1/13, since there are 4 aces in the deck. If you consider just one suit, there is only 1 ace out of 13 cards, so the probability of getting an ace given

that the card is a diamond, for instance, is 1/13. Since the probabilities are the same, getting an ace is independent of the suit.

37. Movies, final showing
   a) Yes. The events of being under 30 and over 50 cannot happen together.
   b) No. These events are disjoint. If one event occurs, the other event cannot occur and so they are not independent.
   c) No. Some customers under 30 are buying comedies and so a customer selected at random can be both a customer buying a comedy and under 30. The events are not disjoint.
   d) No. P(buying a comedy) = 104/240 = 0.4333 and P(buying a comedy | under 30) = (9 + 14)/(31 + 38) = 0.33. P(buying a comedy) ≠ P(buying a comedy | under 30)

39. Men's health, again Consider the two-way table from Exercise 25 Men's health.

	Blood Pressure		
Cholesterol	High	OK	Total
High	0.11	0.21	0.32
OK	0.16	0.52	0.68
Total	0.27	0.73	1.00

High blood pressure and high cholesterol are not independent events. 28.8% of men with OK blood pressure have high cholesterol, while 40.7% of men with high blood pressure have high cholesterol (or compare 40.7% with the unconditional 32% of men who have high cholesterol). If having high blood pressure and high cholesterol were independent, these percentages would be the same.

41. Television
   a) P(Foreign programming | Canadian channels) = 1 – P(Canadian programming | Canadian channels) = 1 – 0.46 = 0.54
   b) Not independent. The proportion of Canadian programming is not the same on Canadian and foreign channels.

43. Luggage Organize the information using a tree diagram.

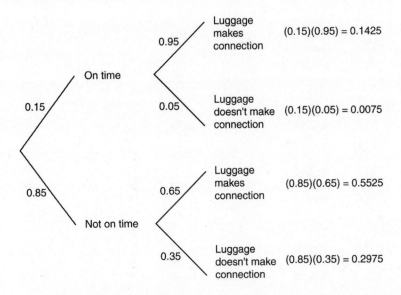

a) No, the flight leaving on time and the luggage making the connection are not independent events. The probability that the luggage makes the connection is dependent on whether or not the flight is on time. The probability is 0.95 if the flight is on time, and only 0.65 if it is not on time.

b)

$$P(\text{luggage}) = P(\text{on time and luggage}) + P(\text{not on time and luggage})$$
$$= (0.15)(0.95) + (0.85)(0.65)$$
$$= 0.695$$

45. Late luggage Refer to the tree diagram constructed for Exercise 43 Luggage.

$$P(\text{not on time}|\text{no lug.}) = \frac{P(\text{not on time and no lug.})}{P(\text{no lug.})} = \frac{(0.85)(0.35)}{(0.15)(0.05)+(0.85)(0.35)} \approx 0.975$$

If you pick Leah up at the Denver airport and her luggage is not there, the probability that her first flight was delayed is 0.975.

47. Absenteeism Organize the information in a tree diagram.

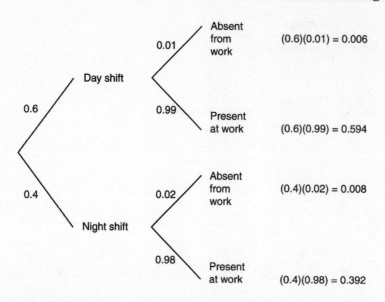

a) No, absenteeism is not independent of shift worked. The rate of absenteeism for the night shift is 2%, while the rate for the day shift is only 1%. If the two were independent, the percentages would be the same.

b)

$P(\text{absent}) = P(\text{day and absent}) + P(\text{night and absent}) = (0.6)(0.01) + (0.4)(0.02)=0.014$

The overall rate of absenteeism at this company is 1.4%.

49. Absenteeism, part II Refer to the tree diagram constructed for Exercise 47 Absenteeism.

$$P(\text{night}|\text{absent})=\frac{P(\text{night and absent})}{P(\text{absent})}=\frac{(0.4)(0.02)}{(0.6)(0.01)+(0.4)(0.02)} \approx 0.571$$

Approximately 57.1% of the company's absenteeism occurs on the night shift.

51. **Drunks** Organize the information into a tree diagram.

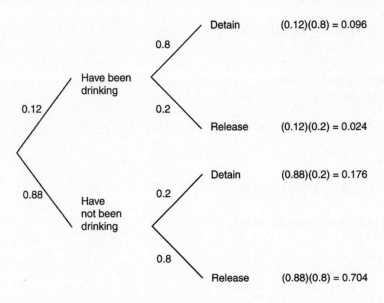

a) $P(\text{detain}|\text{not drinking}) = 0.2$

b)

$P(\text{detain})$

$= P(\text{drinking and det.})$

$\quad + P(\text{not drinking and det.})$

$= (0.12)(0.8) + (0.88)(0.2)$

$= 0.272$

c)

$$P(\text{drunk}|\text{det.}) = \frac{P(\text{drunk and det.})}{P(\text{detain})}$$

$$= \frac{(0.12)(0.8)}{(0.12)(0.8) + (0.88)(0.2)}$$

$$\approx 0.353$$

d)

$$P(\text{drunk}|\text{release}) = \frac{P(\text{drunk and release})}{P(\text{release})}$$

$$= \frac{(0.12)(0.2)}{(0.12)(0.2) + (0.88)(0.8)}$$

$$\approx 0.033$$

53. Dishwashers Organize the information in a tree diagram.

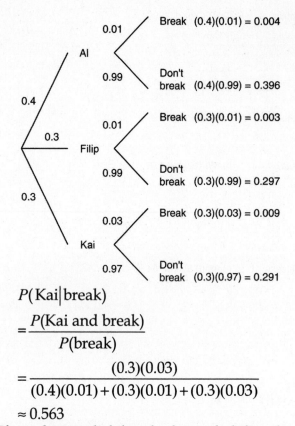

$P(\text{Kai}|\text{break})$

$$= \frac{P(\text{Kai and break})}{P(\text{break})}$$

$$= \frac{(0.3)(0.03)}{(0.4)(0.01)+(0.3)(0.01)+(0.3)(0.03)}$$

≈ 0.563

If you hear a dish break, the probability that Kai is on the job is approximately 0.563.

Chapter 14: Random Variables

1. **Expected value**
   a) $\mu = E(Y) = 10(0.3) + 20(0.5) + 30(0.2) = 19$
   b) $\mu = E(Y) = 2(0.3) + 4(0.4) + 6(0.2) + 8(0.1) = 4.2$

3. **Spinning the wheel**
   a) If you win $100 half the time, you would expect to win an average of $50 per spin.
   b) $E(\text{winnings}) = \$0(0.50) + \$100(0.50) = \$50$

5. **Pick a card, any card**
   a)

Win	$0	$5	$10	$30
P(amount won)	$\dfrac{26}{52}$	$\dfrac{13}{52}$	$\dfrac{12}{52}$	$\dfrac{1}{52}$

   b) $\mu = E(\text{amount won}) = \$0\left(\dfrac{26}{52}\right) + \$5\left(\dfrac{13}{52}\right) + \$10\left(\dfrac{12}{52}\right) + \$30\left(\dfrac{1}{52}\right) \approx \$4.13$

   c) Answers may vary. In the long run, the expected payoff of this game is $4.13 per play. Any amount less than $4.13 would be a reasonable amount to pay to play. Your decision should depend on how long you intend to play. If you are only going to play a few times, you should risk less.

7. **Kids**
   a)

Kids	1	2	3
P(Kids)	0.5	0.25	0.25

   b) $\mu = E(\text{Kids}) = 1(0.5) + 2(0.25) + 3(0.25) = 1.75$ kids

Boys	0	1	2	3
P(boys)	0.5	0.25	0.125	0.125

   c) $\mu = E(\text{Boys}) = 0(0.5) + 1(0.25) + 2(0.125) + 3(0.125) = 0.875$ boys

9. **Software** Since the contracts are awarded independently, the probability that the company will get both contracts is $(0.3)(0.6) = 0.18$. Organize the disjoint events in a Venn diagram.

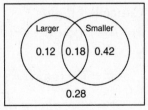

Profit	larger only $50 000	smaller only $20 000	both $70 000	neither $0
P(profit)	0.12	0.42	0.18	0.28

$\mu = E(\text{profit}) = \$50\ 000(0.12) + \$20\ 000(0.42) + \$70\ 000(0.18)$
$\qquad\qquad\quad = \$27\ 000$

11. Variation 1

a)

$$\sigma^2 = Var(Y) = (10-19)^2(0.3) + (20-19)^2(0.5) + (30-19)^2(0.2) = 49$$

$$\sigma = SD(Y) = \sqrt{Var(Y)} = \sqrt{49} = 7$$

b)

$$\sigma^2 = Var(Y) = (2-4.2)^2(0.3) + (4-4.2)^2(0.4) + (6-4.2)^2(0.2) + (8-4.2)^2(0.1) = 3.56$$

$$\sigma = SD(Y) = \sqrt{Var(Y)} = \sqrt{3.56} \approx 1.89$$

13. Pick another card Answers may vary slightly (due to rounding of the mean).

$$\sigma^2 = Var(\text{Won}) = (0-4.13)^2\left(\frac{26}{52}\right) + (5-4.13)^2\left(\frac{13}{52}\right)$$

$$+ (10-4.13)^2\left(\frac{12}{52}\right) + (30-4.13)^2\left(\frac{1}{52}\right) \approx 29.5396$$

$$\sigma = SD(\text{Won}) = \sqrt{Var(\text{Won})} = \sqrt{29.5396} \approx \$5.44$$

15. Kids again

$$\sigma^2 = Var(\text{Kids}) = (1-1.75)^2(0.5) + (2-1.75)^2(0.25) + (3-1.75)^2(0.25) = 0.6875$$

$$\sigma = SD(\text{Kids}) = \sqrt{Var(\text{Kids})} = \sqrt{0.6875} \approx 0.83 \text{ kids}$$

17. Repairs

a) $\mu = E(\text{Number of Repair Calls}) = 0(0.1) + 1(0.3) + 2(0.4) + 3(0.2) = 1.7$ calls

b)

$$\sigma^2 = Var(\text{Calls}) = (0-1.7)^2(0.1) + (1-1.7)^2(0.3) + (2-1.7)^2(0.4) + (3-1.7)^2(0.2) = 0.81$$

$$\sigma = SD(\text{Calls}) = \sqrt{Var(\text{Calls})} = \sqrt{0.81} = 0.9 \text{ calls}$$

19. Defects The percentage of cars with *no* defects is 61%.

$$\mu = E(\text{Defects}) = 0(0.61) + 1(0.21) + 2(0.11) + 3(0.07) = 0.64 \text{ defects}$$

$$\sigma^2 = Var(\text{Defects}) = (0-0.64)^2(0.61) + (1-0.64)^2(0.21)$$

$$+ (2-0.64)^2(0.11) + (3-0.64)^2(0.07) \approx 0.8704$$

$$\sigma = SD(\text{Defects}) = \sqrt{Var(\text{Defects})} \approx \sqrt{0.8704} \approx 0.93 \text{ defects}$$

21. Cancelled flights

a) $\mu = E(\text{gain}) = (-150)(0.20) + 100(0.80) = \$50$

b)

$$\sigma^2 = Var(\text{gain}) = (-150-50)^2(0.20) + (100-50)^2(0.80) = 10,000$$

$$\sigma = SD(\text{gain}) = \sqrt{Var(\text{gain})} \approx \sqrt{10,000} = \$100$$

23. Contest

a) The two games are not independent. The probability that you win the second depends on whether or not you win the first.

b)

$$P(\text{losing both games}) = P(\text{losing the first})\,P(\text{losing the second}\,|\,\text{first was lost})$$
$$= (0.6)(0.7) = 0.42$$

c)

$$P(\text{winning both games}) = P(\text{winning the first})\,P(\text{winning the second}\,|\,\text{first was won})$$
$$= (0.4)(0.2) = 0.08$$

d)

X	0	1	2
$P(X = x)$	0.42	0.50	0.08

e)

$$\mu = E(X) = 0(0.42) + 1(0.50) + 2(0.08) = 0.66 \text{ games}$$
$$\sigma^2 = Var(X) = (0 - 0.66)^2(0.42) + (1 - 0.66)^2(0.50) + (2 - 0.66)^2(0.08) = 0.3844$$
$$\sigma = SD(X) = \sqrt{Var(X)} = \sqrt{0.3844} = 0.62 \text{ games}$$

25. Batteries

a)

Number good	0	1	2
$P(\text{number good})$	$\left(\dfrac{3}{10}\right)\left(\dfrac{2}{9}\right) = \dfrac{6}{90}$	$\left(\dfrac{3}{10}\right)\left(\dfrac{7}{9}\right) + \left(\dfrac{7}{10}\right)\left(\dfrac{3}{9}\right) = \dfrac{42}{90}$	$\left(\dfrac{7}{10}\right)\left(\dfrac{6}{9}\right) = \dfrac{42}{90}$

b) $\mu = E(\text{number good}) = 0\left(\dfrac{6}{90}\right) + 1\left(\dfrac{42}{90}\right) + 2\left(\dfrac{42}{90}\right) = 1.4 \text{ batteries}$

c)

$$\sigma^2 = Var(\text{number good}) = (0 - 1.4)^2\left(\frac{6}{90}\right) + (1 - 1.4)^2\left(\frac{42}{90}\right) + (2 - 1.4)^2\left(\frac{42}{90}\right) \approx 0.3733$$
$$\sigma = SD(\text{number good}) = \sqrt{Var(\text{number good})} \approx \sqrt{0.3733} \approx 0.61 \text{ batteries.}$$

27. Random variables

a)
$$\mu = E(3X) = 3(E(X)) = 3(10) = 30$$
$$\sigma = SD(3X) = 3(SD(X)) = 3(2) = 6$$

b)
$$\mu = E(Y + 6) = E(Y) + 6 = 20 + 6 = 26$$
$$\sigma = SD(Y + 6) = SD(Y) = 5$$

c)
$$\mu = E(X + Y) = E(X) + E(Y) = 10 + 20 = 30$$
$$\sigma = SD(X + Y) = \sqrt{Var(X) + Var(Y)}$$
$$= \sqrt{2^2 + 5^2} \approx 5.39$$

d)

$$\mu = E(X-Y) = E(X) - E(Y) = 10 - 20 = -10$$

$$\sigma = SD(X-Y) = \sqrt{Var(X) + Var(Y)}$$

$$= \sqrt{2^2 + 5^2} \approx 5.39$$

e)

$$\mu = E(X_1 + X_2) = E(X) + E(X) = 10 + 10 = 20$$

$$\sigma = SD(X_1 + X_2) = \sqrt{Var(X) + Var(X)}$$

$$= \sqrt{2^2 + 2^2} \approx 2.83$$

29. Random variables III

a)

$$\mu = E(0.8Y) = 0.8(E(Y)) = 0.8(300) = 240$$

$$\sigma = SD(0.8Y) = 0.8(SD(Y)) = 0.8(16) = 12.8$$

b)

$$\mu = E(2X - 100) = 2(E(X)) - 100 = 140$$

$$\sigma = SD(2X - 100) = 2(SD(X)) = 2(12) = 24$$

c)

$$\mu = E(X + 2Y) = E(X) + 2(E(Y))$$

$$= 120 + 2(300) = 720$$

$$\sigma = SD(X + 2Y) = \sqrt{Var(X) + 2^2 Var(Y)}$$

$$= \sqrt{12^2 + 2^2(16^2)} \approx 34.18$$

d)

$$\mu = E(3X - Y) = 3(E(X)) - E(Y)$$

$$= 3(120) - 300 = 60$$

$$\sigma = SD(3X - Y) = \sqrt{3^2 Var(X) + Var(Y)}$$

$$= \sqrt{3^2(12^2) + 16^2} \approx 39.40$$

e)

$$\mu = E(Y_1 + Y_2 + Y_3 + Y_4) = E(Y) + E(Y) + E(Y) + E(Y) = 300 + 300 + 300 + 300 = 1200$$

$$\sigma = SD(Y_1 + Y_2 + Y_3 + Y_4) = \sqrt{Var(Y) + Var(Y) + Var(Y) + Var(Y)} = \sqrt{16^2 + 16^2 + 16^2 + 16^2} = 32$$

31. Eggs

a) $\mu = E(\text{Broken eggs in 3 dozen}) = 3(E(\text{Broken eggs in 1 dozen})) = 3(0.6) = 1.8$ eggs

b) $\sigma = SD(\text{Broken eggs in 3 dozen}) = \sqrt{0.5^2 + 0.5^2 + 0.5^2} \approx 0.87$ eggs

c) The cartons of eggs must be independent of each other. This assumption might not be warranted because the cartons might be transported together, for example, so that they share a common history of exposure to breakage. Thus, observing a large number of broken eggs in one carton might imply a large number of broken eggs in the other cartons.

33. Repair calls

$\mu = E(\text{calls in 8 hours}) = 8(E(\text{calls in 1 hour}) = 8(1.7) = 13.6 \text{ calls}$

$\sigma = SD(\text{calls in 8 hours}) = \sqrt{8(Var(\text{calls in 1 hour}))} = \sqrt{8(0.9)^2} \approx 2.55 \text{ calls}$

This is only valid if the hours are independent of one another.

35. Tickets

a)

$\mu = E(\text{tickets for 18 trucks}) = 18(E(\text{tickets for one truck})) = 18(1.3) = 23.4 \text{ tickets}$

$\sigma = SD(\text{tickets for 18 trucks}) = \sqrt{18(Var(\text{tickets for one truck})} = \sqrt{18(0.7)^2} \approx 2.97 \text{ tickets}$

b) We are assuming that trucks are ticketed independently.

c) An unusually bad month might represent a number of tickets that is more than two standard deviations higher than average. This would be $23.4 + 2(2.97) = 29.34$, so an unusually bad month might be one in which the company got 30 or more tickets.

37. Fire!

a) The standard deviation is large because the profits on insurance are highly variable. Although there will be many small gains, there will occasionally be large losses when the insurance company has to pay a claim.

b)

$\mu = E(\text{two policies}) = 2(E(\text{one policy})) = 2(150) = \$300$

$\sigma = SD(\text{two policies}) = \sqrt{2(Var(\text{one policy}))} = \sqrt{2(6000^2)} \approx \$8485.28$

c)

$\mu = E(10\ 000 \text{ policies}) = 10\ 000(E(\text{one policy})) = 10\ 000(150) = \$1\ 500\ 000$

$\sigma = SD(10\ 000 \text{ policies}) = \sqrt{10\ 000(Var(\text{one policy}))} = \sqrt{10\ 000(6000^2)} = \$600\ 000$

d) If the company sells 10 000 policies, they are likely to be successful. A profit of $0 is 2.5 standard deviations below the expected profit. This is unlikely to happen. However, if the company sells fewer policies, then the likelihood of turning a profit decreases. In an extreme case, where only two policies are sold, a profit of $0 is more likely, since it is only a small fraction of a standard deviation below the mean.

e) This analysis depends on each of the policies being independent from each other. This assumption of independence may be violated if there are many fire insurance claims as a result of a forest fire or other natural disaster.

39. Bernoulli

a) These are not Bernoulli trials. The possible outcomes are 1, 2, 3, 4, 5, and 6. There are more than two possible outcomes.

b) These may be considered Bernoulli trials. There are only two possible outcomes, Type A and not Type A. Assuming the 120 donors are representative of the population, the probability of having Type A blood is 43%. The trials are not independent, because the population is finite, but the 120 donors represent less than 10% of all possible donors.

c) These are not Bernoulli trials. The probability of getting a heart changes as cards are dealt without replacement.

d) These are not Bernoulli trials. We are sampling without replacement, so the trials are not independent. Samples without replacement may be considered Bernoulli trials if the sample size is less than 10% of the population, but 500 is more than 10% of 3000.

e) These may be considered Bernoulli trials. There are only two possible outcomes, sealed properly and not sealed properly. The probability that a package is unsealed is constant, at about 10%, as long as the packages checked are a representative sample of all packages. Finally, the trials are not independent, since the total number of packages is finite, but the 24 packages checked probably represent less than 10% of the packages.

41. Crosby again

a) Answers will vary. A component is the simulation of the picture in one box of cereal. One possible way to model this component is to generate random digits 0–9. Let 0 and 1 represent Sidney Crosby and 2–9 a picture of another sports star. Each run will consist of generating five random numbers. The response variable will be the number of 0s and 1s in the five random numbers.

b) Answers will vary.

c) Answers will vary. To construct your simulated probability model, start by calculating the simulated probability that you get no pictures of Sidney Crosby in the five boxes. This is the number of trials in which neither 0 nor 1 were generated divided by the total number of trials. Perform similar calculations for the simulated probability that you would get one picture, two pictures, etc.

d) Let X = the number of Sidney Crosby pictures in five boxes.

X	0	1	2	3	4	5
$P(X)$	$(0.20)^0(0.80)^5$ ≈ 0.33	$\binom{5}{1}(0.20)^1(0.80)^4$ ≈ 0.41	$\binom{5}{2}(0.20)^2(0.80)^3$ ≈ 0.20	$\binom{5}{3}(0.20)^3(0.80)^2$ ≈ 0.05	$\binom{5}{4}(0.20)^4(0.80)^1$ ≈ 0.01	$(0.20)^5(0.80)^0$ ≈ 0.0

e) Answers will vary.

43. On time

These departures cannot be considered Bernoulli trials. Departures from the same airport during a two-hour period may not be independent. They all might be affected by weather and delays.

45. Coins and intuition

a) Intuitively, we expect 50 heads.

b) $E(\text{heads}) = np = 100(0.5) = 50$ heads.

47. Lefties.

a) The selection of lefties (and also righties) can be considered Bernoulli trials. Since our group consists of 12 people, and we are considering the righties, use $Binom(12, 0.87)$.

Let Y = the number of righties among $n = 12$.
$E(Y) = np = 12(0.87) = 10.44$ righties
$SD(Y) = \sqrt{npq} = \sqrt{12(0.87)(0.13)} \approx 1.16$ righties

b)

$$P(\text{not all righties}) = 1 - P(\text{all righties})$$
$$= 1 - P(Y = 12)$$
$$= 1 - \binom{12}{12}(0.87)^{12}(0.13)^{0}$$
$$\approx 0.812$$

c)

$$P(\text{no more than 10 righties}) = P(Y \le 10)$$
$$= P(Y = 0) + P(Y = 1) + P(Y = 2) + \ldots + P(Y = 10)$$
$$= \binom{12}{0}(0.87)^{0}(0.13)^{12} + \binom{12}{1}(0.87)^{1}(0.13)^{11} + \ldots + \binom{12}{10}(0.87)^{10}(0.13)^{2}$$
$$\approx 0.475$$

d)

$$P(\text{exactly six of each}) = P(Y = 6)$$
$$= \binom{12}{6}(0.87)^{6}(0.13)^{6}$$
$$\approx 0.00193$$

e)

$$P(\text{majority righties}) = P(Y \ge 7)$$
$$= P(Y = 7) + P(Y = 8) + P(Y = 9) + \ldots + P(Y = 12)$$
$$= \binom{12}{7}(0.87)^{7}(0.13)^{5} + \binom{12}{8}(0.87)^{8}(0.13)^{4} + \ldots + \binom{12}{12}(0.87)^{12}(0.13)^{0}$$
$$\approx 0.998$$

49. **Vision** The vision tests can be considered Bernoulli trials. There are only two possible outcomes, nearsighted or not. The probability of any child being nearsighted is given as $p = 0.12$. Finally, since the population of children is finite, the trials are not independent. However, 169 is certainly less than 10% of all children, and we will assume that the children in this district are representative of all children in relation to nearsightedness. Use $Binom(169, 0.12)$.

$$\mu = E(\text{nearsighted}) = np = 169(0.12) = 20.28 \text{ children.}$$

$$\sigma = SD(\text{nearsighted}) = \sqrt{npq} = \sqrt{169(0.12)(0.88)} \approx 4.22 \text{ children.}$$

51. **Tennis, anyone?** The first serves can be considered Bernoulli trials. There are only two possible outcomes, successful and unsuccessful. The probability of any first serve being good is given as $p = 0.70$. Finally, we are assuming that each serve is independent of the others. Since she is serving six times, use $Binom(6, 0.70)$.

Let $X =$ the number of successful serves in $n = 6$ first serves.

a)

$$P(\text{all six serves in}) = P(X = 6)$$

$$= \binom{6}{6}(0.70)^6(0.30)^0$$

$$\approx 0.118$$

b)

$$P(\text{exactly four serves in}) = P(X = 4)$$

$$= \binom{6}{4}(0.70)^4(0.30)^2$$

$$\approx 0.324$$

c)

$$P(\text{at least four serves in}) = P(X = 4) + P(X = 5) + P(X = 6)$$

$$= \binom{6}{4}(0.70)^4(0.30)^2 + \binom{6}{5}(0.70)^5(0.30)^1 + \binom{6}{6}(0.70)^6(0.30)^0$$

$$\approx 0.744$$

d)

$$P(\text{no more than four serves in}) = P(X = 0) + P(X = 1) + P(X = 2) + P(X = 3) + P(X = 4)$$

$$= \binom{6}{0}(0.70)^0(0.30)^6 + \binom{6}{1}(0.70)^1(0.30)^5 + \binom{6}{2}(0.70)^2(0.30)^4$$

$$+ \binom{6}{3}(0.70)^3(0.30)^3 + \binom{6}{4}(0.70)^4(0.30)^2$$

$$\approx 0.580$$

53. And more tennis The first serves can be considered Bernoulli trials. There are only two possible outcomes, successful and unsuccessful. The probability of any first serve being good is given as $p = 0.70$. Finally, we are assuming that each serve is independent of the others. Since she is serving 80 times, use *Binom*(80, 0.70).

Let X = the number of successful serves in $n = 80$ first serves.

a) $E(X) = np = 80(0.70) = 56$ first serves in.

$SD(X) = \sqrt{npq} = \sqrt{80(0.70)(0.30)} \approx 4.10$ first serves in.

b) Since $np = 56$ and $nq = 24$ are both greater than 10, *Binom*(80, 0.70) may be approximated by the Normal model, $N(56, 4.10)$.

c) According to the Normal model, in matches with 80 serves, she is expected to make between 51.9 and 60.1 first serves approximately 68% of the time, between 47.8 and 64.2 first serves approximately 95% of the time, and between 43.7 and 68.3 first serves approximately 99.7% of the time.

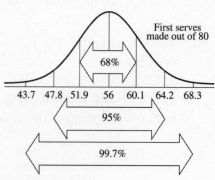

d) Using *Binom*(80, 0.70):

P(at least 65 first serves) = $P(X \geq 65)$

$$= P(X = 65) + P(X = 66) + \ldots + P(X = 80)$$

$$= \binom{80}{65}(0.70)^{65}(0.30)^{15} + \binom{80}{66}(0.70)^{66}(0.30)^{14} + \ldots + \binom{80}{80}(0.70)^{80}(0.30)^{0}$$

$$\approx 0.0161$$

According to the Binomial model, the probability that she makes at least 65 first serves out of 80 is approximately 0.0161.

Using $N(56, 4.10)$:

$$z = \frac{x - \mu}{\sigma}$$

$$z = \frac{65 - 56}{4.10}$$

$$z \approx 2.195$$

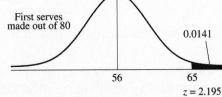

First serves made out of 80

0.0141

56   65

$z = 2.195$

According to the Normal model, the probability that she makes at least 65 first serves out of 80 is approximately 0.0141.

$$P(X \geq 65) \approx P(z > 2.195) \approx 0.0141$$

55. Apples

a) A Binomial model and a Normal model are both appropriate for modelling the number of cider apples that may come from the tree.

Let X = the number of cider apples found in the $n = 300$ apples from the tree.

The quality of the apples may be considered Bernoulli trials. There are only two possible outcomes, cider apple or not a cider apple. The probability that an apple must be used for a cider apple is constant, given as $p = 0.06$. The trials are not independent, since the population of apples is finite, but the apples on the tree are undoubtedly less than 10% of all the apples that the farmer has ever produced, so model with *Binom*(300, 0.06).

$E(X) = np = 300(0.06) = 18$ cider apples

$SD(X) = \sqrt{npq} = \sqrt{300(0.06)(0.94)} \approx 4.11$ cider apples

Since $np = 18$ and $nq = 282$ are both greater than 10, *Binom*(300, 0.06) may be approximated by the Normal model, $N(18, 4.11)$.

b) Using *Binom*(300, 0.06):

P(at most 12 cider apples) = $P(X \leq 12)$

$$= P(X = 0) + \ldots + P(X = 12)$$

$$= \binom{300}{0}(0.06)^{0}(0.94)^{300} + \ldots + \binom{300}{12}(0.06)^{12}(0.94)^{282}$$

$$\approx 0.085$$

According to the Binomial model, the probability that no more than 12 cider apples come from the tree is approximately 0.085.

Using $N(18, 4.11)$:

$$z = \frac{x - \mu}{\sigma}$$

$$z = \frac{12 - 18}{4.11}$$

$$z = -1.460$$

Cider apples
in a tree of
300 apples

0.072

12   18

$z = -1.460$

$P(X \le 12) \approx P(z < -1.460) \approx 0.07$

According to the Normal model, the probability that no more than 12 apples out of 300 are cider apples is approximately 0.072.

c) It is extremely unlikely that the tree will bear more than 50 cider apples. Using the Normal model, $N(18, 4.11)$, 50 cider apples is approximately 7.8 standard deviations above the mean.

57. Lefties again Let X = the number of righties among a class of $n = 188$ students. Using *Binom*(188, 0.87):

These may be considered Bernoulli trials. There are only two possible outcomes, right-handed and not right-handed. The probability of being right-handed is assumed to be constant at about 87%. The trials are not independent, since the population is finite, but a sample of 188 students is certainly fewer than 10% of all people. Therefore, the number of righties in a class of 188 students may be modelled by *Binom*(188, 0.87).

If there are 171 or more righties in the class, some righties have to use a left-handed desk. $P(\text{at least 171 righties}) = P(X \ge 171)$

$$= P(X = 171) + \ldots + P(X = 188)$$

$$= \binom{188}{171}(0.87)^{171}(0.13)^{17} + \ldots + \binom{188}{188}(0.87)^{188}(0.13)^{0}$$

$$\approx 0.061$$

According to the binomial model, the probability that a right-handed student has to use a left-handed desk is approximately 0.061.

Using $N(163.56, 4.61)$:
$E(X) = np = 188(0.87) = 163.56$ righties
$SD(X) = \sqrt{npq} = \sqrt{188(0.87)(0.13)} \approx 4.61$ righties
Since $np = 163.56$ and $nq = 24.44$ are both greater than 10, *Binom*(188, 0.87) may be approximated by the Normal model, $N(163.56, 4.61)$.

$$z = \frac{x - \mu}{\sigma}$$

$$z = \frac{171 - 163.56}{4.61}$$

$$z \approx 1.614$$

Righties in
the class
of 188

0.0533

163.56   171

$z = 1.614$

$P(X \ge 171) \approx P(z > 1.614) \approx 0.05$

According to the Normal model, the probability that there are at least 171 righties in the class of 188 is approximately 0.0533.

59. Annoying phone calls Let X = the number of sales made after making n = 200 calls.
Using *Binom*(200, 0.12):
These may be considered Bernoulli trials. There are only two possible outcomes, making a sale and not making a sale. The telemarketer was told that the probability of making a sale is constant at about p = 0.12. The trials are not independent, since the population is finite, but 200 calls is fewer than 10% of all calls. Therefore, the number of sales made after making 200 calls may be modelled by *Binom*(200, 0.12).

$P(\text{at most } 10) = P(X \le 10)$

$$= P(X = 0) + \ldots + P(X = 10)$$

$$= \binom{200}{0}(0.12)^0(0.88)^{200} + \ldots + \binom{200}{10}(0.12)^{10}(0.88)^{190}$$

$$\approx 0.0006$$

According to the Binomial model, the probability that the telemarketer would make at most 10 sales is approximately 0.0006.

Using N(24, 4.60):
$E(X) = np = 200(0.12) = 24$ sales
$SD(X) = \sqrt{npq} = \sqrt{200(0.12)(0.88)} \approx 4.60$ sales
Since $np = 24$ and $nq = 176$ are both greater than 10, *Binom*(200,0.12) may be approximated by the Normal model, N(24, 4.60).

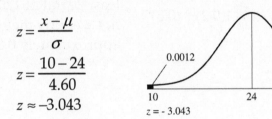

$z = \dfrac{x - \mu}{\sigma}$

$z = \dfrac{10 - 24}{4.60}$

$z \approx -3.043$

$P(X \le 10) \approx P(z < -3.043) \approx 0.0012$

According to the Normal model, the probability that the telemarketer would make at most 10 sales is approximately 0.0012.

Since the probability that the telemarketer made 10 sales, given that the 12% of calls result in sales is so low, it is likely that he was misled about the true success rate.

61. Hurricanes, redux
a) $E(X) = \lambda = 2.7$

$$P(\text{no hurricanes next year}) = \frac{e^{-2.7} 2.7^0}{0!} \approx 0.0672$$

b) $P(\text{“exactly one hurricane in next 2 years”})$
$= P(\text{“hurricane yr 1”}) \, P(\text{“no hurricane yr 2”}) \,\, {}_+ P(\text{“no hurricane yr 1”}) \, P(\text{“hurricane yr 2})\text{”}$

$$= \left(\frac{e^{-2.7}(2.7)^1}{1!}\right)\left(\frac{e^{-2.7}(2.7)^0}{0!}\right) + \left(\frac{e^{-2.7}(2.7)^0}{0!}\right)\left(\frac{e^{-2.7}(2.7)^1}{1!}\right) \approx 0.0244$$

63. HIV again
a) $E(X) = \lambda = np = 8000(0.0005) = 4$ cases

b)

$$P(\text{at least one new case}) = 1 - P(\text{no new cases})$$

$$= 1 - \frac{e^{-4}(4)^0}{0!} \approx 0.9817$$

65. **Seatbelts II** These stops may be considered Bernoulli trials. There are only two possible outcomes, belted or not belted. Police estimate that the probability that a driver is buckled is 80%. (The probability of not being buckled is therefore 20%.) Provided the drivers stopped are representative of all drivers, we can consider the probability constant. The trials are not independent since the population of drivers is finite, but the police will not stop more than 10% of all drivers.

a) $P(\text{The first ten drivers are wearing seatbelts}) = (0.8)^{10} \approx .107$

b) Let Y = the number of drivers wearing their seatbelts in 30 cars.
Use $Binom(30, 0.8)$.
$E(Y) = np = 30(0.8) = 24$ drivers
$SD(Y) = \sqrt{npq} = \sqrt{30(0.8)(0.2)} \approx 2.19$ drivers

c) Let W = the number of drivers not wearing their seatbelts in 120 cars.
Using $Binom(120, 0.2)$:
$P(\text{at least } 20) = P(W \geq 20)$

$$= P(W = 20) + \ldots + P(W = 120)$$

$$= \binom{120}{20}(0.2)^{20}(0.8)^{100} + \ldots + \binom{120}{120}(0.2)^{120}(0.8)^{0}$$

$$\approx 0.848$$

According to the Binomial model, the probability that at least 20 out of 120 drivers are not wearing their seatbelts is approximately 0.848.

Using $N(24, 4.38)$:
$E(W) = np = 120(0.2) = 24$ drivers
$SD(W) = \sqrt{npq} = \sqrt{120(0.2)(0.8)} \approx 4.38$ drivers
Since $np = 24$ and $nq = 96$ are both greater than 10, $Binom(120, 0.2)$ may be approximated by the Normal model, $N(24, 4.38)$.

$$z = \frac{w - \mu}{\sigma}$$

$$z = \frac{20 - 24}{4.38}$$

$$z \approx -0.913$$

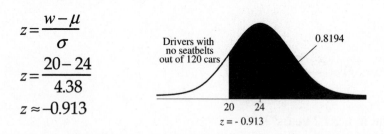

$P(W \geq 120) \approx P(z > -0.913) \approx 0.81$

According to the Normal model, the probability that at least 20 out of 120 drivers stopped are not wearing their seatbelts is approximately 0.8194.

67. **ESP** Choosing symbols may be considered Bernoulli trials. There are only two possible outcomes, correct or incorrect. Assuming that ESP does not exist, the probability of a correct identification from a randomized deck is constant, at $p = 0.20$. The trials are

independent as long as the deck is shuffled after each attempt. Since 100 trials will be performed, use *Binom*(100, 0.2).

Let X = the number of symbols identified correctly out of 100 cards.

$E(X) = np = 100(0.2) = 20$ correct identifications

$SD(X) = \sqrt{npq} = \sqrt{100(0.2)(0.8)} = 4$ correct identifications

Answers may vary. To be convincing, the "mind reader" would have to identify at least 32 out of 100 cards correctly, since 32 is three standard deviations above the mean. Identifying fewer cards than 32 could happen too often, simply due to chance.

69. Hot hand A streak like this is not unusual. The probability that he makes four in a row with a 55% free throw percentage is $(0.55)(0.55)(0.55)(0.55) \approx 0.09$. We can expect this to happen nearly one in ten times for every set of four shots that he makes. One out of ten times is not that unusual.

71. Hotter hand The shots may be considered Bernoulli trials. There are only two possible outcomes, make or miss. The probability of success is constant at 55%, and the shots are independent of one another. Therefore, we can model this situation with *Binom*(32, 0.55).

Let X = the number of free throws made out of 40.

$E(X) = np = 40(0.55) = 22$ free throws made

$SD(X) = \sqrt{npq} = \sqrt{40(0.55)(0.45)} \approx 3.15$ free throws

Answers may vary. The player's performance seems to have increased. Making 32 free throws is $(32 - 22)/3.15 \approx 3.17$ standard deviations above the mean, an extraordinary feat, unless his free throw percentage has increased. This does NOT mean that the sneakers are responsible for the increase in free throw percentage. Some other variable may account for the increase. The player would need to set up a controlled experiment to determine what effect, if any, the sneakers had on his free throw percentage.

73. Continuous uniform model

a)

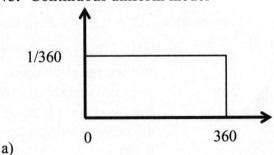

b) Shade area under flat distribution between 20 and 110.

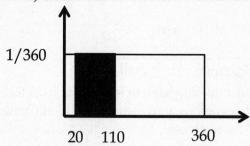

$$P(\ 20 < X < 110\ ) = \text{Shaded area} = \text{height} \times \text{width} = (1/360) \times 90 = 0.25$$

75.   Cereal

a) $E(\text{large bowl} - \text{small bowl}) = E(\text{large bowl}) - E(\text{small bowl}) = 2.5 - 1.5 = 1$ ounce

b)

$$\sigma = SD(\text{large bowl} - \text{small bowl}) = \sqrt{Var(\text{large}) + Var(\text{small})} = \sqrt{0.4^2 + 0.3^2} = 0.5 \text{ ounces}$$

c)

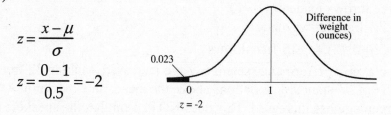

$$z = \frac{x - \mu}{\sigma}$$

$$z = \frac{0 - 1}{0.5} = -2$$

The small bowl will contain more cereal than the large bowl when the difference between the amounts is less than 0. According to the Normal model, the probability of this occurring is approximately 0.023.

d)

$$\mu = E(\text{large bowl} + \text{small bowl}) = E(\text{large bowl}) + E(\text{small bowl}) = 2.5 + 1.5 = 4 \text{ ounces}$$

$$\sigma = SD(\text{large bowl} + \text{small bowl}) = \sqrt{Var(\text{large}) + Var(\text{small})} = \sqrt{0.4^2 + 0.3^2} = 0.5 \text{ ounces}$$

e)

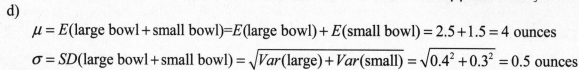

$$z = \frac{x - \mu}{\sigma}$$

$$z = \frac{4.5 - 4}{0.5} = 1$$

According to the Normal model, the probability that the total weight of cereal in the two bowls is more than 4.5 ounces is approximately 0.159.

f)

$$\mu = E(\text{box} - \text{large} - \text{small}) = E(\text{box}) - E(\text{large}) - E(\text{small}) = 16.3 - 2.5 - 1.5 = 12.3 \text{ ounces}$$

$$\sigma = SD(\text{box} - \text{large} - \text{small}) = \sqrt{Var(\text{box}) + Var(\text{large}) + Var(\text{small})}$$

$$= \sqrt{0.2^2 + 0.3^2 + 0.4^2} \approx 0.54 \text{ ounces}$$

77. **More cereal**

a)
$$\mu = E(\text{box} - \text{large} - \text{small}) = E(\text{box}) - E(\text{large}) - E(\text{small}) = 16.2 - 2.5 - 1.5 = 12.2 \text{ ounces}$$

b)
$$\sigma = SD(\text{box} - \text{large} - \text{small}) = \sqrt{Var(\text{box}) + Var(\text{large}) + Var(\text{small})}$$
$$= \sqrt{0.1^2 + 0.3^2 + 0.4^2} \approx 0.51 \text{ ounces}$$

c)

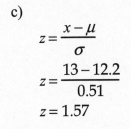

$$z = \frac{x - \mu}{\sigma}$$
$$z = \frac{13 - 12.2}{0.51}$$
$$z = 1.57$$

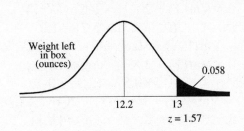

According to the Normal model, the probability that the box contains more than 13 ounces is about 0.058.

79. **Medley**

a)
$$\mu = E(\#1 + \#2 + \#3 + \#4) = E(\#1) + E(\#2) + E(\#3) + E(\#4)$$
$$= 50.72 + 55.51 + 49.43 + 44.91 = 200.57 \text{ seconds}$$

$$\sigma = SD(\#1 + \#2 + \#3 + \#4) = \sqrt{Var(\#1) + Var(\#2) + Var(\#3) + Var(\#4)}$$
$$= \sqrt{0.24^2 + 0.22^2 + 0.25^2 + 0.21^2} \approx 0.46 \text{ seconds}$$

b)

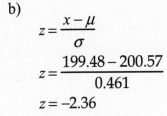

$$z = \frac{x - \mu}{\sigma}$$
$$z = \frac{199.48 - 200.57}{0.461}$$
$$z = -2.36$$

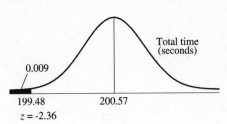

The team is not likely to swim faster than their best time. According to the Normal model, they are only expected to swim that fast or faster about 0.9% of the time.

81. **Farmers' market**

a) Let A = price of a pound of apples, and let P = price of a pound of potatoes.
Profit = $100A + 50P - 2$

b) $\mu = E(100A + 50P - 2) = 100(E(A)) + 50(E(P)) - 2 = 100(0.5) + 50(0.3) - 2 = \$63$

c)
$$\sigma = SD(100A + 50P - 2) = \sqrt{100^2(Var(A)) + 50^2(Var(P))} = \sqrt{100^2(0.2^2) + 50^2(0.1^2)} \approx \$20.62$$

d) No assumptions are necessary to compute the mean. To compute the standard deviation, independent market prices must be assumed.

83. **Geometric model**

a) Yes.
b) No, since we are not counting the number of events in a fixed number of trials.
c) $P(\text{TTTTH}) = 0.5^5 = 0.03125$
d) $P(3) = 0.5 \times 0.5^{3-1} = 0.125$

e) Very skewed since as you increase x, the power of q increases. Since q is less than 1, this part of the distribution formula decreases (geometrically), so the probabilities always decrease as x increases.

85. Correlated RVs.

a) $\text{Var}(W) = 0.5^2\text{Var}(X) + 0.5^2\text{Var}(Y) = \text{Var}(W) = 0.5^2 \times 4 + 0.5^2 \times 4 = 2.0$

b) $\text{Var}(W) = \text{Var}(\frac{1}{2}X + \frac{1}{2}Y) = \text{Var}(\frac{1}{2}X) + \text{Var}(\frac{1}{2}Y) + 2\text{Cov}(\frac{1}{2}X, \frac{1}{2}Y)$
$\qquad = 0.5^2\text{Var}(X) + 0.5^2\text{Var}(Y) + 2 \times \frac{1}{2} \times \frac{1}{2}\text{Cov}(X, Y)$
$\qquad = 0.5^2 \times 4 + 0.5^2 \times 4 + 2(0.5)(0.5) \times (-2) = 1.0$

c) $\text{Var}(W) = \text{Var}(\frac{1}{2}X + \frac{1}{2}Y) = \text{Var}(\frac{1}{2}X) + \text{Var}(\frac{1}{2}Y) + 2\text{Cov}(\frac{1}{2}X, \frac{1}{2}Y)$
$\qquad = 0.5^2\text{Var}(X) + 0.5^2\text{Var}(Y) + 2 \times \frac{1}{2} \times \frac{1}{2}\text{Cov}(X, Y)$
$\qquad = 0.5^2 \times 4 + 0.5^2 \times 4 + 2(0.5)(0.5) \times (2) = 3.0$

d) $\text{Var}(W) = \text{Var}(\frac{1}{2}X + \frac{1}{2}Y) = \text{Var}(\frac{1}{2}X) + \text{Var}(\frac{1}{2}Y) + 2\text{Cov}(\frac{1}{2}X, \frac{1}{2}Y)$
$\qquad = 0.5^2\text{Var}(X) + 0.5^2\text{Var}(Y) + 2 \times \frac{1}{2} \times \frac{1}{2}\text{Cov}(X,Y)$
$\qquad = 0.5^2 \times 4 + 0.5^2 \times 4 + 2(0.5)(0.5) \times (4) = 4.0$

e) The smallest variance occurs when the investments are negatively correlated.

Part IV Review: Randomness and Probability

1. **Quality Control** Construct a Venn diagram of the disjoint outcomes.

   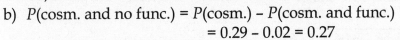

   a) $P(\text{defect}) = P(\text{cosm.}) + P(\text{func.}) - P(\text{cosm. and func.})$
   $$= 0.29 + 0.07 - 0.02 = 0.34$$
   Or, from the Venn: $0.27 + 0.02 + 0.05 = 0.34$

   b) $P(\text{cosm. and no func.}) = P(\text{cosm.}) - P(\text{cosm. and func.})$
   $$= 0.29 - 0.02 = 0.27$$
   Or, from the Venn: 0.27 (the region inside Cosmetic circle, yet outside Functional circle).

   c) $P(\text{func.} \mid \text{cosm.}) = \dfrac{P(\text{func. and cosm.})}{P(\text{cosm.})} = \dfrac{0.02}{0.29} \approx 0.069$

   From the Venn, consider only the region inside the Cosmetic circle. The probability that the car has a functional defect is 0.02 out of a total of 0.29 (the entire Cosmetic circle).

   d) The two kinds of defects are not disjoint events, since 2% of cars have both kinds.

   e) Approximately 6.9% of cars with cosmetic defects also have functional defects. Overall, the probability that a car has a cosmetic defect is 7%. The probabilities are estimates, so these are probably close enough to say that the two types of defects are independent.

3. **Airfares**

   a) Let C = the price of a ticket to China
   Let F = the price of a ticket to France
   Total price of airfare = $3C + 5F$

   b) $\mu = E(3c + 5F) = 3E(C) + 5E(F) = 3(1200) + 5(900) = \$8100$

   $\sigma = SD(3C + 5F) = \sqrt{3^2(Var(C)) + 5^2(Var(F))} = \sqrt{3^2(150^2) + 5^2(100^2)} \approx \$672.68$

   c) $\mu = E = (C - F) = E(C) - E(F) = 1200 - 900 = \$300$

   $\sigma = SD(C - F) = \sqrt{Var(C) + Var(F)}$
   $$= \sqrt{150^2 + 100^2} \approx \$180.28$$

   d) No assumptions are necessary when calculating means. When calculating standard deviations, we must assume that ticket prices are independent of each other for different countries but all tickets to the same country are at the same price.

5. **A game**

X	$0	$2	– $2
P(X)	0.10	0.40	0.50

a) Let X = net amount won

$\mu = E(X) = 0(0.10) + 2(0.40) - 2(0.50) = -\$0.20$

$\sigma^2 = Var(X) = (0 - (-0.20))^2(0.10) + (2 - (-0.20))^2(0.40) + (-2 - (-0.20))^2(0.50) = 3.56$

$\sigma = SD(X) = \sqrt{Var(X)} = \sqrt{3.56} \approx \$1.89$

b) $X + X$ = the total winnings for two plays

$\mu = E(X + X) = E(X) + E(X) = (-0.20) + (-0.20) = -\$0.40$

$\sigma = SD(X + X) = \sqrt{Var(X) + Var(X)}$

$\qquad\qquad\quad = \sqrt{3.56 + 3.56} \approx \$2.67$

7. Facebook

The selection of these people can be considered Bernoulli trials. There are two possible outcomes, on Facebook or not on Facebook. As long as the people selected are representative of the population of all people, then $p = 0.50$ for the adults, and $p = 0.75$ for the teenagers. The trials are not independent since the population of all people is finite, but 10 people are fewer than 10% of the population of all people.

Let X = the number of Facebook users from $n = 10$ adults

Let Y = the number of Facebook users from $n = 10$ teenagers

a) Use $Binom(10, 0.50)$

$$\begin{aligned}
P(\text{at least one not on Facebook}) &= 1 - P(\text{all use Facebooks}) \\
&= 1 - P(X=10) \\
&= 1 - \binom{10}{10}0.5^{10}0.5^{0} \\
&= 0.999
\end{aligned}$$

b) Use $Binom(10, 0.75)$

$$\begin{aligned}
P(\text{at least one not on Facebook}) &= 1 - P(\text{all use Facebooks}) \\
&= 1 - P(Y=10) \\
&= 1 - \binom{10}{10}0.75^{10}0.25^{0} \\
&= 0.944
\end{aligned}$$

c) Use $Binom(5, 0.5)$ and $Binom(5, 0.75)$

$$\begin{aligned}
P(\text{at least one not on Facebook}) &= 1 - P(\text{all use Facebooks})P(\text{all teenagers on Facebook}) \\
&= 1 - P(X = 5)\, P(Y = 5) \\
&= 1 - \left(\binom{5}{5}0.5^{5}0.5^{0}\right)\left(\binom{5}{5}0.75^{5}0.25^{0}\right) \\
&= 0.993
\end{aligned}$$

9. **More Facebook** In Exercise 7, it was determined that these were Bernoulli trials. Use *Binom*(5, 0.75).

Let X = the number of teenagers on Facebook from $n = 5$ teenagers

a)

P(all on Facebook)	$=$	$P(X=5)$
	$=$	$\binom{5}{5}0.75^5 0.25^0$
	$=$	0.237

b)

P(exactly 1 on Facebook)	$=$	$P(X=1)$
	$=$	$\binom{5}{1}0.75^1 0.25^4$
	$=$	0.015

c)

P(at least 3 on Facebook)	$=$	$P(X = 3)+P(X = 4)+P(X = 5)$
	$=$	
		$\binom{5}{3}0.75^3 0.25^2 +\binom{5}{4}0.75^4 0.25^1 \binom{5}{5}0.75^5 0.25^0$
	$=$	0.896

11. **Friend me?**

In Exercise 7, it was determined that these were Bernoulli trials.
Use *Binom*(158, 0.75).
Let X = the number of teenagers on Facebook from $n = 158$ teenagers

a) $E(X) = np = 158(0.75) = 118.5$ teenagers

$SD(X) = \sqrt{npq} = \sqrt{158(0.75)(0.25)} = 5.44$ teenagers

b) Since $np = 118.5$ and $nq = 39.5$ are both greater than 10, the Success/Failure condition is met and *Binom*(158, 0.75) may be approximated by $N(118.5, 5.44)$.

c) Using *Binom*(158, 0.75):

P(no more than 110) = $\qquad\qquad P(X \leq 110)$

$= \qquad P(X = 0)+P(X = 1) +...+ P(X = 110)$

$= \binom{158}{0}0.75^0 0.25^{158} +\binom{158}{1}0.75^1 0.25^{157} +\cdots+\binom{158}{110}0.75^{110}0.25^{48}$

$= \qquad 0.073$

According to the Binomial model, the probability that no more than 110 teenagers are on Facebook is approximately 0.073.

Using $N(118.5, 5.44)$:

$$z = \frac{x - \mu}{\sigma}$$

$$= \frac{110 - 118.5}{5.44}$$

$$= -1.5625 \qquad P(X \le 110) \approx P(z \le -1.5625) = 0.059$$

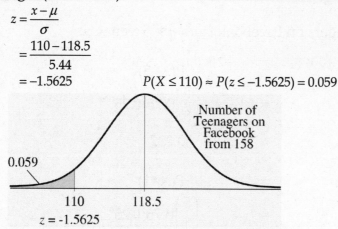

According to the Normal model, the probability that no more than 110 of 158 teenagers are on Facebook is 0.059.

13. **Language** Assuming that the first-year composition class consists of 25 randomly selected people, these may be considered Bernoulli trials. There are only two possible outcomes, having a specified language centre or not having the specified language centre. The probabilities of the specified language centres are constant at 80%, 10%, and 10%, for right, left, and two-sided language centre, respectively. The trials are not independent, since the population of people is finite, but we will select fewer than 10% of all people.

a) Let L be the number of people with left-brain language control from $n = 25$ people

Use $Binom(25, 0.80)$.
$$P(\text{no more than } 15) = P(L \le 15)$$
$$= P(L = 0) + \ldots + P(L = 15)$$
$$= {}_{25}C_0(0.80)^0(0.20)^{25} + \ldots + {}_{25}C_{15}(0.80)^{15}(0.20)^{10}$$
$$\approx 0.0173$$

According to the Binomial model, the probability that no more than 15 students in a class of 25 will have left-brain language centres is approximately 0.0173.

b) Let T = the number of people with two-sided language control from $n = 5$ people
Use $Binom(5, 0.10)$.
$$P(\text{none have two-sided language control}) = P(T = 0)$$
$$= {}_5C_0(0.10)^0(0.90)^5$$
$$\approx 0.590$$

c) Using Binomial models:

$E(\text{left}) = np_L = 1200(0.80) = 960$ people

$E(\text{right}) = np_R = 1200(0.10) = 120$ people

$E(\text{two} - \text{sided}) = np_T = 1200(0.10) = 120$ people

d) Let R = the number of people with right-brain language control

$E(R) = np_R = 1200(0.10) = 120$ people

$SD(R) = \sqrt{np_R q_R} = \sqrt{1200(0.10)(0.90)} \approx 10.39$ people

e) Since $np_R = 120$ and $nq_R = 1080$ are both greater than 10, the Normal model, $N(120, 10.39)$, may be used to approximate $Binom(1200, 0.10)$. According to the Normal model, about 68% of randomly selected groups of 1200 people could be expected to have between 109.61 and 130.39 people with right-brain language control. About 95% of randomly selected groups of 1200 people could be expected to have between 99.22 and 140.78 people with right-brain language control. About 99.7% of randomly selected groups of 1200 people could be expected to have between 88.83 and 151.17 people with right-brain language control.

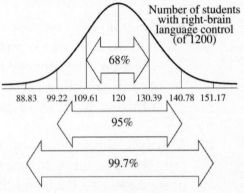

15. **Beanstalks**

a) The greater standard deviation for men's heights indicates that men's heights are more variable than women's heights.

b) Admission to a Beanstalk Club is based upon extraordinary height for both men and women, but men are slightly more likely to qualify. The qualifying height for women is about 2.4 standard deviations above the mean height of women, while the qualifying height for men is about 1.75 standard deviations above the mean height for men.

c) Let M = the height of a randomly selected man from $N(69.1, 2.8)$

Let W = the height of a randomly selected woman from $N(64.0, 2.5)$

$M - W$ = the difference in height of a randomly selected man and woman

d) $E(M - W) = E(M) - E(W) = 69.1 - 64.0 = 5.1$ inches

e) $SD(M - W) = \sqrt{Var(M) + Var(W)} = \sqrt{2.8^2 + 2.5^2} \approx 3.75$ inches

f)   Since each distribution is described by a Normal model, the distribution of the difference in height between a randomly selected man and woman is $N(5.1, 3.75)$.

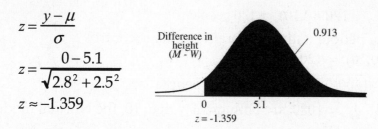

$$z = \frac{y - \mu}{\sigma}$$

$$z = \frac{0 - 5.1}{\sqrt{2.8^2 + 2.5^2}}$$

$$z \approx -1.359$$

According to the Normal model, the probability that a randomly selected man is taller than a randomly selected woman is approximately 0.913.

g) If people chose spouses independent of height, we would expect 91.3% of married couples to consist of a taller husband and shorter wife. The 92% that was seen in the survey is close to 91.3%, and the difference may be due to natural sampling variability. Unless this survey is very large, there is not sufficient evidence of association between height and choice of partner.

17. **Multiple choice** Guessing at questions can be considered Bernoulli trials. There are only two possible outcomes, correct or incorrect. If you are guessing, the probability of success is $p = 0.25$, and the questions are independent. Use $Binom(50, 0.25)$ to model the number of correct guesses on the test.

a) Let X = the number of correct guesses
$$P(at \text{ least } 30 \text{ of } 50 \text{ correct}) = P(X \geq 30)$$
$$= P(X = 30) + ... + P(X = 50)$$
$$= \binom{50}{30}(0.25)^{30}(0.75)^{20} + ... + \binom{50}{50}(0.25)^{50}(0.75)^{0}$$
$$\approx 0.00000016$$

You are very unlikely to pass by guessing on every question.

b) Use $Binom(50, 0.70)$.
$$P(at \text{ least } 30 \text{ of } 50 \text{ correct}) = P(X \geq 30)$$
$$= P(X = 30) + ... + P(X = 50)$$
$$= \binom{50}{30}(0.70)^{30}(0.30)^{20} + ... + \binom{50}{50}(0.70)^{50}(0.30)^{0}$$
$$\approx 0.952$$

According to the Binomial model, your chances of passing are about 95.2%.

19. **Insurance** The company is expected to pay $100 000 only 2.6% of the time, while always gaining $520 from every policy sold. When they pay, they actually only pay $99 480.
$$E(profit) = \$520(0.974) - \$99\,480(0.026) = -\$2080$$
The expected profit is actually a loss of $2080 per policy. The company had better raise its premiums if it hopes to stay in business.

21. **Passing stats** Organize the information in a tree diagram.

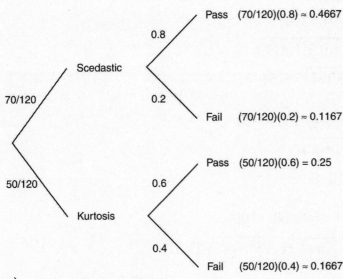

a)

   P(Passing Statistics)

   $= P$(Scedastic and Pass)

   $\quad + P$(Kurtosis and Pass)

   $\approx 0.4667 + 0.25$

   ≈ 0.717

b)

   P(Kurtosis | Fail)

   $= \dfrac{P(\text{Kurtosis and Fail})}{P(\text{Fail})}$

   $\approx \dfrac{0.1667}{0.1167 + 0.1667} \approx 0.588$

23. **Random variables**

   a)

   $\mu = E(X + 50) = E(X) + 50 = 50 + 50 = 100$

   $\sigma = SD(X + 50) = SD(X) = 8$

   b)

   $\mu = E(10Y) = 10E(Y) = 10(100) = 1000$

   $\sigma = SD(10Y) = 10SD(Y) = 60$

c)
$$\mu = E(X + 0.5Y) = E(X) + 0.5E(Y)$$
$$= 50 + 0.5(100) = 100$$
$$\sigma = SD(X + 0.5Y) = \sqrt{Var(X) + 0.5^2 Var(Y)}$$
$$= \sqrt{8^2 + 0.5^2(6^2)} \approx 8.54$$

d)
$$\mu = E(X - Y) = E(X) - E(Y) = 50 - 100 = -50$$
$$\sigma = SD(X - Y) = \sqrt{Var(X) + Var(Y)}$$
$$= \sqrt{8^2 + 6^2} = 10$$

e)
$$\mu = E(X_1 + X_2) = E(X) + E(X) = 50 + 50 = 100$$
$$\sigma = SD(X_1 + X_2) = \sqrt{Var(X) + Var(X)} = \sqrt{8^2 + 8^2} \approx 11.31$$

25. Youth survey
   a) Many boys play computer games and use e-mail, so the probabilities can total more than 100%. There is no evidence that there is a mistake in the report.
   b) Playing computer games and using e-mail are not disjoint. If they were, the probabilities would total 100% or less.
   c) E-mailing friends and being a boy or girl are not independent. 76% of girls e-mailed friends in the last week, but only 65% of boys e-mailed. If e-mailing were independent of being a boy or girl, the probabilities would be the same.

27. Travel to Kyrgyzstan
   a) Spending an average of 4237 soms per day, you can stay about
   $$\frac{90\ 000}{4237} \approx 21 \text{ days.}$$
   b) Assuming that your daily spending is independent, the standard deviation is the square root of the sum of the variances for 21 days.
   $$\sigma = \sqrt{21(360)^2} \approx 1649.73 \text{ soms}$$
   c) The standard deviation in your total expenditures is about 1650 soms, so if you don't think you will exceed your expectation by more than 2 standard deviations, bring an extra 3300 soms. This gives you a cushion of about 157 soms for each of the 21 days.

29. Home sweet home 2011 Since the homes are randomly selected, these can be considered Bernoulli trials. There are only two possible outcomes, owning the home or not owning the home. The probability of any randomly selected resident home being owned by the current resident is 0.67. The trials are not independent, since the population is finite, but as long as the city has more than 8200 homes, we are not sampling more than 10% of the population. The Binomial model, *Binom*(820,

0.67), can be used to model the number of homeowners among the 820 homes surveyed.

Let H = the number of homeowners found in n = 820 homes

$E(H) = np = 820(0.67) = 549.4$ homes

$SD(H) = \sqrt{npq} = \sqrt{820(0.67)(0.33)} = 13.46$ homes

The 523 homeowners found in the candidate's survey represent a number of homeowners that is about 2 standard deviations below the expected number of homeowners. It is somewhat unusual to be 2 standard deviations below the mean. There is some support for the candidate's claim of a low level of home ownership.

31. Who's the boss 2011?

a) P(first three owned by women) = 0.17^3 = 0.005

b) P(none of the first four owned by women) = 0.83^4 = 0.4746

c) P(sixth firm is owned by a woman | none of the first five were) = 0.17
   Since the firms are chosen randomly, the fact that the first five firms were owned by men has no bearing on the ownership of the sixth firm.

33. When to stop?

a) If 6s are not allowed, the mean of each die roll is $\dfrac{1+2+3+4+5}{5} = 3$. You would
   expect to get 15 if you rolled 5 times.

b) $P(5 \text{ rolls without a } 6) = \left(\dfrac{5}{6}\right)^5 \approx 0.402$

35. Technology on campus 2005 Construct a Venn diagram of the disjoint outcomes.

a) P(neither tech.) = 1 – P(either tech.) = 1 – [P(calculator) + P(computer) + P(both)]
   = 1 – [0.51 + 0.21 + 0.10] = 0.38
   Or, from the Venn: 0.38 (the region outside both circles). This is MUCH easier.
   38% of students use neither type of technology.

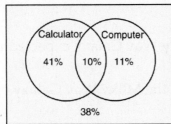

b) P(calc. and no comp.) = P(calc.) – P(calc. and comp.) = 0.51 – 0.10 = 0.41
   Or, from the Venn: 0.41 (region inside the Calculator circle, outside the Computer circle)
   41% of students use calculators, but not computers.

c) P("computer | calculator") = P(comp.&calc.)/P(calc.) = 0.10/0.51 = 0.196
   About 19.6% of calculator users have computer assignments.

d) 19.6% of calculator users were computer users. In classes where calculators were not used, 22.4% of the classes used computers. Since the percentages are different, there is evidence of an association between computer use and calculator use. Calculator classes are less likely to use computers.

37. O-rings
  a) A Poisson model would be used.
  b) If the probability of failure for one O-ring is 0.01, then the mean number of failures for 10 O-rings is $E(X) = \lambda = np = 10(0.01) = 0.1$ O-ring. We are able to calculate this because the Poisson model scales according to sample size.
  c) $P(\text{one failed O - ring}) = \dfrac{e^{-0.1}(0.1)^1}{1!} \approx 0.090$
  d)
  $$P(\text{at least one failed O - ring}) = 1 - P(\text{no failures})$$
  $$= 1 - \frac{e^{-0.1}(0.1)^0}{0!} \approx 0.095$$

39. Socks Since we are sampling without replacement, use conditional probabilities throughout.

  a) $P(\text{2 blue}) = \left(\dfrac{4}{12}\right)\left(\dfrac{3}{11}\right) = \dfrac{12}{132} = \dfrac{1}{11}$

  b) $P(\text{no grey}) = \left(\dfrac{7}{12}\right)\left(\dfrac{6}{11}\right) = \dfrac{42}{132} = \dfrac{7}{22}$

  c) $P(\text{at least one black}) = 1 - P(\text{no black}) = 1 - \left(\dfrac{9}{12}\right)\left(\dfrac{8}{11}\right) = \dfrac{60}{132} = \dfrac{5}{11}$

  d) $P(\text{green}) = 0$ (There aren't any green socks in the drawer.)

  e) $P(\text{match}) = P(\text{2 blue}) + P(\text{2 grey}) + P(\text{2 black}) = \left(\dfrac{4}{12}\right)\left(\dfrac{3}{11}\right) + \left(\dfrac{5}{12}\right)\left(\dfrac{4}{11}\right) + \left(\dfrac{3}{12}\right)\left(\dfrac{2}{11}\right) = \dfrac{19}{66}$

41. The Drake equation
  a) $N \cdot f_p$ represents the number of stars in the Milky Way Galaxy expected to have planets.
  b) $N \cdot f_p \cdot n_e \cdot f_i$ represents the number of planets in the Milky Way Galaxy expected to have intelligent life.
  c) $f_l \cdot f_i$ is the probability that a planet has a suitable environment and has intelligent life.
  d) $f_l = P(\text{life} \mid \text{suitable environment})$. This is the probability that life develops, if a planet has a suitable environment.
  $f_i = P(\text{intelligence} \mid \text{life})$ This is the probability that the life develops intelligence, if a planet already has life.

$f_c = P(\text{communication} \mid \text{intelligence})$ This is the probability that radio communication develops, if a planet already has intelligent life.

43. **Pregnant?** Organize the information in a tree diagram.

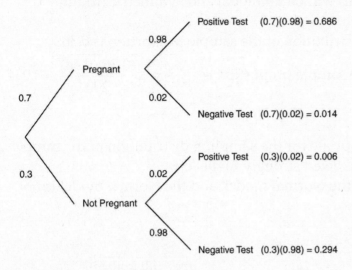

$P(\text{pregnant} \mid \text{positive test})$

$$= \frac{P(\text{pregnant and positive test})}{P(\text{positive test})}$$

$$= \frac{0.686}{0.686 + 0.006}$$

$$\approx 0.991$$

Chapter 15: Sampling Distribution Models

1. **Web site**
   a) Since the sample is drawn at random, and assuming that 200 investors is a small portion of their customers, the sampling distribution for the proportion of 200 investors that use smartphones will be unimodal and symmetric (roughly Normal).
   b) The centre of the sampling distribution of the sample proportion is 0.36.
   c) The standard deviation of the sample proportion is $\sqrt{\dfrac{pg}{n}} = \sqrt{\dfrac{(0.36)(0.64)}{200}} \approx 0.034$.

3. **Sample Maximum**
   a) A Normal model is not appropriate for the sampling distribution of the sample maximum. The histogram is skewed strongly to the left.
   b) No. The 95% rule is based on the Normal model, and the Normal model is not appropriate here.

5. **Market research**
   a) The standard deviation of the sample proportion is $\sqrt{\dfrac{pq}{n}} = \sqrt{\dfrac{(0.15)(0.85)}{100}} \approx 0.0357$.
   b) To reduce the standard deviation by half, she needs a sample 4 times as large, or 400 people.

7. **Send money** All of the histograms are centred around $p = 0.05$. As n gets larger, the shape of the histograms get more unimodal and symmetric, approaching a Normal model, while the variability in the sample proportions decreases.

9. **Send more money**
   a)

n	Observed mean	Theoretical mean	Observed st. dev.	Theoretical standard deviation
20	0.0497	0.05	0.0479	$\sqrt{(0.05)(0.95)/20} \approx 0.0487$
50	0.0516	0.05	0.0309	$\sqrt{(0.05)(0.95)/50} \approx 0.0308$
100	0.0497	0.05	0.0215	$\sqrt{(0.05)(0.95)/100} \approx 0.0218$
200	0.0501	0.05	0.0152	$\sqrt{(0.05)(0.95)/200} \approx 0.0154$

   b) All of the values seem very close to what we would expect from theory.
   c) The histogram for $n = 200$ looks quite unimodal and symmetric. We should be able to use the Normal model.
   d) The Success/Failure Condition requires np and nq to both be at least 10, which is not satisfied until $n = 200$ for $p = 0.05$. The theory supports the choice in part c).

11. Coin tosses

    a) The histogram of these proportions is expected to be symmetric, but not because of the Central Limit Theorem. The sample of 16 coin flips is not large. The distribution of these proportions is expected to be symmetric because the probability that the coin lands heads is the same as the probability that the coin lands tails.

    b) The histogram is expected to have its centre at 0.5, the probability that the coin lands heads.

    c) The standard deviation of data displayed in this histogram should be approximately equal to the standard deviation of the sampling distribution model, $\sqrt{\dfrac{pq}{n}} = \sqrt{\dfrac{(0.5)(0.5)}{16}} = 0.125$.

    d) The expected number of heads, $np = 16(0.5) = 8$, which is less than 10. The Success/Failure Condition is not met. The Normal model is not appropriate in this case.

13.   **More coins**

    a) $\mu_{\hat{p}} = p = 0.5$ and $\sigma(\hat{p}) = \sqrt{\dfrac{pq}{n}} = \sqrt{\dfrac{(0.5)(0.5)}{25}} = 0.1$

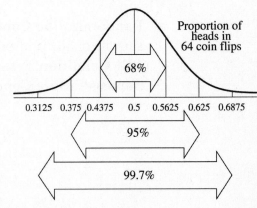

    About 68% of the sample proportions are expected to be between 0.4 and 0.6, about 95% are expected to be between 0.3 and 0.7, and about 99.7% are expected to be between 0.2 and 0.8.

    b) First of all, coin flips are independent of one another. There is no need to check the 10% Condition. Second, $np = nq = 12.5$, so both are greater than 10. The Success/Failure Condition is met, so the sampling distribution model is $N(0.5, 0.1)$.

    c) $\mu_{\hat{p}} = p = 0.5$ and $\sigma(\hat{p}) = \sqrt{\dfrac{pq}{n}} = \sqrt{\dfrac{(0.5)(0.5)}{64}} = 0.0625$

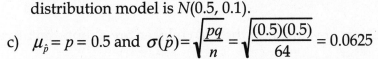

    About 68% of the sample proportions are expected to be between 0.4375 and 0.5625, about 95% are expected to be between 0.375 and 0.625, and about 99.7% are expected to be between 0.3125 and 0.6875.

    Coin flips are independent of one another, and $np = nq = 32$, so both are greater than 10. The Success/Failure Condition is met, so the sampling distribution model is $N(0.5, 0.0625)$.

d) As the number of tosses increases, the sampling distribution model will still be Normal and centred at 0.5, but the standard deviation will decrease. The sampling distribution model will be less spread out.

15. **Just (un)lucky?** For 200 flips, the sampling distribution model is Normal with

$\mu_{\hat{p}} = p = 0.5$ and $\sigma(\hat{p}) = \sqrt{\dfrac{pq}{n}} = \sqrt{\dfrac{(0.5)(0.5)}{200}} \approx 0.0354$. Her sample proportion of

$\hat{p} = 0.42$ is about 2.26 standard deviations below the expected proportion, which is unusual, but not extraordinary. According to the Normal model, we expect sample proportions this low or lower about 1.2% of the time.

17. **Speeding**

a) $\mu_{\hat{p}} = p = 0.70$

$\sigma(\hat{p}) = \sqrt{\dfrac{pq}{n}} = \sqrt{\dfrac{(0.7)(0.3)}{80}} \approx 0.051$.

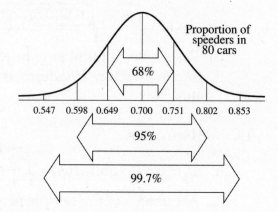

About 68% of the sample proportions are expected to be between 0.649 and 0.751, about 95% are expected to be between 0.598 and 0.802, and about 99.7% are expected to be between 0.547 and 0.853.

b) **Randomization Condition:** The sample may not be representative. If the flow of traffic is very fast, the speed of the other cars around may have some effect on the speed of each driver. Likewise, if traffic is slow, the police may find a smaller proportion of speeders than they expect.
10% Condition: 80 cars represent less than 10% of all cars.
Success/Failure Condition: $np = 56$ and $nq = 24$ are both greater than 10.
The Normal model may not be appropriate. Use caution. (And don't speed!)

19. **Vision**

a) **Randomization Condition:** Assume that the 170 children are a representative sample of all children.
10% Condition: A sample of this size is less than 10% of all children.
Success/Failure Condition: $np = 20.4$ and $nq = 149.6$ are both greater than 10.

Therefore, the sampling distribution model for the proportion of 170 children who are nearsighted is $N(0.12, 0.025)$.

b) The Normal model is shown below.

c) They might expect that the proportion of nearsighted students to be within 2 standard deviations of the mean. According to the Normal model, this means they might expect between 7% and 17% of the students to be nearsighted, or between about 12 and 29 students.

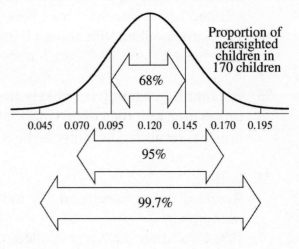

21. **Loans**

a) $\mu_{\hat{p}} = p = 7\%$

$$\sigma(\hat{p}) = \sqrt{\frac{pq}{n}} = \sqrt{\frac{(0.07)(0.93)}{200}} \approx 1.8\%$$

b) **Randomization Condition**: Assume that the 200 people are a representative sample of all loan recipients.
   10% Condition: A sample of this size is less than 10% of all loan recipients.
   Success/Failure Condition: $np = 14$ and $nq = 186$ are both greater than 10. Therefore, the sampling distribution model for the proportion of 200 loan recipients who will not make payments on time is $N(0.07, 0.018)$.

c) According to the Normal model, the probability that over 10% of these clients will not make timely payments is approximately 0.048.

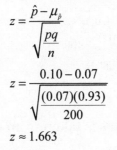

$$z = \frac{\hat{p} - \mu_{\hat{p}}}{\sqrt{\frac{pq}{n}}}$$

$$z = \frac{0.10 - 0.07}{\sqrt{\frac{(0.07)(0.93)}{200}}}$$

$$z \approx 1.663$$

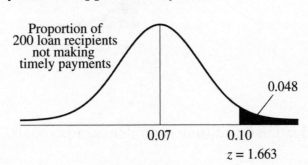

23. **Big families**

10% Condition: 400 is less than 10% of all families.
Success/Failure Condition: $np = 33.2$ and $nq = 366.8$ are both greater than 10. Therefore, the sampling distribution model for the proportion is *Normal*, with:

$$\mu_{\beta} = p = 0.083$$

$$\sigma(\beta) = \sqrt{\frac{pq}{n}} = \sqrt{\frac{(0.083)(0.917)}{400}} = 0.0138$$

There is roughly a 68% chance that between 6.9% and 9.7% of the 400 families selected has five or more members, about a 95% chance that between 5.5% and 11.1% are families with at least five members, and about a 99.7% chance that between 4.2% and 12.4% are families with at least five members.

25. **Big families, again** It is unlikely that rural families compose a SRS. However, proceeding as if they do, we find $\sigma(\hat{p}) = 1.10\%$ and so 16% would be a very unusual result, as it is outside the 3 $\sigma(\hat{p})$ range by the 68–95–99.7 Rule.

27. **Polling**
Randomization Condition: We must assume that the 400 voters were polled randomly.
10% Condition: 400 voters polled represent less than 10% of potential voters.
Success/Failure Condition: $np = 192$ and $nq = 208$ are both greater than 10. Therefore, the sampling distribution model for $\hat{p}$ is Normal, with:

$$\mu_{\hat{p}} = p = 0.48$$

$$\sigma(\hat{p}) = \sqrt{\frac{pq}{n}} = \sqrt{\frac{(0.48)(0.52)}{400}} \approx 0.025$$

According to the Normal model, the probability that the newspaper's sample will lead them to predict victory (that is, predict sovereignty support above 50%) is approximately 0.212.

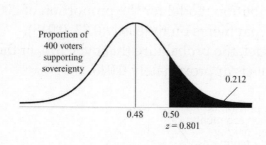

$$z = \frac{\hat{p} - \mu_{\hat{p}}}{\sqrt{\frac{pq}{n}}}$$

$$z = \frac{0.50 - 0.48}{\sqrt{\frac{(0.48)(0.52)}{400}}}$$

$$z \approx 0.801$$

29. **Apples**
Randomization Condition: A random sample of 150 apples is taken from each truck.
10% Condition: 150 is less than 10% of all apples.
Success/Failure Condition: $np = 12$ and $nq = 138$ are both greater than 10. Therefore, the sampling distribution model for $\hat{p}$ is Normal, with:

$$\mu_{\hat{p}} = p = 0.08$$

$$\sigma(\hat{p}) = \sqrt{\frac{pq}{n}} = \sqrt{\frac{(0.08)(0.92)}{150}} \approx 0.0222$$

According to the Normal model, the probability that less than 5% of the apples in the sample are unsatisfactory is approximately 0.088.

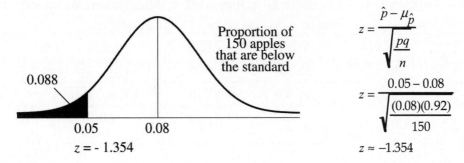

$$z = \frac{\hat{p} - \mu_{\hat{p}}}{\sqrt{\dfrac{pq}{n}}}$$

$$z = \frac{0.05 - 0.08}{\sqrt{\dfrac{(0.08)(0.92)}{150}}}$$

$$z \approx -1.354$$

31. **Nonsmokers**

Randomization Condition: We will assume that the 120 customers (to fill the restaurant to capacity) are representative of all customers.

10% Condition: 120 customers represent less than 10% of all potential customers.

Success/Failure Condition: $np = 72$ and $nq = 48$ are both greater than 10.

Therefore, the sampling distribution model for $\hat{p}$ is Normal, with:

$$\mu_{\hat{p}} = p = 0.60$$

$$\sigma(\hat{p}) = \sqrt{\frac{pq}{n}} = \sqrt{\frac{(0.60)(0.40)}{120}} \approx 0.0447$$

Answers may vary. We will use 3 standard deviations above the expected proportion of customers who demand nonsmoking seats to be "very sure."

$$\mu_{\hat{p}} + 3\left(\sqrt{\frac{pq}{n}}\right) \approx 0.60 + 3(0.0447) \approx 0.734$$

Since $120(0.734) = 88.08$, the restaurant needs at least 89 seats in the nonsmoking section.

33. **Sampling**

a) The sampling distribution model for the sample mean is $N\left(\mu, \dfrac{\sigma}{\sqrt{n}}\right)$.

b) If we choose a larger sample, the mean of the sampling distribution model will remain the same, but the standard deviation will be smaller.

35. **Waist size**

a) The distribution of waist size of 250 men in Utah is unimodal and slightly skewed to the right. A typical waist size is approximately 36 inches, and the standard deviation in waist sizes is approximately 4 inches.

b) All of the histograms show distributions of sample means centred near 36 inches. As n gets larger, the histograms approach the Normal model in shape and the variability in the sample means decreases. The histograms are fairly Normal by the time the sample reaches size 5.

37. Waist size revisited

a)

n	Observed mean	Theoretical mean	Observed st. dev.	Theoretical standard deviation
2	36.314	36.33	2.855	$4.019/\sqrt{2} \approx 2.842$
5	36.314	36.33	1.805	$4.019/\sqrt{5} \approx 1.797$
10	36.341	36.33	1.276	$4.019/\sqrt{10} \approx 1.271$
20	36.339	36.33	0.895	$4.019/\sqrt{20} \approx 0.899$

b) The observed values are all very close to the theoretical values.

c) For samples as small as 5, the sampling distribution of sample means is unimodal and symmetric. The Normal model would be appropriate.

d) The distribution of the original data is nearly unimodal and symmetric, so it doesn't take a very large sample size for the distribution of sample means to be approximately Normal.

39. GPAs

Randomization Condition: Assume that the students are randomly assigned to seminars.
Independence Assumption: It is reasonable to think that high school averages for randomly selected students are mutually independent.
10% Condition: The 25 students in the seminar certainly represent less than 10% of the population of students.
Large Enough Sample Condition: The distribution of high school averages is roughly unimodal and symmetric, so the sample of 25 students is large enough.

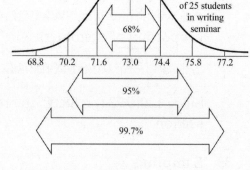

The mean of the population is $\mu = 73$, with $\sigma = 7$. The mean of a sample of 25 will be the same ($\mu\,(\bar{y}) = 73$) but the standard deviation will be tighter $\sigma(\bar{y}) = \dfrac{\sigma}{\sqrt{n}} = 1.4$.

41. The trial of the pyx

a) The mean of 100 coins varies less from sample to sample than the individual coins.

b) $SD(\bar{y}) = \dfrac{\sigma}{\sqrt{n}} = \dfrac{0.09}{\sqrt{100}} = 0.009$

The limit on the standard deviation of the mean should have been 0.009 grains.

43. Pregnancy

a)

$$z = \frac{y - \mu}{\sigma}$$

$$z = \frac{270 - 266}{16}$$

$$z = 0.25$$

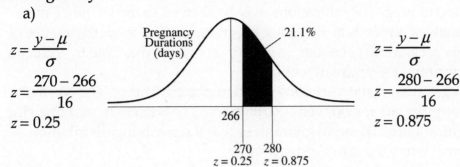

270 280
$z = 0.25$ $z = 0.875$

$$z = \frac{y - \mu}{\sigma}$$

$$z = \frac{280 - 266}{16}$$

$$z = 0.875$$

According to the Normal model, approximately 21.1% of all pregnancies are expected to last between 270 and 280 days.

b)

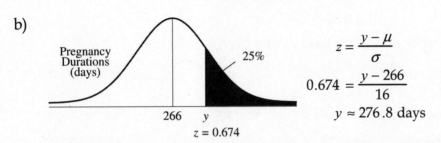

$z = 0.674$

$$z = \frac{y - \mu}{\sigma}$$

$$0.674 = \frac{y - 266}{16}$$

$$y \approx 276.8 \text{ days}$$

According to the Normal model, the longest 25% of pregnancies are expected to last approximately 276.8 days or more.

c) **Randomization Condition**: Assume that the 60 women the doctor is treating can be considered a representative sample of all pregnant women.
Independence Assumption: It is reasonable to think that the durations of the patients' pregnancies are mutually independent.
10% Condition: The 60 women that the doctor is treating certainly represent less than 10% of the population of all women.
Large Enough Sample Condition: The sample of 60 women is large enough. In this case, any sample would be large enough, since the distribution of pregnancies is Normal.
The mean duration of the pregnancies was $\mu = 266$ days, with standard deviation $\sigma = 16$ days. Since the distribution of pregnancy durations is Normal, we can model the sampling distribution of the mean pregnancy duration with a Normal model, with $\mu_{\bar{y}} = 266$ days and standard deviation

$$\sigma(\bar{y}) = \frac{16}{\sqrt{60}} \approx 2.07 \text{ days}.$$

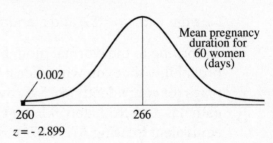

d) According to the Normal model, the probability that the mean pregnancy duration is less than 260 days is 0.002.

45. Pregnant again

a) The distribution of pregnancy durations may be skewed to the left since there are more premature births than very long pregnancies. The modern practice of medicine stops pregnancies at about 2 weeks past normal due date by inducing labour or performing a Caesarean section.

b) We can no longer answer the questions posed in parts a) and b). The Normal model is not appropriate for skewed distributions. The answer to part c) is still valid. The Central Limit Theorem guarantees that the sampling distribution model is Normal when the sample size is large.

47. Dice and dollars

a) Let X = the number of dollars won in one play.

$$\mu = E(X) = 0\left(\frac{3}{6}\right) + 1\left(\frac{2}{6}\right) + 10\left(\frac{1}{6}\right) = \$2$$

$$\sigma^2 = Var(X) = (0-2)^2\left(\frac{3}{6}\right) + (1-2)^2\left(\frac{2}{6}\right) + (10-2)^2\left(\frac{1}{6}\right) = 13$$

$$\sigma = SD(X) = \sqrt{Var(X)} = \sqrt{13} \approx \$3.61$$

b) $X_1 + X_2$ = the total winnings for two plays.

$$\mu = E(X_1 + X_2) = E(X_1) + E(X_2) = 2 + 2 = \$4$$

$$\sigma = SD(X_1 + X_2) = \sqrt{Var(X_1) + Var(X_2)} = \sqrt{13 + 13} \approx \$5.10$$

c) In order to win at least \$100 in 40 plays, you must average at least $\frac{100}{40} = \$2.50$ per play.

The expected value of the winnings is $\mu = \$2$, with standard deviation $\sigma = \$3.61$. Rolling a die is random and the outcomes are mutually independent, so the Central Limit Theorem guarantees that the sampling distribution model is Normal with $\mu_{\bar{x}} = \$2$ and standard deviation $\sigma(\bar{x}) = \frac{\$3.61}{\sqrt{40}} \approx \$0.571$.

According to the Normal model, the probability that you win at least \$100 in 40 plays (or equivalently, \$2.5 on average per game) is approximately 0.191. (This is equivalent to using $N(2, 0.571)$ to model your total winnings.)

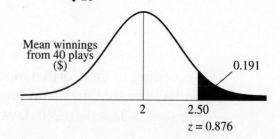

49. Canadian families

a) $\mu = 1(0.276) + 2(0.341) + 3(0.156) + 4(0.143) + 5(0.054) + 6(0.029) = 2.442$

$\sigma = \sqrt{((1-2.442)^2(0.276)+(2-2.442)^2(0.341)+\ldots+(6-2.442)^2(0.029))} = 1.325$

This calculation is based on a rounded mean.

b) They would not be expected to follow a Normal Distribution. They would follow the frequency distribution given in the table.

c) By the Central Limit Theorem, the mean of consecutive large random samples would follow an approximate normal distribution with mean 2.442 and standard deviation 0.210.

51. Canadian families, again Assuming farm families are a typical Canadian Family, we have $\mu = 2.442$ and $\sigma = 1.325$. For $n = 36$, $\mu(\bar{y}) = 2.442$ and $\sigma(\bar{y}) = 1.325/\sqrt{36}$ $=0.2208$. From the Normal Distribution of the mean, $z = (x - \mu(\bar{y})/\sigma(\bar{y}) = (3 - 2.442)/0.2208 = 2.53$. From the table in the Appendix, this corresponds to an area of 0.0057 beyond that z-score. Thus, the probability of finding a group of 36 families that large is 0.0057. This tells us that farm families are probably not representative of the rest of the country. This makes sense intuitively, because farms can always use more workers to help out.

53. Pollution

a) **Randomization Condition:** Assume that the 80 cars can be considered a representative sample of all cars of this type.

Independence Assumption: It is reasonable to think that the CO emissions for these cars are mutually independent.

10% Condition: The 80 cars in the fleet certainly represent less than 10% of all cars of this type.

Large Enough Sample Condition: A sample of 80 cars is large enough.
The mean CO level was $\mu = 2.9$ gm/mi, with standard deviation $\sigma = 0.4$ gm/mi. Since the conditions are met, the CLT allows us to model the sampling distribution of the $\bar{y}$ with a Normal model, with $\mu_{\bar{y}} = 2.9$ gm/mi and standard deviation $\sigma(\bar{y}) = \dfrac{0.4}{\sqrt{80}} = 0.045$ gm/mi

b) According to the Normal model, the probability that $\bar{y}$ is between 3.0 and 3.1 gm/mi is approximately 0.0131.

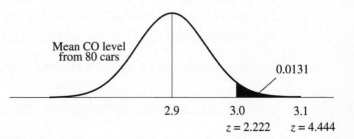

c) According to the Normal model, there is only a 5% chance that the fleet's mean CO level is greater than approximately 2.97 gm/mi.

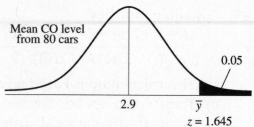

55. Tips

a) Since the distribution of tips is skewed to the right, we can't use the Normal model to determine the probability that a given party will tip at least $20.

b) No. A sample of four parties is probably not a large enough sample for the CLT to allow us to use the Normal model to estimate the distribution of averages.

c) A sample of 10 parties may not be large enough to allow the use of a Normal model to describe the distribution of averages. It would be risky to attempt to estimate the probability that his next 10 parties tip an average of $15. However, since the distribution of tips has $\mu = \$9.60$, with standard deviation $\sigma = \$5.40$, we still know that the mean of the sampling distribution model is $\mu_{\bar{y}} = \$9.60$ with standard deviation $\sigma(\bar{y}) = \dfrac{5.40}{\sqrt{10}} \approx \$1.71$. We don't know the exact shape of the distribution, but we can still assess the likelihood of specific means. A mean tip of $15 is over 3 standard deviations above the expected mean tip for 10 parties. That's not very likely to happen.

57. More tips

a) **Randomization Condition:** Assume that the tips from 40 parties can be considered a representative sample of all tips.
Independence Assumption: It is reasonable to think that the tips are mutually independent, unless the service is particularly good or bad during this weekend.

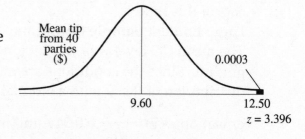

10% Condition: The tips of 40 parties certainly represent less than 10% of all tips.
Large Enough Sample Condition: The sample of 40 parties is large enough.
The mean tip is $\mu = \$9.60$, with standard deviation $\sigma = \$5.40$. Since the conditions are satisfied, the CLT allows us to model the sampling distribution of $\bar{y}$ with a Normal model, with $\mu_{\bar{y}} = \$9.60$ and standard deviation $\sigma(\bar{y}) = \dfrac{5.40}{\sqrt{40}} \approx \$0.8538$.

To earn at least $500, the waiter would have to average $\dfrac{500}{40} = \$12.50$ per party.

According to the Normal model, the probability that the waiter earns at least $500 in tips in a weekend is approximately 0.0003.

b) According to the Normal model, the waiter can expect to have a mean tip of about $10.6942, which corresponds to about $427.77 for 40 parties, in the best 10% of such weekends.

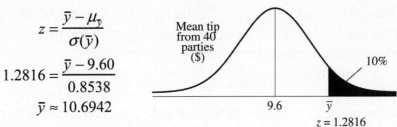

$$z = \frac{\bar{y} - \mu_{\bar{y}}}{\sigma(\bar{y})}$$

$$1.2816 = \frac{\bar{y} - 9.60}{0.8538}$$

$$\bar{y} \approx 10.6942$$

59. Big stats course

a) Let X denote the test score of this student. Then $X \sim N(\mu = 72, \sigma = 16)$.
    $z = (80 - 72)/16 = 0.5$ and $P(X > 80) = 1 - 0.6915 = 0.3085$.

b) Let Y denote the test score of the student from Xiu Xiu's tutorial. We want $P(X - Y \geq 7)$.

$$X - Y \sim N\left(72 - 72 \sqrt{16^2 + 16^2} = 22.63\right)$$

$$P(X - Y \geq 7) = P(Z \geq \frac{7-0}{22.63}) = P(Z \geq 0.31) = 1 - 0.6217 = 0.3783.$$

c) Let $\bar{X}$ denote the average test score in Sabri's tutorial. $\bar{X} \sim N\left(72, \frac{16}{\sqrt{30}} = 2.92\right)$

$$P(\bar{X} \geq 80) = P(Z \geq \frac{80-72}{2.92} = 2.74) = 1 - 0.9969 = 0.0031$$

d) Let $\bar{Y}$ denote the average test score in Xiu Xiu's tutorial. $\bar{Y}$ also has the same distribution as $\bar{X}$.

$$\bar{Y} \sim N\left(72\frac{16}{\sqrt{30}} = 2.92\right)$$

$$\bar{X} - \bar{Y} \sim N\left(72 - 72, \sqrt{\frac{16^2}{30} + \frac{16^2}{30}} = 4.13\right)$$

We want $P(\bar{X} - \bar{Y} \geq 7) = P\left(Z \geq \frac{7-0}{4.13} = 1.69\right) = 1 - 0.9545 = 0.0455.$

61. Mean or median? Answers may vary.

63. Central Limit Theorem Answers may vary.

65. Central Limit Theorem, finis Answers may vary.

1. **Lying about age**

   a) This means that 49% of the 799 teens in the sample saidthey have misrepresented their age online. This is our best estimate of p, the proportion of all U.S. teens who would say they have done.

   b) $SE(\hat{p}) = \sqrt{\dfrac{(0.49)(0.51)}{799}} \approx 0.018$

   c) Because we don't know p, we use $\hat{p}$ to estimate the standard deviation of the sampling distribution. So the standard error is our estimate of the amount of variation in the sample proportion we expect to see from sample to sample when we ask 799 teens whether they've misrepresented their age online.

3. **Lying about ages again**

   a) We are 95% confident that, if we were to ask all teens whether they have misrepresented their age online, between 45.6% and 52.5% of them would say they have.

   b) If we were to collect many random samples of 799 teens, about 95% of the confidence intervals we construct would contain the proportion of all teens who admit to misrepresenting their age online.

5. **Margin of error** He believes the true proportion of voters with a certain opinion is within 4% of his estimate, with some degree of confidence, perhaps 95% confidence.

7. **Conditions**

   a) *Population* – all cars; *sample* – 134 cars actually stopped at the checkpoint; p – proportion of all cars with safety problems; $\hat{p}$ – proportion of cars in the sample that actually have safety problems (10.4%).

   Plausible Independence Condition: There is no reason to believe that the safety problems of cars are related to each other.

   Randomization Condition: This sample is not random, so hopefully the cars stopped are representative of cars in the area.

   10% Condition: The 134 cars stopped represent a small fraction of all cars, certainly less than 10%.

   Success/Failure Condition: $n\hat{p} = 14$ and $n\hat{q} = 120$ are both greater than 10, so the sample is large enough.

   A one-proportion z-interval can be created for the proportion of all cars in the area with safety problems.

b) *Population* – the general public; *sample* – 602 viewers that logged on to the Web site; *p* – proportion of the general public that support prayer in school; $\hat{p}$ – proportion of viewers that logged on to the Web site and voted to support prayer in schools (81.1%).
Randomization Condition: This sample is not random, but biased by voluntary response.
It would be very unwise to attempt to use this sample to infer anything about the opinion of the general public related to school prayer.

c) *Population* – parents at the school; *sample* – 380 parents who returned surveys; *p* – proportion of all parents in favour of uniforms; $\hat{p}$ – proportion of those who responded that are in favour of uniforms (60%).
Randomization Condition: This sample is not random, but rather biased by non-response. There may be lurking variables that affect the opinions of parents who return surveys (and the children who deliver them!).
It would be very unwise to attempt to use this sample to infer anything about the opinion of the parents about uniforms.

d) *Population* – all first year enrollees at the college (not just one year); *sample* – 1632 first year enrollees during the specified year; $\hat{p}$ – proportion of all students who will graduate on time; *p* – proportion of students from that year who graduate on time (85.05%).
Plausible Independence Condition: It is reasonable to think that the abilities of students to graduate on time are mutually independent.
Randomization Condition: This sample is not random, but this year's first year class is probably representative of first year classes in other years.
10% Condition: The 1632 students in this year's first year class represent less than 10% of all possible students.
Success/Failure Condition: $n\hat{p}$ = 1388 and $n\hat{q}$ = 244 are both greater than 10, so the sample is large enough.
A one-proportion *z*-interval can be created for the proportion of first year enrollees that graduate on time from this college.

9. **Conclusions**
   a) Not correct. This statement implies certainty. There is no level of confidence in the statement.
   b) Not correct. Different samples will give different results. Many fewer than 95% of samples are expected to have *exactly* 88% on-time orders.
   c) Not correct. A confidence interval should say something about the unknown population proportion, not the sample proportion in different samples.
   d) Not correct. We *know* that 88% of the orders arrived on time. There is no need to make an interval for the sample proportion.
   e) Not correct. The interval should be about the proportion of on-time orders, not the days.

11. Confidence intervals

a) False. For a given sample size, higher confidence means a *larger* margin of error.

b) True. Larger samples lead to smaller standard errors, which lead to smaller margins of error.

c) True. Larger samples are less variable, which makes us more confident that a given confidence interval succeeds in catching the population proportion.

d) False. The margin of error decreases as the square root of the sample size increases. Halving the margin of error requires a sample four times as large as the original.

13. Cars We are 90% confident that between 29.9% and 47.0% of cars are made in Japan.

15. Mislabelled seafood

a) $\hat{p} \pm z^* \sqrt{\frac{\hat{p}\hat{q}}{n}} = \left(\frac{42}{190}\right) \pm 1.960 \sqrt{\frac{\left(\frac{42}{190}\right)\left(\frac{148}{190}\right)}{190}} = (0.162, 0.280)$

b) We are 95% confident that between 16.2% and 28.0% of all seafood packages sold in these three states are mislabelled.

c) The size of the population is irrelevant. If *Consumer Reports* had a random sample, 95% of intervals genereated by studies like this are expected to capture the true proportion of seafood packages that are mislabelled.

17. Canadian pride

a) The margin of error is $1.645^* \sqrt{\frac{0.58^* 0.42}{1000}} \approx 0.0257$. For a more cautious answer, we could use the more conservative proportion estimate of 0.5 when finding the standard error. The margin of error is then $1.645^* \sqrt{\frac{0.50^* 0.50}{1000}} \approx 0.0260$.

b) We are 90% sure that the true proportion of Canadians who are proud of health care is within ±2.6% of 58%.

c) The margin of error would have to be larger, if we wanted to be more certain of capturing the true proportion.

d) The margin of error is $2.576^* \sqrt{\frac{0.58^* 0.42}{1000}} \approx 0.040$. A more conservative answer would be $2.576^* \sqrt{\frac{0.5^* 0.5}{1000}} \approx 0.041$

e) Smaller margins of error will produce less confidence in the interval.

f) Our new interval (55.4%, 60.6%) does not include the old measurement of 50%. Even with the wider margins associated with the 99% CI, we still don't include 50%. Thus, we have strong evidence that the real proportion has changed.

19. **Contributions please**
    a) **Randomization Condition**: Letters were sent to a random sample of 100 000 potential donors.
    10% Condition: We assume that the potential donor list has more than 1 000 000 names.
    Success/Failure Condition: $n\hat{p} = 4781$ and $n\hat{q} = 95\,219$ are both much greater than 10, so the sample is large enough.

    $$\hat{p} \pm z^* \sqrt{\frac{\hat{p}\hat{q}}{n}} = \left(\frac{4781}{100\,000}\right) \pm 1.960\sqrt{\frac{\left(\frac{4781}{100\,000}\right)\left(\frac{95\,219}{100\,000}\right)}{100\,000}} = (0.0465, 0.0491)$$

    We are 95% confident that the between 4.65% and 4.91% of potential donors would donate.
    b) The confidence interval gives the set of plausible values with 95% confidence. Since 5% is outside the interval, it seems to be a bit optimistic.

21. **Teenage drivers**
    a) **Independence Assumption**: There is no reason to believe that accidents selected at random would be related to one another.
    Randomization Condition: The insurance company randomly selected 582 accidents.
    10% Condition: 582 accidents represent less than 10% of all accidents.
    Success/Failure Condition: $n\hat{p} = 91$ and $n\hat{q} = 491$ are both greater than 10, so the sample is large enough.
    Since the conditions are met, we can use a one-proportion z-interval to estimate the percentage of accidents involving teenagers.

    $$\hat{p} \pm z^* \sqrt{\frac{\hat{p}\hat{q}}{n}} = \left(\frac{91}{582}\right) \pm 1.960\sqrt{\frac{\left(\frac{91}{582}\right)\left(\frac{491}{582}\right)}{582}} = (12.7\%, 18.6\%)$$

    b) We are 95% confident that between 12.7% and 18.6% of all accidents involve teenagers.
    c) About 95% of random samples of size 582 will produce intervals that contain the true proportion of accidents involving teenagers.
    d) Our confidence interval contradicts the assertion of the politician. The figure quoted by the politician, 1 out of every 5, or 20%, is outside the interval.

23. **Safe food** The grocer can conclude nothing about the opinions of all his customers from this survey. Those customers who bothered to fill out the survey represent a voluntary response sample, consisting of people who felt strongly one way or another about irradiated food. The random condition was not met.

25. **Death penalty, again**
    a) There may be a leading question bias present here.

b) Combining the samples give $(0.48 + 0.58)/2 = 0.53$ of people believing the death penalty was fairly applied. The margin of error of the 95% confidence interval is $1.96^* \sqrt{\dfrac{0.53^* 0.47}{1020}} \approx 0.03$. The 95% CI is $0.53 \pm 0.03 = (0.50, 0.56)$.

c) The margin of error is smaller. This is expected because the sample size is effectively larger when using the pooled data.

27. Rickets

a) **Independence Assumption**: It is reasonable to think that the randomly selected children are mutually independent in regards to vitamin D deficiency.
Randomization Condition: The 2700 children were chosen at random.
10% Condition: 2700 children are less than 10% of all English children.
Success/Failure Condition: $n\hat{p} = (2700)(0.20) = 540$ and $n\hat{q} = (2700)(0.80) = 2160$ are both greater than 10, so the sample is large enough.
Since the conditions are met, we can use a one-proportion z-interval to estimate the proportion of the English children with vitamin D deficiency.

$$\hat{p} \pm z^* \sqrt{\frac{\hat{p}\hat{q}}{n}} = (0.20) \pm 2.326 \sqrt{\frac{(0.20)(0.80)}{2700}} = (18.2\%, 21.8\%)$$

b) We are 98% confident that between 18.2% and 21.8% of English children are deficient in vitamin D.

c) About 98% of random samples of size 2700 will produce confidence intervals that contain the true proportion of English children that are deficient in vitamin D.

29. Studying poverty

$n = 300$, $\hat{p} = 45/300 = 0.15$

The margin of error of the 95% confidence interval =

$$z^* \sqrt{\frac{\hat{p}(1-\hat{p})}{n}} = 1.96 \times \sqrt{\frac{0.15(1-0.15)}{300}} = 1.96 \times 0.02061552813 \approx 0.04.$$

The 95% CI is $0.15 \pm 0.04 = (0.11, 0.19)$.

The method used above assumes independence of observations. This is unlikely to be true when observing individuals within a block. Some blocks may be poor neighbourhoods and some rich neighbourhoods. In this case, all individuals we observe from the same block can be poor (if the block is in a poor neighbourhood) or all the individuals can be rich (if the block is in a rich neighbourhood).

If our randomly selected blocks happened to be from poor neighbourhoods, then all the 300 individuals can be poor, or if all the selected blocks happened to be from rich neighbourhoods, then all the 300 individuals can be rich. This can make the variability of the sample proportion very high (ranging from 0 to 1). The methods we used (assuming independence) underestimates the variability of the sample proportion and also the SE of the sample proportion.

31. **Deer ticks**
    a) **Independence Assumption**: Deer ticks are parasites. A deer carrying the parasite may spread it to others. Ticks may not be distributed evenly throughout the population.
    Randomization Condition: The sample is not random and may not represent all deer.
    10% Condition: 153 deer are less than 10% of all deer.
    Success/Failure Condition: $n\hat{p} = 32$ and $n\hat{q} = 121$ are both greater than 10, so the sample is large enough.
    The conditions are not satisfied, so we should use caution when a one-proportion z-interval is used to estimate the proportion of deer carrying ticks.

    $$\hat{p} \pm z^* \sqrt{\frac{\hat{p}\hat{q}}{n}} = \left(\frac{32}{153}\right) \pm 1.645 \sqrt{\frac{\left(\frac{32}{153}\right)\left(\frac{121}{153}\right)}{153}} = (15.5\%, 26.3\%)$$

    We are 90% confident that between 15.5% and 26.3% of deer have ticks.
    b) In order to cut the margin of error in half, they must sample four times as many deer.
    4(153) = 612 deer.
    c) The incidence of deer ticks is not plausibly independent, and the sample may not be representative of all deer, since females and young deer are usually not hunted.

33. **Graduation**
    a)
    $$ME = z^* \sqrt{\frac{\hat{p}\hat{q}}{n}}$$
    $$0.06 = 1.645 \sqrt{\frac{(0.25)(0.75)}{n}}$$
    $$n = \frac{(1.645)^2(0.25)(0.75)}{(0.06)^2}$$
    $$n \approx 141 \text{ people}$$

    In order to estimate the proportion of non-graduates in the 25-to 30-year-old age group to within 6% with 90% confidence, we would need a sample of at least 141 people. All decimals in the final answer must be rounded up to the next integer.
    (For a more cautious answer, let $\hat{p} = \hat{q} = 0.5$. This method results in a required sample of 188 people.)

    b)
    $$ME = z^* \sqrt{\frac{\hat{p}\hat{q}}{n}}$$
    $$0.04 = 1.645 \sqrt{\frac{(0.25)(0.75)}{n}}$$
    $$n = \frac{(1.645)^2(0.25)(0.75)}{(0.04)^2}$$
    $$n \approx 318 \text{ people}$$

    In order to estimate the proportion of non-graduates in the 25-to 30-year-old age group to within 4% with 90% confidence, we would need a sample of at least 318 people. All decimals in the final answer must be rounded up to the next integer.
    (For a more cautious answer, let $\hat{p} = \hat{q} = 0.5$. This method results in a required sample of 423 people.) Alternatively, the margin of error is now 2/3 of the original, so the sample size must be increased by a factor of 9/4. 141(9/4) ≈ 318 people.

c)

$$ME = z^* \sqrt{\frac{\hat{p}\hat{q}}{n}}$$

$$0.03 = 1.645 \sqrt{\frac{(0.25)(0.75)}{n}}$$

$$n = \frac{(1.645)^2 (0.25)(0.75)}{(0.03)^2}$$

$$n \approx 564 \text{ people}$$

In order to estimate the proportion of non-graduates in the 25-to 30-year-old age group to within 3% with 90% confidence, we would need a sample of at least 564 people. All decimals in the final answer must be rounded up to the next integer.

(For a more cautious answer, let $\hat{p} = \hat{q} = 0.5$. This method results in a required sample of 752 people.)

Alternatively, the margin of error is now half that of the original, so the sample size must be increased by a factor of 9/4. $141(9/4) \approx 318$ people.

35. Graduation, again

$$ME = z^* \sqrt{\frac{\hat{p}\hat{q}}{n}}$$

$$0.02 = 1.960 \sqrt{\frac{(0.25)(0.75)}{n}}$$

$$n = \frac{(1.960)^2 (0.25)(0.75)}{(0.02)^2}$$

$$n \approx 1801 \text{ people}$$

In order to estimate the proportion of non-graduates in the 25-to 30-year-old age group to within 2% with 95% confidence, we would need a sample of at least 1801 people. All decimals in the final answer must be rounded up, to the next integer.

(For a more cautious answer, let $\hat{p} = \hat{q} = 0.5$. This method results in a required sample of 2401 people.)

37. Pilot study

$$ME = z^* \sqrt{\frac{\hat{p}\hat{q}}{n}}$$

$$0.03 = 1.645 \sqrt{\frac{(0.15)(0.85)}{n}}$$

$$n = \frac{(1.645)^2 (0.15)(0.85)}{(0.03)^2}$$

$$n \approx 384 \text{ cars}$$

Use $\hat{p} = \frac{9}{60} = 0.15$ from the pilot study as an estimate.

In order to estimate the percentage of cars with faulty emissions systems to within 3% with 90% confidence, the province's environmental agency will need a sample of at least 384 cars. All decimals in the final answer must be rounded up to the next integer.

39. Approval rating

$$ME = z^* \sqrt{\frac{\hat{p}\hat{q}}{n}}$$

$$0.025 = z^* \sqrt{\frac{(0.65)(0.35)}{972}}$$

$$z^* = \frac{0.025}{\sqrt{\frac{(0.65)(0.35)}{972}}}$$

$$z^* \approx 1.634$$

Since $z^* \approx 1.634$, which is close to 1.645, the pollsters were probably using 90% confidence. The slight difference in the z^* values is due to rounding of the premier's approval rating.

41. **Bad countries**

    a) $n = 2001$, $\hat{p} = 0.52$.

    The standard error of $\hat{p}$ and the standard error

    $$= \sqrt{\frac{\hat{p}(1-\hat{p})}{n}} = \sqrt{\frac{0.52(1-0.52)}{2001}} = 0.01116860023$$

    and the margin of error of the 99% confidence interval

    $$= z^* \sqrt{\frac{\hat{p}(1-\hat{p})}{n}} = 2.576 \times \sqrt{\frac{0.52(1-0.52)}{2001}} = 0.03.$$

    So, a 99% CI for the true population proportion is 0.52 ± 0.03 or $(0.49, 0.55)$.

    b) The margin of error increases as the sample size decreases. The sample size for British Columbia must be smaller than the sample size for Canada and so, we should expect an increase in margin of error.

    The margin of error of the confidence interval for population proportions also depends on the sample proportion ($\hat{p}$). The value of $\hat{p}(1-\hat{p})$ takes its maximum possible value when $\hat{p} = 0.5$ and decreases as $\hat{p}$ moves away from 0.5 (either increases or decreases). It is most likely that the increase in the margin of error due to the decrease in sample size will dominate the decrease due to $\hat{p}$ moving away from 0.5 (since it only moves from 0.52 to 0.58, but the decrease in the sample size can be substantial).

    c) The full sample had 2001 and 25% of this is approximately 500. The margin of error of the 99% confidence interval is:

    $$z^* \sqrt{\frac{\hat{p}(1-\hat{p})}{n}} = 2.576 \times \sqrt{\frac{0.063(1-0.063)}{500}} = 0.0280$$

    d) To reduce the margin of error by one half, the sample size should increase four times (because the sample size is inside the square root in the formula for margin of error). This means the sample size should increase from 2001 to 8004.

43. **Simulations**

    a) The confidence interval depends on the sample simulated. The 90% confidence interval based on the sample simulated is $(0.689751, 0.830249)$ and this includes the true population proportion (i.e., 0.8).

    b) The 90% confidence intervals based on the 100 samples (each of size 100) is given in the table below (these confidence intervals depend on the samples simulated). 92% of the confidence intervals contain the true population proportion (i.e., 0.8). For 100 simulated samples, if X = number of CIs that contain 0.8, then X should have a binomial distribution with $n = 100$ and $p = 0.9$.

Sample	Lower limit Upper limit	Is p in CI? (1 = yes, 0 = no)
1	0.689751 0.830249	1
2	0.656975 0.803025	1
3	0.791267 0.908733	1
4	0.734206 0.865794	1
5	0.734206 0.865794	1
6	0.756807 0.883193	1
7	0.711863 0.848137	1
8	0.745472 0.874528	1
9	0.723004 0.856996	1
10	0.689751 0.830249	1
11	0.814683 0.925317	0
12	0.745472 0.874528	1
13	0.711863 0.848137	1
14	0.768214 0.891786	1
15	0.779699 0.900301	1
16	0.779699 0.900301	1
17	0.678776 0.821224	1
18	0.802926 0.917074	0
19	0.768214 0.891786	1
20	0.734206 0.865794	1
21	0.756807 0.883193	1
22	0.689751 0.830249	1
23	0.734206 0.865794	1
24	0.667851 0.812149	1
25	0.768214 0.891786	1
26	0.756807 0.883193	1
27	0.734206 0.865794	1
28	0.768214 0.891786	1
29	0.723004 0.856996	1
30	0.814683 0.925317	0
31	0.734206 0.865794	1
32	0.646146 0.793854	0
33	0.700779 0.839221	1
34	0.689751 0.830249	1
35	0.711863 0.848137	1
36	0.779699 0.900301	1
37	0.791267 0.908733	1
38	0.814683 0.925317	0
39	0.745472 0.874528	1
40	0.723004 0.856996	1
41	0.802926 0.917074	0

Sample	Lower limit Upper limit	Is p in CI? (1 = yes, 0 = no)
42	0.711863 0.848137	1
43	0.700779 0.839221	1
44	0.734206 0.865794	1
45	0.723004 0.856996	1
46	0.635363 0.784637	0
47	0.667851 0.812149	1
48	0.779699 0.900301	1
49	0.700779 0.839221	1
50	0.711863 0.848137	1
51	0.756807 0.883193	1
52	0.745472 0.874528	1
53	0.768214 0.891786	1
54	0.723004 0.856996	1
55	0.745472 0.874528	1
56	0.745472 0.874528	1
57	0.779699 0.900301	1
58	0.768214 0.891786	1
59	0.768214 0.891786	1
60	0.667851 0.812149	1
61	0.734206 0.865794	1
62	0.656975 0.803025	1
63	0.756807 0.883193	1
64	0.814683 0.925317	0
65	0.779699 0.900301	1
66	0.779699 0.900301	1
67	0.779699 0.900301	1
68	0.711863 0.848137	1
69	0.745472 0.874528	1
70	0.700779 0.839221	1
71	0.711863 0.848137	1
72	0.756807 0.883193	1
73	0.745472 0.874528	1
74	0.723004 0.856996	1
75	0.734206 0.865794	1
76	0.723004 0.856996	1
77	0.723004 0.856996	1
78	0.711863 0.848137	1
79	0.768214 0.891786	1
80	0.779699 0.900301	1
81	0.700779 0.839221	1
82	0.745472 0.874528	1

Sample	Lower limit	Upper limit	Is p in CI? (1 = yes, 0 = no)
83	0.768214	0.891786	1
84	0.756807	0.883193	1
85	0.723004	0.856996	1
86	0.745472	0.874528	1
87	0.667851	0.812149	1
88	0.711863	0.848137	1
89	0.711863	0.848137	1
90	0.745472	0.874528	1
91	0.756807	0.883193	1
92	0.756807	0.883193	1
93	0.678776	0.821224	1
94	0.723004	0.856996	1
95	0.689751	0.830249	1
96	0.723004	0.856996	1
97	0.734206	0.865794	1
98	0.656975	0.803025	1
99	0.734206	0.865794	1
100	0.689751	0.830249	1

45. Dogs

a) $n = 42$, $\hat{p} = 5/42 = 0.12$

$n\hat{p} = 5 < 10$ and so the sample size is not large enough for the distribution of the sample proportion to be approximately Normal. We cannot use a Normal approximation based confidence interval.

b) $\tilde{p} = \dfrac{5+2}{42+4} = 0.15$ and the 95% confidence interval for the population proportion

(p) is $\tilde{p} \pm z^* \sqrt{\dfrac{\tilde{p}(1-\tilde{p})}{\tilde{n}}} = 0.15 \pm 1.96 \sqrt{\dfrac{0.15(1-0.15)}{42+4}} = 0.15 \pm 0.10 = (0.05 - 0.25)$.

We are 95% confident that between 5% to 25% of all puppies in this population have early hip dysplasia.

47. Departures 2011

a) **Independence Assumption :** Since there is no time trend, and also the timeplot shows little relationship between one month and the next month, the monthly on-time departure rates appear to be nearly independent. This is not a random sample, but should be representative of future flights.

10% Condition: These months represent fewer than 10% of all possible months.
Sample Size Assumption: The histogram looks unimodal and slightly skewed to the left. Since the sample size is 201, the sample mean should be approximately Normal, by the CLT, and we can plug in s for σ in the confidence interval.

b) The on-time departure rates in the sample had a mean of 80.752%, and a standard deviation in of 4.594%. Since the conditions have been satisfied, construct a one-sample *z*-interval, at 90% confidence.

$$\bar{y} \pm Z^* \frac{s}{\sqrt{n}} = 80.752 \pm 1.645 \frac{4.594}{\sqrt{20}_1} = (80.22, 81.29)$$

c) On-time departures average somewhere between 80.22% and 81.29% of flights per month (with 90% confidence). Without more data, this is the only interval we can construct, but unless the number of flights vary greatly from month to month, this will also be a good estimate of the probability that your flight will depart on time (of course, you might also want to condition on your particular airline!).

49. Speed of light

a) The standard deviation is $79.0/\sqrt{(100)} = 7.9$. Adding a constant does not change the SD or variance of a random variable.
b) 99% CI: $852.4 \pm 2.576*7.9 = (832.0, 872.8)$ in km/sec coded
c) Michelson is 99% sure that this interval captures the true speed of light.
d) You would need to check that the data was not severely skewed or with outliers, as your use of the z^* critical values (i.e., 2.576) depends on the Central Limit Theorem for sample means. Also, the measurements would need to be independent, and unbiased for our analysis to hold.
e) This value is well outside his 99% confidence interval. He was either very unlucky, or some of our assumptions about the normality and impartiality of his data are not correct. However, the result is still remarkably close for 1897!

51. Hot dog margin of error

To achieve a reduction of 50%, the sample size would need to increase by a factor of 2^2 or four times. For a ME of 4.0, set $1.645*32/\sqrt{n} = 4.0$. Then, $n = (1.645*32/4)^2 = 173.18$. We must round up to 174 to guarantee our confidence level.

1. **Better than aspirin?**

   a) The new drug is not more effective than aspirin, and reduces the risk of heart attack by 44%. (p = 0.44)

   b) The new drug is more effective than aspirin, and reduces the risk of heart attack by more than 44%. (p > 0.44)

3. **Better than aspirin II?**

   a) Since the P-value of 0.28 is big, there is insufficient evidence to conclude that the new drug is better than aspirin.

   b) Since the P-value of 0.004 is very small, there is evidence that the new drug is more effective than aspirin.

5. **Psychic again (you should have seen this coming)**
   The alternative hypothesis would be one-sided, because the only evidence that would support the friend's claim is guessing more than 25% of the suits correctly.

7. **Hypotheses**
   a) H_0: The governor's "negatives" are 30%. ($p = 0.30$)
      H_A: The governor's "negatives" are less than 30%. ($p < 0.30$)
   b) H_0: The proportion of heads is 50%. ($p = 0.50$)
      H_A: The proportion of heads is not 50%. ($p \neq 0.50$)
   c) H_0: The proportion of people who quit smoking is 20%. ($p = 0.20$)
      H_A: The proportion of people who quit smoking is greater than 20%. ($p > 0.20$)

9. **Negatives** Statement d is the correct interpretation of a P-value.

11. **Relief** It is *not* reasonable to conclude that the new formula and the old one are equally effective. Furthermore, our inability to make that conclusion has nothing to do with the *P*-value. We cannot prove the null hypothesis (that the new formula and the old formula are equally effective), but can only fail to find evidence that would cause us to reject it. All we can say about this P-value is that there is a 27% chance of seeing the observed effectiveness from natural sampling variation if the new formula and the old one are equally effective.

13. He cheats!
  a) Two losses in a row aren't convincing. There is a 25% chance of losing twice in a row, and that is not unusual.
  b) If the process is fair, three losses in a row can be expected to happen about 12.5% of the time. $(0.5)(0.5)(0.5) = 0.125$.
  c) Three losses in a row is still not a convincing occurrence. We'd expect that to happen about once every eight times we tossed a coin three times.
  d) Answers may vary. Maybe five times would be convincing. The chances of five losses in a row are only one in 32, which seems unusual.

15. Smartphones
Null and alternative hypotheses should involve p, not $\hat{p}$.

The question is about *failing* to meet the goal. H_A should be $p < 0.96$.

The student failed to check $nq = (200)(0.04) = 8$. Since $nq < 10$, the Success/Failure condition is violated. Similarly, the 10% Condition is not verified.

$$SD(\hat{p}) = \sqrt{\frac{pq}{n}} = \sqrt{\frac{(0.96)(0.04)}{200}} \approx 0.014.$$ The student used $\hat{p}$ and $\hat{q}$.

Value of z is incorrect. The correct value is $z = \dfrac{0.94 - 0.96}{0.014} \approx -1.43$.

P-value is incorrect. $P = P(z < -1.43) = 0.076$

For the P-value given, an incorrect conclusion is drawn. A P-value of 0.12 provides no evidence that the new system has failed to meet the goal. The correct conclusion for the corrected *P*-value is: Since the *P*-value of 0.076 is fairly low, there is weak evidence that the new system has failed to meet the goal.

17. Dowsing
  a) H_0: The percentage of successful wells drilled by the dowser is 30%. ($p = 0.30$)
     H_A: The percentage of successful wells drilled by the dowser is greater than 30%. ($p > 0.30$)
  b) **Independence Assumption:** There is no reason to think that finding water in one well will affect the probability that water is found in another, unless the wells are close enough to be fed by the same underground water source.
     Randomization Condition: This sample is not random, so hopefully the customers you check with are representative of all of the dowser's customers.
     10% Condition: The 80 customers sampled may be considered less than 10% of all possible customers.
     Success/Failure Condition: $np = (80)(0.30) = 24$ and $nq = (80)(0.70) = 56$ are both greater than 10, so the sample is large enough.
  c) The sample of customers may not be representative of all customers, so we will proceed cautiously. A Normal model can be used to model the sampling

distribution of the proportion, with $\mu_{\hat{p}} = p = 0.30$ and

$$\sigma(\hat{p}) = \sqrt{\frac{pq}{n}} = \sqrt{\frac{(0.30)(0.70)}{80}} \approx 0.0512.$$

We can perform a one-proportion z-test.

The observed proportion of successful wells is $\hat{p} = \frac{27}{80} = 0.3375$.

d) If his dowsing has the same success rate as standard drilling methods, there is more than a 23% chance of seeing results as good as those of the dowser, or better, by natural sampling variation.

e) With a P-value of 0.233, we fail to reject the null hypothesis. There is no evidence to suggest that the dowser has a success rate any higher than 30%.

$$z = \frac{\hat{p} - p_0}{\sqrt{\frac{pq}{n}}}$$

$$z = \frac{0.3375 - 0.30}{\sqrt{\frac{(0.30)(0.70)}{80}}}$$

$$z \approx 0.73$$

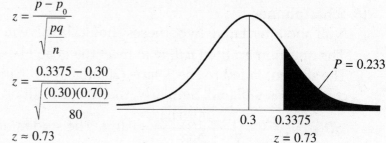

$P = 0.233$

$0.3 \quad 0.3375$

$z = 0.73$

19. Educated

There is no need to conduct any hypothesis testing since the figures (i.e., 64.1% in 2013 and 60.7% in 2006) are already the parameters (recall that it is a census and is based on the entire population).

21. Contributions, please, part II

a) H_0: The contribution rate is 5%. ($p = 0.05$)
   H_A: The contribution rate is less than 5%. ($p < 0.05$)

b) **Independence Assumption:** There is no reason to believe that one randomly selected potential donor's decision will affect another's decision.
   Randomization Condition: The sample was 100 000 randomly selected potential donors.
   10% Condition: We will assume that the entire mailing list has over 1 000 000 names.
   Success/Failure Condition: $np = 5000$ and $nq = 95\ 000$ are both greater than 10, so the sample is large enough.
   The conditions have been satisfied, so a Normal model can be used to model the sampling distribution of the proportion, with $\mu_{\hat{p}} = p = 0.05$ and

$$\sigma(\hat{p}) = \sqrt{\frac{pq}{n}} = \sqrt{\frac{(0.05)(0.95)}{100,000}} \approx 0.0007.$$

We can perform a one-proportion z-test. The observed contribution rate is $\hat{p} = \frac{4781}{100\ 000} = 0.04781.$

c)

$$z = \frac{\hat{p} - p}{\sqrt{\dfrac{pq}{n}}}$$

$$z = \frac{0.04781 - 0.05}{\sqrt{\dfrac{(0.05)(0.95)}{100000}}}$$

$$z = -3.18$$

The corresponding one sided *p*-value would be 0.0007. Since the *p*-value is low, we reject the null hypothesis. There is strong evidence that contribution rate for all potential donors is lower than 5%.

23. Law school

a) H_0: The law school acceptance rate for the training program is 29.6%. ($p = 0.296$)
H_A: The law school acceptance rate for the training program is greater than 29.6%. ($p > 0.296$)

b) **Randomization Condition:** These 240 students may be considered representative of the population of law school applicants.
10% Condition: There are certainly more than 2400 law school applicants.
Success/Failure Condition: $np = 71.04$ and $nq = 168.96$ are both greater than 10, so the sample is large enough.
The conditions have been satisfied, so a Normal model can be used to model the sampling distribution of the proportion, with $\mu_{\hat{p}} = p = 0.296$ and

$$\sigma(\hat{p}) = \sqrt{\frac{pq}{n}} = \sqrt{\frac{(0.296)(0.704)}{240}} \approx 0.0295$$

We can perform a one-proportion *z*-test. The observed success rate is

$$\hat{p} = \frac{83}{240} = 0.3458$$

This gives the corresponding z-value as:

$$z = \frac{\hat{p} - p}{\sqrt{\dfrac{pq}{n}}}$$

$$z = \frac{0.3458 - 0.296}{\sqrt{\dfrac{(0.296)(0.704)}{240}}}$$

$$z = 1.69$$

and the corresponding *p*-value is 0.046.

c) Since the P-value is fairly low, we can reject the null hypothesis. There is weak evidence that the law school acceptance rate is higher for applicants from that training program. Candidates should decide whether they can afford the time and expense.

25. Pollution

H_0: The percentage of cars with faulty emissions is 20%. ($p = 0.20$)

H_A: The percentage of cars with faulty emissions is greater than 20%. ($p > 0.20$)

Two conditions are not satisfied. 22 is greater than 10% of the population of 150 cars, and $np = (22)(0.20) = 4.4$, which is not greater than 10. It's probably not a good idea to proceed with a hypothesis test.

27. Twins

H_0: The percentage of twin births to teenage girls is 3%. ($p = 0.03$)

H_A: The percentage of twin births to teenage girls differs from 3%. ($p \neq 0.03$)

Independence Assumption: One mother having twins will not affect another. Observations are plausibly independent.

Randomization Condition: This sample may not be random, but it is reasonable to think that this hospital has a representative sample of teen mothers, with regards to twin births.

10% Condition: The sample of 469 teenage mothers is less than 10% of all such mothers.

Success/Failure Condition: $np = (469)(0.03) = 14.07$ and $nq = (469)(0.97) = 454.93$ are both greater than 10, so the sample is large enough.

The conditions have been satisfied, so a Normal model can be used to model the sampling distribution of the proportion, with $\mu_{\hat{p}} = p = 0.03$ and

$$\sigma(\hat{p}) = \sqrt{\frac{pq}{n}} = \sqrt{\frac{(0.03)(0.97)}{469}} \approx 0.0079.$$

We can perform a one-proportion z-test. The observed proportion of twin births to teenage mothers is $\hat{p} = \dfrac{7}{469} \approx 0.01493$.

Since the P-value = 0.0556 is fairly low, we reject the null hypothesis. There is some evidence that the proportion of twin births for teenage mothers at this large city hospital is lower than the proportion of twin births for all mothers.

$$z = \frac{\hat{p} - p_0}{\sqrt{\dfrac{pq}{n}}}$$

$$z = \frac{0.015 - 0.03}{\sqrt{\dfrac{(0.03)(0.97)}{469}}}$$

$$z \approx -1.91$$

P = 0.0556

0.015   0.03   0.045

$z = -1.91$ $z = 1.91$

29. WebZine

H_0: The percentage of readers interested in an online edition is 25%. ($p = 0.25$).

H_A: The percentage of readers interested in an online edition is greater than 25%. ($p > 0.25$)

Independence Assumption: Interest of one reader should not affect interest of other readers.

Randomization Condition: The magazine conducted an SRS of 500 current readers.

10% Condition: 500 readers are less than 10% of all potential subscribers.

Success/Failure Condition: $np = (500)(0.25) = 125$ and $nq = (500)(0.75) = 375$ are both greater than 10, so the sample is large enough.

The conditions have been satisfied, so a Normal model can be used to model the sampling distribution of the proportion, with $\sigma_p = p = 0.25$ and

$$\sigma(\hat{p}) = \sqrt{\frac{pq}{n}} = \sqrt{\frac{(0.25)(0.75)}{500}} \approx 0.0194 .$$

We can perform a one-proportion z-test. The observed proportion of interested readers is $\hat{p} = \dfrac{137}{500} = 0.274$.

Since the P-value = 0.1075 is high, we fail to reject the null hypothesis. There is little evidence to suggest that the proportion of interested readers is greater than 25%. The magazine should not publish the online edition.

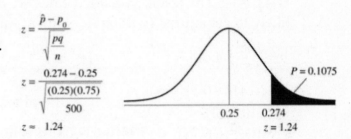

$$z = \frac{\hat{p} - p_0}{\sqrt{\dfrac{pq}{n}}}$$

$$z = \frac{0.274 - 0.25}{\sqrt{\dfrac{(0.25)(0.75)}{500}}}$$

$$z \approx 1.24$$

$P = 0.1075$

$0.25 \quad 0.274$

$z = 1.24$

31. Female executives

H_0: The proportion of female executives is similar to the overall proportion of female employees at the company. ($p = 0.40$)

H_A: The proportion of female executives is lower than the overall proportion of female employees at the company. ($p < 0.40$)

Independence Assumption: It is reasonable to think that executives at this company were chosen independently.

Randomization Condition: The executives were not chosen randomly, but it is reasonable to think of these executives as representative of all potential executives over many years.

10% Condition: 43 executives are less than 10% of all possible executives at the company.

Success/Failure Condition: $np = (43)(0.40) = 17.2$ and $nq = (43)(0.60) = 25.8$ are both greater than 10, so the sample is large enough.

The conditions have been satisfied, so a Normal model can be used to model the sampling distribution of the proportion, with $\mu_{\hat{p}} = p = 0.40$ and

$$\sigma(\hat{p}) = \sqrt{\frac{pq}{n}} = \sqrt{\frac{(0.40)(0.60)}{43}} \approx 0.0747 \,.$$

We can perform a one-proportion z-test. The observed proportion is $\hat{p} = \dfrac{13}{43} \approx 0.302$.

Since the P-value = 0.0951 is high, we fail to reject the null hypothesis. There is little evidence to suggest proportion of female executives is any different from the overall proportion of 40% female employees at the company.

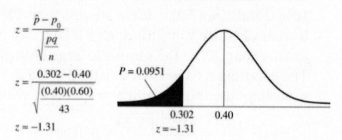

$$z = \frac{\hat{p} - p_0}{\sqrt{\dfrac{pq}{n}}}$$

$$z = \frac{0.302 - 0.40}{\sqrt{\dfrac{(0.40)(0.60)}{43}}}$$

$$z \approx -1.31$$

$P = 0.0951$

$0.302 \qquad 0.40$

$z = -1.31$

33. **Hockey injuries and relative age**
    a) H_0: p = 0.5

    H_A: p ≠ 0.5 The test is two-sided since we don't know which age group is more likely to get injured (without looking at the data)

    $$\hat{p} = \frac{156}{377} = 4.414 \qquad\qquad \sigma\left(\hat{p}\right) = \sqrt{\frac{pq}{n}} = \sqrt{\frac{0.5^{*}0.5}{377}} = 0.02575$$

    $$z = \frac{(0.414 - 0.5)}{0.02575} = -3.34 \qquad\qquad \text{P-value} = 0.0008$$

    p-value $= 2 \times P(Z < -3.34) \approx 0.0008$. It is highly unlikely for the sample proportion of 0.414 (in a sample of size 377) to occur by chance when the population proportion is 0.5.
    The sample does provide evidence to conclude that an injured Atom player is not equally likely to be from either quartile.

    b) For the Bantam level injured players,
    $$\hat{p} = \frac{433}{(433 + 646)} = 0.4012974977 \,. \; n = 433 + 646 = 1079 \text{ and}$$

    $$z = \frac{\hat{p} - p_0}{\sqrt{\dfrac{p_0 q_0}{n}}} = \frac{0.4012974977 - 0.5}{\sqrt{\dfrac{0.5 \times (1 - 0.5)}{1079}}} = --6.48 \,.$$

    p-value $= 2 \times P(Z < -6.48) \approx 0$. It is highly unlikely for the sample proportion of 0.4012974977 (in a sample of size 1079) to occur by chance when the population proportion is 0.5. The sample does provide evidence to conclude that an injured youth is not equally likely to be from either quartile.

c) We really should be looking at proportions injured of those in each age group, since there are more older players actually selected to play for the teams. We could try to obtain ages or age quintiles for all players, then compare the proportion injured in one age group with the proportion injured in the other age group.

35. Lost luggage

H_0: The proportion of lost luggage returned the next day is 90%. ($p = 0.90$)
H_A: The proportion of lost luggage returned the next day is lower than 90%. ($p < 0.90$)

Independence Assumption: It is reasonable to think that the people surveyed were independent with regards to their luggage woes.

Randomization Condition: Although not stated, we will hope that the survey was conducted randomly, or at least that these air travellers are representative of all air travellers for that airline.

10% Condition: 122 air travellers are less than 10% of all air travellers on the airline.

Success/Failure Condition: $np = (122)(0.90) = 109.8$ and $nq = (122)(0.10) = 12.2$ are both greater than 10, so the sample is large enough.

The conditions have been satisfied, so a Normal model can be used to model the sampling distribution of the proportion, with $\mu_{\hat{p}} = p = 0.90$ and

$$\sigma(\hat{p}) = \sqrt{\frac{pq}{n}} = \sqrt{\frac{(0.90)(0.10)}{122}} \approx 0.0272.$$

We can perform a one-proportion z-test. The observed proportion of lost luggage returned the next day is $\hat{p} = \dfrac{103}{122} \approx 0.8443$.

Since the P-value = 0.0201 is low, we reject the null hypothesis. There is evidence that the proportion of lost luggage returned the next day is lower than the 90% claimed by the airline.

$$z = \frac{\hat{p} - p_0}{\sqrt{\dfrac{pq}{n}}}$$

$$z = \frac{0.844 - 0.90}{\sqrt{\dfrac{(0.90)(0.10)}{122}}}$$

$$z \approx -2.05$$

$P = 0.0201$

0.844   0.90

$z = -2.05$

37. John Wayne

a) H_0: The death rate from cancer for people working on the film was similar to that predicted by cancer experts, 30 out of 220.
   H_A: The death rate from cancer for people working on the film was higher than the rate predicted by cancer experts.

   The conditions for inference are not met, since this is not a random sample. We will assume that the cancer rates for people working on the film are similar to those predicted by the cancer experts, and a Normal model can be used to model the sampling distribution of the rate, with $\mu_{\hat{p}} = p = 30/220$ and

$$\sigma(\hat{p}) = \sqrt{\frac{pq}{n}} = \sqrt{\frac{\left(\frac{30}{220}\right)\left(\frac{190}{220}\right)}{220}} \approx 0.0231 \,.$$

We can perform a one-proportion z-test. The observed cancer rate is

$$\hat{p} = \frac{46}{220} \approx 0.209 \,.$$

$$z = \frac{\hat{p} - p_0}{\sigma(\hat{p})}$$

$$z = \frac{\frac{46}{220} - \frac{30}{220}}{\sqrt{\frac{\left(\frac{30}{220}\right)\left(\frac{190}{220}\right)}{220}}}$$

$$z = 3.14$$

Since the P-value = 0.0008 is very low, we reject the null hypothesis. There is strong evidence that the cancer rate is higher than expected among the workers on the film.

b) This does not prove that exposure to radiation may increase the risk of cancer. This group of people may be atypical for reasons that have nothing to do with the radiation.

39. Obama or Harper?

H_0: $p = 0.5$

H_A: $p \neq 0.5$ The test is two-sided (given)

$$\hat{p} = \frac{21}{47} = 0.4468 \qquad\qquad \sigma\left(\hat{p}\right) = \sqrt{\frac{pq}{n}} = \sqrt{\frac{0.5^*0.5}{470}} = 0.023$$

$$z = \frac{(0.4468 - 0.5)}{0.023} = -2.31 \qquad\qquad \text{P-value} = 0.0208$$

We have moderately strong evidence that more Canadians would choose Obama as their most admired figure.

41. Body temperature

a)

H_0: $\mu = 37.0°C$

H_A: $\mu \neq 37.0°C$ The test is two-sided

$$\bar{x} = 36.83 \qquad\qquad \sigma(\bar{x}) = \frac{\sigma}{\sqrt{n}} = \frac{0.38}{\sqrt{64}} = 0.0475$$

$$z = \frac{(36.83 - 37)}{0.0475} = -3.58 \qquad\qquad \text{P-value} = 0.0003$$

We have strong evidence that "normal" body temperature is not $37.0°C$ but lower.

b) 95% CI: $\bar{x} \pm z^* \sigma(\bar{x}) = 36.83 \pm 1.96(0.0475) = (36.74, 36.92)$ Since $37.0°C$ falls above the 95% CI for body temperature, we can conclude that "normal" body temperature is not $37.0°C$ but lower.

43. Speed of light

H_0: $\mu = 299710.5$

H_A: $\mu \neq 299710.5$ The test is two-sided since he doesn't know in what direction he has erred (not that it will matter)

$\bar{x} = 299852.4$ $\qquad\qquad$ $\sigma(\bar{x}) = \dfrac{\sigma}{\sqrt{n}} = \dfrac{79.0}{\sqrt{100}} = 7.9$

$z = \dfrac{(299852.4 - 299710.5)}{7.9} = 17.96$ $\qquad\qquad$ P-value < 0.000001

This indicates that Michelson's measurements were biased. The chance of him having done the experiment perfectly and getting a number this far off is basically zero.

1. **One-sided or two?**
   a) Two sided. Let p be the percentage of students who prefer Diet Coke.
      H_0: 50% of students prefer Diet Coke. ($p = 0.50$)
      H_A: The percentage of students who prefer Diet Coke is not 50%. ($p \neq 0.50$)
   b) One sided. Let p be the percentage of teenagers who prefer the new formulation.
      H_0: 50% of students prefer the new formulation. ($p = 0.50$)
      H_A: More than 50% of students prefer the new formulation. ($p > 0.50$)
   c) One sided. Let p be the percentage of people who plan to vote for the override.
      H_0: 2/3 of the residents intend to vote for the override. ($p = 2/3$)
      H_A: More than 2/3 of the residents intend to vote for the override. ($p > 2/3$)
   d) Two sided. Let p be the percentage of days the market goes up.
      H_0: The market goes up on 50% of days. ($p = 0.50$)
      H_A: The percentage of days the market goes up is not 50%. ($p \neq 0.50$)

3. **P-value** If the effectiveness of the new poison ivy treatment is the same as the effectiveness of the old treatment, the chance of observing an effectiveness this large or larger in a sample of the same size is 4.7% by natural sampling variation alone.

5. **Alpha** Since the null hypothesis was rejected at $a = 0.05$, the P-value for the researcher's test must have been less than 0.05. He would have made the same decision at $a = 0.10$, since the P-value must also be less than 0.10. We can't be certain whether or not he would have made the same decision at $a = 0.01$, since we only know that the P-value was less than 0.05. It may have been less than 0.01, but we can't be sure.

7. **Significant?**
   a) If 90% of children have really been vaccinated, there is only a 1.1% chance of observing 89.4% of children (in a sample of 13 000) vaccinated by natural sampling variation alone.
   b) We conclude that the proportion of children who have been vaccinated is below 90%, but a 95% confidence interval would show that the true proportion is between 88.9% and 89.9%. Most likely a decrease from 90% to 89.9% would not be considered important. The 90% figure was probably an approximate figure anyway.

9. **Groceries**
   a) **Randomization Condition:** We will assume that the Yahoo! survey was conducted randomly.
      10% Condition: 2400 men represent less than 10% of all men.
      Success/Failure Condition: $n\hat{p} = 1224$ and $n\hat{q} = 1176$ are both greater than 10, so the sample is large enough.

Since the conditions are met, we can use a one-proportion z-interval to estimate the percentage of men who identify themselves as the primary grocery shopper in their household.

$$\hat{p} \pm z^* \sqrt{\frac{\hat{p}\hat{q}}{n}} = \left(\frac{1224}{2400}\right) \pm 2.326 \sqrt{\frac{\left(\frac{1224}{2400}\right)\left(\frac{1176}{2400}\right)}{2400}} = (48.6\%, 53.4\%)$$

We are 98% confident that between 48.6% and 53.4% of all men identify themselves as the primary grocery shopper in their household.

b) Since 45% is not in the confidencem interval, there is strong evidence that the percentage of all men that identify themselves as the primary grocery shopper in their household is not 45%.

c) The significance level of this test is $\alpha = 0.02$. It's a two-tailed test based on a 98% confidence interval.

11. Rural med school applicants, again

a) 95% CI for the population proportion is given by $\hat{p} \pm z^* \sqrt{\frac{\hat{p}\hat{q}}{n}}$.

Substituting $z^* = 1.96$ and $\hat{p} = 360/4948 = 0.072757$ and $n = 4948$,

$$\hat{p} \pm z^* \sqrt{\frac{\hat{p}\hat{q}}{n}} = 0.072757 \pm 1.96 \sqrt{\frac{0.072757(1 - 0.072757)}{4948}} = (0.065520, 0.079994).$$

b) The hypothesized value 0.13 is not in the above interval (in fact this interval is entirely below 0.13) and so we reject the null hypothesis $H_0 : p = 0.13$ against the two sided alternative ($H_A : p \neq 0.13$). Since all values in the 95% CI lie below 0.13, we have some reasons to suspect a possible lack of rural applicants.

13. Loans

a) The bank has made a Type II error. The person was not a good credit risk, and the bank failed to notice this.

b) The bank has made a Type I error. The person was a good credit risk, and the bank was convinced that he/she was not.

c) By making it easier to get a loan, the bank has reduced the alpha level. It takes less evidence to grant the person the loan.

d) The risk of Type I error is decreased and the risk of Type II error has increased.

15. Second loan

a) Power is the probability that the bank denies a loan that could not have been repaid.

b) To increase power, the bank could raise the cutoff score.

c) If the bank raised the cutoff score, a larger number of trustworthy people would be denied credit, and the bank would lose the opportunity to collect the interest on these loans.

17 Obesity in First Nations youth

a) The null hypothesis is that the obesity rate in First Nations youth has not decreased and the alternative hypothesis is that the obesity rate in First Nations youth has decreased.

b) Type I error in this situation is the error in concluding that the obesity rate in First Nations youth has decreased when in fact it has not decreased.

c) Type II error in this situation is the error in concluding that the obesity rate in First Nations youth has not decreased when in fact it has decreased.

d) If we make a type I error (i.e., if we conclude that the obesity rate in First Nations youth has decreased when in fact it has not decreased), the federal and provincial governments will continue the program that does not help reduce obesity rate. This harms the federal and provincial governments.

  If we make a type II error (i.e., if we conclude that the obesity rate in First Nations youth has not decreased when in fact it has decreased), the federal and provincial governments will stop the program that does help reduce obesity rate. This harms the First Nations children as they are losing a program effective in reducing obesity rate.

e) The power of the test represents the probability of concluding that the obesity rate in First Nations youth has decreased when in fact it has decreased.

19. Testing cars

H_0: The shop is meeting the emissions standards.

H_A: The shop is not meeting the emissions standards.

a) Type I error is when the regulators decide that the shop is not meeting standards when they actually are meeting the standards.

b) Type II error is when the regulators certify the shop when they are not meeting the standards.

c) Type I would be more serious to the shop owners. They would lose their certification, even though they are meeting the standards.

d) Type II would be more serious to environmentalists. Shops are allowed to operate, even though they are allowing polluting cars to operate.

21. Cars again

a) The power of the test is the probability of detecting that the shop is not meeting standards when they are not.

b) The power of the test will be greater when 40 cars are tested. A larger sample size increases the power of the test.

c) The power of the test will be greater when the level of significance is 10%. There is a greater chance that the null hypothesis will be rejected.

d) The power of the test will be greater when the shop is out of compliance "a lot." Larger problems are easier to detect.

23. Equal opportunity?

H_0: The company is not discriminating against minorities.
H_A: The company is discriminating against minorities.

a) This is a one-tailed test. They wouldn't sue if "too many" minorities were hired.
b) Type I error would be deciding that the company is discriminating against minorities when they are not discriminating.
c) Type II error would be deciding that the company is not discriminating against minorities when they actually are discriminating.
d) The power of the test is the probability that discrimination is detected when it is actually occurring.
e) The power of the test will increase when the level of significance is increased from 0.01 to 0.05.
f) The power of the test is lower when the lawsuit is based on 37 employees instead of 87. Lower sample size leads to less power.

25. Dropouts

a) The test is one-tailed. We are testing to see if a decrease in the dropout rate is associated with the software.
b) H_0: The dropout rate does not change following the use of the software. ($p = 0.13$)
   H_A: The dropout rate decreases following the use of the software. ($p < 0.13$)
c) The professor makes a Type I error if he buys the software when the dropout rate has not actually decreased.
d) The professor makes a Type II error if he doesn't buy the software when the dropout rate has actually decreased.
e) The power of the test is the probability of buying the software when the dropout rate has actually decreased.

27. Wind power again

a) We advise to buy a turbine, when in truth it will be a losing proposition financially.
b) We advise not to buy a turbine, when in truth it would have been a money-saving investment.
c) Decrease it so that there is less of a chance of recommending a bad investment in a turbine.
d) Power is the probability of advising to purchase a turbine when in fact it will be a money-saving investment.
e) Increase.
f) Make more measurements, say every 3 hours.
g) No, the small p-value indicates very strong and clear evidence that the wind speed exceeds 8 mph, but the wind speed might not exceed 8 mph by much. This

can happen in this particular situation because of the large sample size, so it would be useful to look at a confidence interval as well to see how large the average wind speed might be.

29. Dropouts, part II

a) H_0: The dropout rate does not change following the use of the software. ($p = 0.13$)
H_A: The dropout rate decreases following the use of the software. ($p < 0.13$)
Independence Assumption: One student's decision about dropping out should not influence another's decision.
Randomization Condition: This year's class of 203 students is probably representative of all Stats students.
10% Condition: A sample of 203 students is less than 10% of all students.
Success/Failure Condition: $np = (203)(0.13) = 26.39$ and $nq = (203)(0.87) = 176.61$ are both greater than 10, so the sample is large enough.
The conditions have been satisfied, so a Normal model can be used to model the sampling distribution of the proportion, with $\mu_{\hat{p}} = p = 0.13$ and

$$\sigma(\hat{p}) = \sqrt{\frac{pq}{n}} = \sqrt{\frac{(0.13)(0.87)}{203}} \approx 0.0236 \; .$$

We can perform a one-proportion z-test. The observed proportion of dropouts is $\hat{p} = \dfrac{11}{203} \approx 0.054 \; .$

Since the P-value $= 0.0007$ is very low, we reject the null hypothesis. There is strong evidence that the dropout rate has dropped since use of the software program was implemented. As long as the professor feels confident that this class of students is representative of all potential

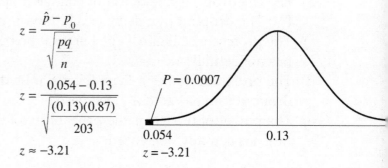

$$z = \frac{\hat{p} - p_0}{\sqrt{\dfrac{pq}{n}}}$$

$$z = \frac{0.054 - 0.13}{\sqrt{\dfrac{(0.13)(0.87)}{203}}}$$

$$z \approx -3.21$$

students, then he should buy the program.
If you used a 95% confidence interval to assess the effectiveness of the program:

$$\hat{p} \pm z^* \sqrt{\frac{\hat{p}\hat{q}}{n}} = \left(\frac{11}{203}\right) \pm 1.960 \sqrt{\frac{\left(\frac{11}{203}\right)\left(\frac{192}{203}\right)}{203}} = (2.3\%, \, 8.5\%)$$

We are 95% confident that the dropout rate is between 2.3% and 8.5%. Since 13% is not contained in the interval, this provides evidence that the dropout rate has changed following the implementation of the software program.

b) The chance of observing 11 or fewer dropouts in a class of 203 is only 0.07% if the dropout rate in the population is really 13%.

31. Two coins

a) The alternative hypothesis is that your coin produces 30% heads.

b) Reject the null hypothesis if the coin comes up tails. Otherwise, fail to reject.

c) There is a 10% chance that the coin comes up tails if the null hypothesis is true, so alpha is 10%.

d) Power is our ability to detect the 30% coin. The coin will come up tails 70% of the time. That's the power of our test.

e) To increase the power and lower the probability of Type I error at the same time, simply flip the coin more times.

33. Hoops

H_0: The player's foul-shot percentage is only 60%. ($p = 0.60$)

H_A: The player's foul-shot percentage is better than 60%. ($p > 0.60$)

a) The player's shots can be considered Bernoulli trials. There are only two possible outcomes, make the shot and miss the shot. The probability of making any shot is constant at $p = 0.60$. Assume that the shots are independent of each other. Use $Binom(10, 0.60)$.

Let X = the number of shots made out of $n = 10$

$$P(\text{makes at least 9 out of 10}) = P(X \geq 9)$$
$$= P(X = 9) + P(X = 10)$$
$$= {}_{10}C_9 (0.60)^9 (0.40)^1 + {}_{10}C_{10} (0.60)^{10} (0.40)^0$$
$$\approx 0.0464$$

b) The coach made a Type I error.

c) The power of the test can be calculated for specific values of the new probability of success. Each true value of p has a power calculation associated with it. In this case, we are finding the power of the test to detect an 80% foul-shot. Use $Binom(10, 0.80)$.

Let X = the number of shots made out of $n = 10$

$$P(\text{makes at least 9 out of 10}) = P(X \geq 9)$$
$$= P(X = 9) + P(X = 10)$$
$$= {}_{10}C_9 (0.80)^9 (0.20)^1 + {}_{10}C_{10} (0.80)^{10} (0.20)^0$$
$$\approx 0.376$$

The power of the test to detect an increase in foul-shot percentage from 60% to 80% is about 37.6%.

d) The power of the test to detect improvement in foul-shooting can be increased by increasing the number of shots, or by keeping the number of shots at 10 but increasing the level of significance by declaring that 8, 9, or 10 shots made will convince the coach that the player has improved. In other words, the coach can increase the power of the test by lowering the standard of proof.

35. Hoops II

a) We reject the null hypothesis $H_0 : p = 0.6$ in favour of the alternative $H_A : p > 0.6$

if $z = \dfrac{\hat{p} - p_0}{\sqrt{\dfrac{p_0 q_0}{n}}} = \dfrac{\hat{p} - 0.6}{\sqrt{\dfrac{0.6 \times (1 - 0.6)}{80}}} = \dfrac{\hat{p} - 0.6}{0.05477225575} > 1.645$.

That is $\hat{p} > 1.645 \times 0.05477225575 + 0.6 = 0.6901003607$.

b) If p = 0.75,

$$P(\hat{p} > 0.6901003607) = P\left(Z > \dfrac{0.6901003607 - 0.75}{\sqrt{\dfrac{0.75 \times (1 - 0.75)}{80}}}\right) = P(Z > -1.24) = 1 - 0.1075 = 0.8925$$

The power of the test if p = 0.75 is 0.8925.

c) From statistical software, the power of the test when p = 0.75 (and sample size = 80, significance level = 0.05) is 0.892039.
   To raise the power of the test to 0.99 when his true foul-shot percentage has risen to 75%, we need a sample of size 147 (using statistical software).

37. Simulations

a) The simulated sample from a Bernoulli distribution with $p = 0.8$ is shown below (Note: This is a random sample and so the answers can vary.):
   0 1 0 1 0 0 1 1 0 1 1 1 0 1 1 1 1 1 1
   1 1 0 1 0 1 1 0 1 1 0 1 1 1 1 1 1 1 1
   1 1 1 1 1 1 1 1 1 1 1 1 1 1 1 1 1 1 1
   0 1 1 1 0 1 1 1 1 1 1 1 0 1 1 1 1 1

   The number of 1s = 62.
   $\hat{p} = 62 / 75 = 0.826667$
   $n = 75$
   The test statistic to test the null hypothesis that $p = 0.8$ is

   $z = \dfrac{\hat{p} - p_0}{\sqrt{\dfrac{p_0 q_0}{n}}} = \dfrac{0.826667 - 0.8}{\sqrt{\dfrac{0.8 \times (1 - 0.8)}{75}}} = 0.58$

   p-value $= 2 \times P(Z > 0.58) = 0.564 > 0.05$ and so no evidence to reject the null hypothesis p = 0.8 against the two-sided alternative.

b) The results for 100 samples are given below. 94 of these tests do not reject the null hypothesis at the 5% level and 6 of them reject the null hypothesis. We expect 95% percent of tests to produce a correct decision, if we repeated these simulations many times. If X = number wrong decisions, then X has a binomial distribution with $n = 100$ and $p = 0.05$.

Sample	Sample size	Number of successes	Sample proportion	Z-statistic	p-value	Is p-value < 0.05? (1 = yes 0 = no)
1	75	56	0.746667	–1.15470	0.24821	0
2	75	66	0.880000	1.73205	0.08326	0
3	75	62	0.826667	0.57735	0.56370	0
4	75	59	0.786667	–0.28868	0.77283	0
5	75	57	0.760000	–0.86603	0.38648	0
6	75	63	0.840000	0.86603	0.38648	0
7	75	55	0.733333	–1.44338	0.14891	0
8	75	59	0.786667	–0.28868	0.77283	0
9	75	56	0.746667	–1.15470	0.24821	0
10	75	57	0.760000	–0.86603	0.38648	0
11	75	60	0.800000	0.00000	1.00000	0
12	75	57	0.760000	–0.86603	0.38648	0
13	75	59	0.786667	–0.28868	0.77283	0
14	75	60	0.800000	0.00000	1.00000	0
15	75	65	0.866667	1.44338	0.14891	0
16	75	59	0.786667	–0.28868	0.77283	0
17	75	59	0.786667	–0.28868	0.77283	0
18	75	61	0.813333	0.28868	0.77283	0
19	75	63	0.840000	0.86603	0.38648	0
20	75	58	0.773333	–0.57735	0.56370	0
21	75	62	0.826667	0.57735	0.56370	0
22	75	65	0.866667	1.44338	0.14891	0
23	75	57	0.760000	–0.86603	0.38648	0
24	75	61	0.813333	0.28868	0.77283	0
25	75	55	0.733333	–1.44338	0.14891	0
26	75	52	0.693333	–2.30940	0.02092	1
27	75	59	0.786667	–0.28868	0.77283	0
28	75	61	0.813333	0.28868	0.77283	0
29	75	61	0.813333	0.28868	0.77283	0
30	75	60	0.800000	0.00000	1.00000	0
31	75	58	0.773333	–0.57735	0.56370	0
32	75	62	0.826667	0.57735	0.56370	0
33	75	55	0.733333	–1.44338	0.14891	0
34	75	53	0.706667	–2.02073	0.04331	1

Sample	Sample size	Number of successes	Sample proportion	Z-statistic	p-value	Is p-value < 0.05? (1 = yes 0 = no)
35	75	63	0.840000	0.86603	0.38648	0
36	75	56	0.746667	–1.15470	0.24821	0
37	75	60	0.800000	0.00000	1.00000	0
38	75	59	0.786667	–0.28868	0.77283	0
39	75	54	0.720000	–1.73205	0.08326	0
40	75	63	0.840000	0.86603	0.38648	0
41	75	60	0.800000	0.00000	1.00000	0
42	75	58	0.773333	–0.57735	0.56370	0
43	75	60	0.800000	0.00000	1.00000	0
44	75	67	0.893333	2.02073	0.04331	1
45	75	63	0.840000	0.86603	0.38648	0
46	75	59	0.786667	–0.28868	0.77283	0
47	75	62	0.826667	0.57735	0.56370	0
48	75	62	0.826667	0.57735	0.56370	0
49	75	64	0.853333	1.15470	0.24821	0
50	75	62	0.826667	0.57735	0.56370	0
51	75	63	0.840000	0.86603	0.38648	0
52	75	57	0.760000	–0.86603	0.38648	0
53	75	60	0.800000	0.00000	1.00000	0
54	75	58	0.773333	–0.57735	0.56370	0
55	75	60	0.800000	0.00000	1.00000	0
56	75	58	0.773333	–0.57735	0.56370	0
57	75	63	0.840000	0.86603	0.38648	0
58	75	58	0.773333	–0.57735	0.56370	0
59	75	62	0.826667	0.57735	0.56370	0
60	75	58	0.773333	–0.57735	0.56370	0
61	75	60	0.800000	0.00000	1.00000	0
62	75	60	0.800000	0.00000	1.00000	0
63	75	61	0.813333	0.28868	0.77283	0
64	75	62	0.826667	0.57735	0.56370	0
65	75	62	0.826667	0.57735	0.56370	0
66	75	58	0.773333	–0.57735	0.56370	0
67	75	68	0.906667	2.30940	0.02092	1
68	75	65	0.866667	1.44338	0.14891	0

Sample	Sample size	Number of successes	Sample proportion	Z-statistic	p-value	Is p-value < 0.05? (1 = yes 0 = no)
69	75	54	0.720000	–1.73205	0.08326	0
70	75	62	0.826667	0.57735	0.56370	0
71	75	54	0.720000	–1.73205	0.08326	0
72	75	64	0.853333	1.15470	0.24821	0
73	75	59	0.786667	–0.28868	0.77283	0
74	75	62	0.826667	0.57735	0.56370	0
75	75	62	0.826667	0.57735	0.56370	0
76	75	62	0.826667	0.57735	0.56370	0
77	75	60	0.800000	0.00000	1.00000	0
78	75	53	0.706667	–2.02073	0.04331	1
79	75	62	0.826667	0.57735	0.56370	0
80	75	62	0.826667	0.57735	0.56370	0
81	75	59	0.786667	–0.28868	0.77283	0
82	75	63	0.840000	0.86603	0.38648	0
83	75	58	0.773333	–0.57735	0.56370	0
84	75	57	0.760000	–0.86603	0.38648	0
85	75	57	0.760000	–0.86603	0.38648	0
86	75	57	0.760000	–0.86603	0.38648	0
87	75	66	0.880000	1.73205	0.08326	0
88	75	64	0.853333	1.15470	0.24821	0
89	75	56	0.746667	–1.15470	0.24821	0
90	75	60	0.800000	0.00000	1.00000	0
91	75	66	0.880000	1.73205	0.08326	0
92	75	56	0.746667	–1.15470	0.24821	0
93	75	57	0.760000	–0.86603	0.38648	0
94	75	62	0.826667	0.57735	0.56370	0
95	75	57	0.760000	–0.86603	0.38648	0
96	75	53	0.706667	–2.02073	0.04331	1
97	75	63	0.840000	0.86603	0.38648	0
98	75	63	0.840000	0.86603	0.38648	0
99	75	61	0.813333	0.28868	0.77283	0
100	75	62	0.826667	0.57735	0.56370	0

Summary
Sample size: 75.0000
Probability of success: 0.800000
Number of samples: 100.000
Significance level: 0.0500000
The value of p_0: 0.800000
Observed significance level: 0.0600000

39. Still more simulations

a) The simulated sample from a Bernoulli distribution with $p = 0.8$ is shown below (Note: This is a random sample and so the answers can vary):

```
1 1 1 1 1 1 1 1 1 1 1 1 1 0 1 1 1 1 1
1 1 0 1 1 1 1 1 1 1 1 1 1 1 1 1 1 1 1
1 1 1 1 0 1 1 1 1 1 0 1 1 1 1 1 0 1 1
1 1 1 1 1 1 1 1 1 1 1 1 1 1 1 0 0
```

The number of 1s = 68.
$\hat{p} = 68/75 = 0.906667$
$n = 75$
The test statistic to test the null hypothesis that p = 0.3 is

$$z = \frac{\hat{p} - p_0}{\sqrt{\dfrac{p_0 q_0}{n}}} = \frac{0.906667 - 0.7}{\sqrt{\dfrac{0.7 \times (1 - 0.7)}{75}}} = 3.91$$

p-value = $2 \times P(Z > 3.91) = 0.00$ < 0.05 and so we have evidence to reject the null hypothesis $p = 0.7$ against the two-sided alternative.

b) The results for 100 samples are given below. 47 of these tests reject the null hypothesis at the 5% percent level and 53 of them failed to reject the null hypothesis. Type II error is the error of failing to reject the null hypothesis (when it is false). The probability of a type II error based on this simulation is 53/100 = 0.53 and the power = 1 – probability of a type II error = 0.47.
An estimate of power from a simulation based on 1500 samples (each of size 75) was 0.465333.

c) The exact power of the test (using statistical software) is 0.467982.

Sample	Sample size	Number of successes	Sample proportion	Z-statistic	p-value	Is p-value < 0.05? (1 = yes, 0 = no)
1	75	57	0.760000	1.13389	0.256839	0
2	75	59	0.786667	1.63785	0.101454	0
3	75	66	0.880000	3.40168	0.000670	1
4	75	65	0.866667	3.14970	0.001634	1
5	75	51	0.680000	−0.37796	0.705457	0
6	75	66	0.880000	3.40168	0.000670	1
7	75	67	0.893333	3.65366	0.000259	1
8	75	53	0.706667	0.12599	0.899741	0
9	75	61	0.813333	2.14180	0.032210	1
10	75	59	0.786667	1.63785	0.101454	0
11	75	59	0.786667	1.63785	0.101454	0
12	75	58	0.773333	1.38587	0.165787	0
13	75	59	0.786667	1.63785	0.101454	0
14	75	62	0.826667	2.39377	0.016676	1
15	75	57	0.760000	1.13389	0.256839	0
16	75	55	0.733333	0.62994	0.528733	0
17	75	62	0.826667	2.39377	0.016676	1
18	75	56	0.746667	0.88192	0.377822	0
19	75	61	0.813333	2.14180	0.032210	1
20	75	57	0.760000	1.13389	0.256839	0
21	75	56	0.746667	0.88192	0.377822	0
22	75	59	0.786667	1.63785	0.101454	0
23	75	57	0.760000	1.13389	0.256839	0
24	75	59	0.786667	1.63785	0.101454	0
25	75	56	0.746667	0.88192	0.377822	0
26	75	57	0.760000	1.13389	0.256839	0
27	75	56	0.746667	0.88192	0.377822	0
28	75	61	0.813333	2.14180	0.032210	1
29	75	63	0.840000	2.64575	0.008151	1
30	75	57	0.760000	1.13389	0.256839	0
31	75	59	0.786667	1.63785	0.101454	0
32	75	57	0.760000	1.13389	0.256839	0
33	75	60	0.800000	1.88982	0.058782	0
34	75	61	0.813333	2.14180	0.032210	1
35	75	57	0.760000	1.13389	0.256839	0
36	75	61	0.813333	2.14180	0.032210	1
37	75	59	0.786667	1.63785	0.101454	0
38	75	61	0.813333	2.14180	0.032210	1
39	75	60	0.800000	1.88982	0.058782	0

Sample	Sample size	Number of successes	Sample proportion	Z-statistic	p-value	Is p-value < 0.05? (1 = yes, 0 = no)
40	75	59	0.786667	1.63785	0.101454	0
41	75	56	0.746667	0.88192	0.377822	0
42	75	65	0.866667	3.14970	0.001634	1
43	75	57	0.760000	1.13389	0.256839	0
44	75	60	0.800000	1.88982	0.058782	0
45	75	62	0.826667	2.39377	0.016676	1
46	75	57	0.760000	1.13389	0.256839	0
47	75	61	0.813333	2.14180	0.032210	1
48	75	63	0.840000	2.64575	0.008151	1
49	75	59	0.786667	1.63785	0.101454	0
50	75	67	0.893333	3.65366	0.000259	1
51	75	61	0.813333	2.14180	0.032210	1
52	75	59	0.786667	1.63785	0.101454	0
53	75	58	0.773333	1.38587	0.165787	0
54	75	63	0.840000	2.64575	0.008151	1
55	75	61	0.813333	2.14180	0.032210	1
56	75	62	0.826667	2.39377	0.016676	1
57	75	59	0.786667	1.63785	0.101454	0
58	75	65	0.866667	3.14970	0.001634	1
59	75	61	0.813333	2.14180	0.032210	1
60	75	54	0.720000	0.37796	0.705457	0
61	75	63	0.840000	2.64575	0.008151	1
62	75	63	0.840000	2.64575	0.008151	1
63	75	64	0.853333	2.89773	0.003759	1
64	75	57	0.760000	1.13389	0.256839	0
65	75	63	0.840000	2.64575	0.008151	1
66	75	55	0.733333	0.62994	0.528733	0
67	75	56	0.746667	0.88192	0.377822	0
68	75	60	0.800000	1.88982	0.058782	0
69	75	61	0.813333	2.14180	0.032210	1
70	75	57	0.760000	1.13389	0.256839	0
71	75	60	0.800000	1.88982	0.058782	0
72	75	54	0.720000	0.37796	0.705457	0
73	75	53	0.706667	0.12599	0.899741	0
74	75	62	0.826667	2.39377	0.016676	1
75	75	65	0.866667	3.14970	0.001634	1
76	75	59	0.786667	1.63785	0.101454	0
77	75	64	0.853333	2.89773	0.003759	1
78	75	61	0.813333	2.14180	0.032210	1

Sample	Sample size	Number of successes	Sample proportion	Z-statistic	p-value	Is p-value < 0.05? (1 = yes, 0 = no)
79	75	59	0.786667	1.63785	0.101454	0
80	75	59	0.786667	1.63785	0.101454	0
81	75	63	0.840000	2.64575	0.008151	1
82	75	62	0.826667	2.39377	0.016676	1
83	75	65	0.866667	3.14970	0.001634	1
84	75	62	0.826667	2.39377	0.016676	1
85	75	61	0.813333	2.14180	0.032210	1
86	75	58	0.773333	1.38587	0.165787	0
87	75	61	0.813333	2.14180	0.032210	1
88	75	60	0.800000	1.88982	0.058782	0
89	75	60	0.800000	1.88982	0.058782	0
90	75	66	0.880000	3.40168	0.000670	1
91	75	54	0.720000	0.37796	0.705457	0
92	75	58	0.773333	1.38587	0.165787	0
93	75	62	0.826667	2.39377	0.016676	1
94	75	61	0.813333	2.14180	0.032210	1
95	75	64	0.853333	2.89773	0.003759	1
96	75	59	0.786667	1.63785	0.101454	0
97	75	64	0.853333	2.89773	0.003759	1
98	75	61	0.813333	2.14180	0.032210	1
99	75	64	0.853333	2.89773	0.003759	1
100	75	68	0.906667	3.90563	0.000094	1

Summary

Sample size: 75.0000

Probability of success: 0.800000

Number of samples: 100.000

Significance level: 0.0500000

The value of p_0: 0.700000

Observed significance level: 0.470000

41. Multiple tests

Let X = the number of significant tests. Then X is Binomial(10, 0.05), so

$P(X \geq 1) = 1 - P(X = 0) = 1 - 0.95^{10} = 1 - 0.60 = 0.40$. If exactly one centre finds a significant result, this is not very significant in the big picture. A single significant result in 10 trials is easily explained by chance, so we should not reject the null hypothesis.

1. **Online social networking** It is very unlikely that samples would show an observed difference this large if, in fact, there was no real difference in the proportions of teens that have Facebook accounts and teens that have MySpace accounts.

3. **Name recognition** Without knowing anything about how the polls were taken, we should be cautious about drawing conclusions from this study. For example, the second poll could have been taken at a support rally for this candidate, causing a large source of bias. If the same group was polled twice, we do not satisfy the independence assumption. Assuming proper polling techniques, a P-value this small would give us good evidence that the campaign ads are working.

5. **Revealing information** This test is not appropriate for these data, since the responses are not from independent groups, but are from the same individuals. The independent samples condition has been violated.

7. **Gender gap**
   a) This is a stratified random sample, stratified by gender.
   b) We would expect the difference in proportions in the sample to be the same as the difference in proportions in the population, with the percentage of respondents with a favourable impression of the candidate 6% higher among males.
   c) The standard deviation of the difference in proportions is:
   $$\sigma(\hat{p}_M - \hat{p}_F) = \sqrt{\frac{\hat{p}_M \hat{q}_M}{n_M} + \frac{\hat{p}_F \hat{q}_F}{n_F}} = \sqrt{\frac{(0.59)(0.41)}{300} + \frac{(0.53)(0.47)}{300}} \approx 4\%$$

   d)

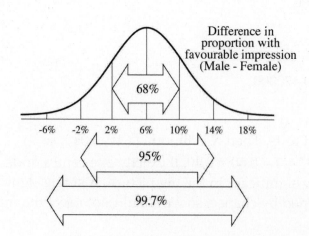

e) The campaign could certainly be misled by the poll. According to the model, a poll showing little difference could occur relatively frequently. That result is only 1.5 standard deviations below the expected difference in proportions.

9. Arthritis

a) **Randomization Condition**: Americans age 65 and older were selected randomly.

10% Condition: 1012 men and 1062 women are less than 10% of all men and women.

Independent Samples Condition: The sample of men and the sample of women were drawn independently of each other.

Success/Failure Condition: $n\hat{p}_M = 411$, $n\hat{q}_M = 601$, $n\hat{p}_F = 535$, and $n\hat{q}_F = 527$ are all greater than 10, so the samples are both large enough.

Since the conditions have been satisfied, we will find a two-proportion z-interval.

b)

$$\left(\hat{p}_F - \hat{p}_M\right) \pm z^* \sqrt{\frac{\hat{p}_F \hat{q}_F}{n_F} + \frac{\hat{p}_M \hat{q}_M}{n_M}} = \left(\frac{535}{1062} - \frac{411}{1012}\right) \pm 1.960 \sqrt{\frac{\left(\frac{535}{1062}\right)\left(\frac{527}{1062}\right)}{1062} + \frac{\left(\frac{411}{1012}\right)\left(\frac{601}{1012}\right)}{1012}} = (0.055,\ 0.140)$$

c) We are 95% confident that the proportion of American women age 65 and older who suffer from arthritis is between 5.5% and 14.0% higher than the proportion of American men the same age who suffer from arthritis.

d) Since the interval for the difference in proportions of arthritis sufferers does not contain 0, there is strong evidence that arthritis is more likely to afflict women than men.

11. Pets

a)

$$SE\left(\hat{p}_{Herb} - \hat{p}_{None}\right) = \sqrt{\frac{\hat{p}_{Herb} \hat{q}_{Herb}}{n_{Herb}} + \frac{\hat{p}_{None} \hat{q}_{None}}{n_{None}}} = \sqrt{\frac{\left(\frac{473}{827}\right)\left(\frac{354}{827}\right)}{827} + \frac{\left(\frac{19}{130}\right)\left(\frac{111}{130}\right)}{130}} = 0.035$$

b) **Randomization Condition**: Assume that the dogs studied were representative of all dogs.

10% Condition: 827 dogs from homes with herbicide used regularly and 130 dogs from homes with no herbicide used are less than 10% of all dogs.

Independent Samples Condition: The samples were drawn independently of each other.

Success/Failure Condition: $n\hat{p}_{Herb} = 473$, $n\hat{q}_{Herb} = 354$, $n\hat{p}_{None} = 19$, and $n\hat{q}_{None} = 111$ are all greater than 10, so the samples are both large enough.

Since the conditions have been satisfied, we will find a two-proportion z-interval.

$$\left(\hat{p}_{Herb} - \hat{p}_{None} \right) \pm z^* \sqrt{\frac{\hat{p}_{Herb}\hat{q}_{Herb}}{n_{Herb}} + \frac{\hat{p}_{None}\hat{q}_{None}}{n_{None}}}$$

$$= 9\left(\frac{473}{827} - \frac{19}{130} \right) \pm 1.960 \sqrt{\frac{\left(\frac{473}{827}\right)\left(\frac{354}{827}\right)}{827} + \frac{\left(\frac{19}{130}\right)\left(\frac{111}{130}\right)}{130}} = (0.356, 0.495)$$

c) We are 95% confident that the proportion of dogs with malignant lymphoma in homes where herbicides are used is between 35.6% and 49.5% higher than the proportion of dogs with lymphoma in homes where no pesticides are used.

13. **Ear infections**
    a) **Randomization Condition**: The babies were randomly assigned to the two treatment groups.
    Independent Samples Condition: The groups were assigned randomly, so the groups are not related.
    Success/Failure Condition: $n\hat{p}_{Vacc}$ = 333, $n\hat{q}_{Vacc}$ = 2122, $n\hat{p}_{None}$ = 499, and $n\hat{q}_{None}$ = 1953 are all greater than 10, so the samples are both large enough.
    Since the conditions have been satisfied, we will find a two-proportion z-interval.

    b) $$\left(\hat{p}_{None} - \hat{p}_{Vacc} \right) \pm z^* \sqrt{\frac{\hat{p}_{None}\hat{q}_{None}}{n_{None}} + \frac{\hat{p}_{Vacc}\hat{q}_{Vacc}}{n_{Vacc}}}$$

    $$= \left(\frac{499}{2452} - \frac{333}{2455} \right) \pm 1.960 \sqrt{\frac{\left(\frac{499}{2452}\right)\left(\frac{1953}{2452}\right)}{2452} + \frac{\left(\frac{333}{2455}\right)\left(\frac{2122}{2455}\right)}{2455}} = \left(0.047, 0.089 \right)$$

    c) We are 95% confident that the proportion of unvaccinated babies who develop ear infections is between 4.7% and 8.9% higher than the proportion of vaccinated babies who develop ear infections. The vaccinations appear to be effective, especially considering the 20% infection rate among the unvaccinated. A reduction of between 5% and 9% is meaningful.

15. **Another ear infection**
    a) H_0: The proportion of vaccinated babies who get ear infections is the same as the proportion of unvaccinated babies who get ear infections.
    $$\left(p_{Vacc} = p_{None} \text{ or } p_{Vacc} - p_{None} = 0 \right)$$
    H_A: The proportion of vaccinated babies who get ear infections is lower than the proportion of unvaccinated babies who get ear infections.
    $$\left(p_{Vacc} < p_{None} \text{ or } p_{Vacc} - p_{None} < 0 \right)$$
    b) Since 0 is not in the confidence interval, reject the null hypothesis. There is evidence that the vaccine reduces the rate of ear infections.

c) If we think that the vaccine really reduces the rate of ear infections and it really does not reduce the rate of ear infections, then we have committed a Type I error.

d) Babies would be given ineffective vaccines.

17. Teen smoking, part I

a) This is a prospective observational study.

b) H_0: The proportion of teen smokers among the group whose parents disapprove of smoking is the same as the proportion of teen smokers among the group whose parents are lenient about smoking.

$$\left(p_{Dis} = p_{Len} \text{ or } p_{Dis} - p_{Len} = 0\right)$$

H_A: The proportion of teen smokers among the group whose parents disapprove of smoking is lower than the proportion of teen smokers among the group whose parents are lenient about smoking.

$$\left(p_{Dis} < p_{Len} \text{ or } p_{Dis} - p_{Len} < 0\right)$$

c) **Randomization Condition**: Assume that the teens surveyed are representative of all teens.

10% Condition: 284 and 41 are both less than 10% of all teens.

Independent Samples Condition: The groups were surveyed independently.

Success/Failure Condition: $n\hat{p}_{Dis} = 54$, $n\hat{q}_{Dis} = 230$, $n\hat{p}_{Len} = 11$, and $n\hat{q}_{Len} = 30$ are all greater than 10, so the samples are both large enough.

Since the conditions have been satisfied, we will model the sampling distribution of the difference in proportion with a Normal model with mean 0 and standard deviation estimated by:

$$SE_{pooled}\left(\hat{p}_{Dis} - \hat{p}_{Len}\right) = \sqrt{\frac{\hat{p}_{pooled}\hat{q}_{pooled}}{n_{Dis}} + \frac{\hat{p}_{pooled}\hat{q}_{pooled}}{n_{Len}}} = \sqrt{\frac{\left(\frac{65}{325}\right)\left(\frac{260}{325}\right)}{284} + \frac{\left(\frac{65}{325}\right)\left(\frac{260}{325}\right)}{41}} = 0.0668$$

d) The observed difference between the proportions is 0.190 – 0.268 = – 0.078.

Since the *P*-value = 0.1211 is high, we fail to reject the null hypothesis. There is little evidence to suggest that parental attitudes influence teens' decisions to smoke.

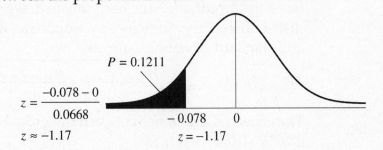

$$z = \frac{-0.078 - 0}{0.0668}$$

$$z \approx -1.17$$

$P = 0.1211$

$-0.078 \quad 0$

$z = -1.17$

e) If there is no difference in the proportions, there is about a 12% chance of seeing the observed difference or larger by natural sampling variation.

f) If teens' decisions about smoking *are* influenced, we have committed a Type II error.

19. Teen smoking, part II

a) Since the conditions have already been satisfied in Exercise 17, we will find a two-proportion z-interval.

$$(\hat{p}_{Dis} - \hat{p}_{Len}) \pm z^* \sqrt{\frac{\hat{p}_{Dis}\,\hat{q}_{Dis}}{n_{Dis}} + \frac{\hat{p}_{Len}\,\hat{q}_{Len}}{n_{Len}}}$$

$$= \left(\frac{54}{284} - \frac{11}{41}\right) \pm 1.960 \sqrt{\frac{\left(\frac{54}{284}\right)\left(\frac{230}{284}\right)}{284} + \frac{\left(\frac{11}{41}\right)\left(\frac{30}{41}\right)}{41}} = (-0.221,\ 0.065)$$

b) We are 95% confident that the proportion of teens whose parents disapprove of smoking who will eventually smoke is between 22.1% less and 6.5% more than for teens with parents who are lenient about smoking.

c) We expect 95% of random samples of this size to produce intervals that contain the true difference between the proportions.

21. Pregnancy

a) No, this is observational data, not an experiment.

b) H_0: The proportion of live births is the same for women under the age of 38 as it is for women 38 or older. $\left(p_{<38} = p_{\geq 38}\ \text{or}\ p_{<38} - p_{\geq 38} = 0\right)$

H_A: The proportion of live births is different for women under the age of 38 than for women 38 or older. $\left(p_{<38} \neq p_{\geq 38}\ \text{or}\ p_{<38} - p_{\geq 38} \neq 0\right)$

Randomization Condition: Assume that the women studied are representative of all women.

10% Condition: 157 and 89 are both less than 10% of all women.

Independent Samples Condition: The groups are not associated.

Success/Failure Condition: $n\hat{p}$(under 38) = 42, $n\hat{q}$(under 38) = 115, $n\hat{p}$(38 and over) = 7, and $n\hat{q}$(38 and over) = 82 are not all greater than 10, since the observed number of live births is only 7. However, the pooled value, $n\hat{p}_{pooled}$ (38 and over) = (89)(0.1992) = 18. All of the samples are large enough.

Since the conditions have been satisfied, we will model the sampling distribution of the difference in proportion with a Normal model with mean 0 and standard deviation estimated by:

$$SE_{pooled}\left(\hat{p}_{<38} - \hat{p}_{\geq 38}\right) = \sqrt{\frac{\hat{p}_{pooled}\,\hat{q}_{pooled}}{n_{<38}} + \frac{\hat{p}_{pooled}\,\hat{q}_{pooled}}{n_{\geq 38}}} = \sqrt{\frac{\left(\frac{49}{246}\right)\left(\frac{197}{246}\right)}{157} + \frac{\left(\frac{49}{246}\right)\left(\frac{197}{246}\right)}{89}} \approx 0.0530$$

The observed difference between the proportions is:

0.2675 – 0.0787 = 0.1888.

Since the P-value = 0.0004 is low, we reject the null hypothesis. There is strong evidence to suggest a difference in the proportion of live births for women under 38 and

$$z = \frac{0.1888 - 0}{0.0530}$$

$$z \approx 3.56$$

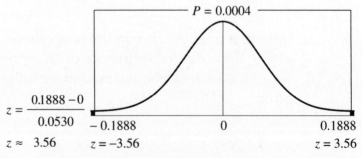

women 38 and over at this clinic. In fact, the evidence suggests that women under 38 have a higher proportion of live births.

c)
$$(\hat{p}_{<38} - \hat{p}_{\geq 38}) \pm z^* \sqrt{\frac{\hat{p}_{<38}\hat{q}_{<38}}{n_{<38}} + \frac{\hat{p}_{\geq 38}\hat{q}_{\geq 38}}{n_{\geq 38}}}$$

$$= \left(\tfrac{42}{157} - \tfrac{7}{89}\right) \pm 1.960 \sqrt{\frac{\left(\tfrac{42}{157}\right)\left(\tfrac{115}{157}\right)}{157} + \frac{\left(\tfrac{7}{89}\right)\left(\tfrac{82}{89}\right)}{89}} = (0.100, 0.278)$$

We are 95% confident that the proportion of live births for patients at this clinic is between 10.0% and 27.8% higher for women under 38 than for women 38 and over. However, the Success/Failure Condition is not met for the older women, so we should be cautious when using this interval. (The expected number of successes from the pooled proportion cannot be used for a condition for a confidence interval. It's based upon an assumption that the proportions are the same. We don't make that assumption in a confidence interval. In fact, we are implicitly assuming a *difference*, by finding an interval for the difference in proportion.)

23. Politics and sex

a) H_0: The proportion of voters in support of the candidate is the same before and after news of his extramarital affair got out.

$$(p_B = p_A \quad \text{or} \quad p_B - p_A = 0)$$

H_A: The proportion of voters in support of the candidate decreased after news of his extramarital affair got out.

$$(p_B > p_A \quad \text{or} \quad p_B - p_A > 0)$$

Randomization Condition: Voters were randomly selected.
10% Condition: 630 and 1010 are both less than 10% of all voters.
Independent Samples Condition: Since the samples were random, the groups are independent.
Success/Failure Condition: $n\hat{p}$(before) = (630)(0.54) = 340, $n\hat{q}$(before) = (630)(0.46) = 290, $n\hat{p}$(after) = (1010)(0.51) = 515, and $n\hat{q}$(after) = (1010)(0.49) = 505 are all greater than 10, so both samples are large enough.
Since the conditions have been satisfied, we will model the sampling distribution of the difference in proportion with a Normal model with mean 0 and standard deviation estimated by:

$$SE_{pooled}(\hat{p}_B - \hat{p}_A) = \sqrt{\frac{\hat{p}_{pooled}\hat{q}_{pooled}}{n_B} + \frac{\hat{p}_{pooled}\hat{q}_{pooled}}{n_A}} = \sqrt{\frac{(0.5215)(0.4785)}{630} + \frac{(0.5215)(0.4785)}{1010}} \approx 0.02536$$

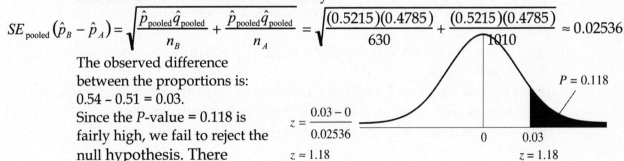

The observed difference between the proportions is:
0.54 − 0.51 = 0.03.
Since the *P*-value = 0.118 is fairly high, we fail to reject the null hypothesis. There

$$z = \frac{0.03 - 0}{0.02536}$$

$$z \approx 1.18$$

$P = 0.118$

$z = 1.18$

is little evidence of a decrease in the proportion of voters in support of the candidate after the news of his extramarital affair got out.

b) If we are wrong, we would be making a Type II error, since we failed to accept the alternative hypothesis when it is actually true.

c) A 95% confidence interval would be

$$0.03 \pm 1.96(0.02536) = 0.03 \pm 0.0497 = (-0.02, 0.08)$$

It is estimated with 95% confidence that the change in support is anywhere from an 8% decrease to a 2% increase. We could not conclude, with 95% confidence, that there was a decrease in support.

25. Pain

a) **Randomization Condition**: The patients were randomly selected AND randomly assigned to treatment groups. If that's not random enough for you, I don't know what is!

10% Condition: 112 and 108 are both less than 10% of all people with joint pain.

Success/Failure Condition: $n\hat{p}(A) = 84$, $n\hat{q}(A) = 28$, $n\hat{p}(B) = 66$, and $n\hat{q}(B) = 42$ are all greater than 10, so both samples are large enough.

Since the conditions are met, we can use a one-proportion z-interval to estimate the percentage of patients who may get relief from medication A.

$$\hat{p} \pm z^* \sqrt{\frac{\hat{p}\hat{q}}{n}} = \left(\frac{84}{112}\right) \pm 1.960 \sqrt{\frac{\left(\frac{84}{112}\right)\left(\frac{28}{112}\right)}{112}} = (67.0\%, \ 83.0\%)$$

We are 95% confident that between 67.0% and 83.0% of patients with joint pain will find medication A to be effective.

b) Since the conditions were met in part a), we can use a one-proportion z-interval to estimate the percentage of patients who may get relief from medication B.

$$\hat{p} \pm z^* \sqrt{\frac{\hat{p}\hat{q}}{n}} = \left(\frac{66}{108}\right) \pm 1.960 \sqrt{\frac{\left(\frac{66}{108}\right)\left(\frac{42}{108}\right)}{108}} = (51.9\%, \ 70.3\%)$$

We are 95% confident that between 51.9% and 70.3% of patients with joint pain will find medication B to be effective.

c) The 95% confidence intervals overlap, which might lead one to believe that there is no evidence of a difference in the proportions of people who find each medication effective. However, if one was led to believe that, one should proceed to part d).

d) Most of the conditions were checked in part a). We only have one more condition to check:

Independent Samples Condition: The groups were assigned randomly, so there is no reason to believe there is a relationship between them.

Since the conditions have been satisfied, we will find a two-proportion z-interval.

$$(\hat{p}_A - \hat{p}_B) \pm z^* \sqrt{\frac{\hat{p}_A \hat{q}_A}{n_A} + \frac{\hat{p}_B \hat{q}_B}{n_B}}$$

$$= \left(\frac{84}{112} - \frac{60}{108}\right) \pm 1.96 \sqrt{\frac{\left(\frac{84}{112}\right)\left(\frac{28}{112}\right)}{112} + \frac{\left(\frac{66}{108}\right)\left(\frac{42}{108}\right)}{108}} = (0.017, 0.261)$$

We are 95% confident that the proportion of patients with joint pain who will find medication A effective is between 1.70% and 26.1% higher than the proportion of patients who will find medication B effective.

e) The interval does not contain zero. There is evidence that medication A is more effective than medication B.

f) The two-proportion method is the proper method. By attempting to use two, separate, confidence intervals, you are adding standard deviations when looking for a difference in proportions. We know from our previous studies that *variances* add when finding the standard deviation of a difference. The two-proportion method does this.

27. **Sensitive men**
H_0: The proportion of 18–24-year-old men who are comfortable talking about their problems is the same as the proportion of 25–34-year old men.
$$\left(p_{Young} = p_{Old} \text{ or } p_{Young} - p_{Old} = 0\right)$$
H_A: The proportion of 18–24-year-old men who are comfortable talking about their problems is higher than the proportion of 25–34-year old men.
$$\left(p_{Young} > p_{Old} \text{ or } p_{Young} - p_{Old} > 0\right)$$
Randomization Condition: We must assume that the respondents were chosen randomly.
10% Condition: 129 and 184 are both less than 10% of all people.
Independent Samples Condition: The groups were chosen independently.
Success/Failure Condition: $n_{\hat{p}}$(young) = 80, $n_{\hat{q}}$(young) = 49, $n_{\hat{p}}$(old) = 98, and $n_{\hat{q}}$(old) = 86 are all greater than 10, so both samples are large enough.
Since the conditions have been satisfied, we will model the sampling distribution of the difference in proportion with a Normal model with mean 0 and standard deviation estimated by:

$$SE_{pooled}\left(\hat{p}_{Young} - \hat{p}_{Old}\right) = \sqrt{\frac{\hat{p}_{pooled}\,\hat{q}_{pooled}}{n_Y} + \frac{\hat{p}_{pooled}\,\hat{q}_{pooled}}{n_O}} = \sqrt{\frac{\left(\frac{178}{313}\right)\left(\frac{135}{313}\right)}{129} + \frac{\left(\frac{178}{313}\right)\left(\frac{135}{313}\right)}{184}} \approx 0.05687$$

The observed difference between the proportions is: 0.620 – 0.533 = 0.087. Since the *P*-value = 0.0619 is high, we fail to reject the null hypothesis. There is little evidence that the proportion of 18–24-year-old men who are

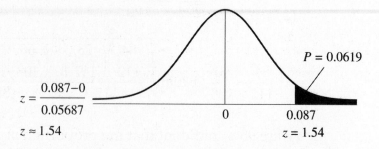

$$z = \frac{0.087 - 0}{0.05687}$$

$$z \approx 1.54$$

$P = 0.0619$

$z = 1.54$

comfortable talking about their problems is higher than the proportion of 25–34-year-old men who are comfortable. *Time* magazine's interpretation is questionable.

29.　Online activity checks

H_0: The proportion of teens who say their parents check to see what Web sites they visited is the same in 2006 as it was in 2004.　$\left(p_{2006} = p_{2004} \text{ or } p_{2006} - p_{2004} = 0\right)$

H_A: The proportion of teens who say their parents check to see what Web sites they visited is higher in 2006 than it was in 2004. $\left(p_{2006} > p_{2004} \text{ or } p_{2006} - p_{2004} > 0\right)$

Randomization Condition: The samples were random.
10% Condition: 811 and 868 are both less than 10% of all teens.
Independent Samples Condition: The samples were taken independently.
Success/Failure Condition: $n\hat{p}(2006) = 333$, $n\hat{q}(2006) = 478$, $n\hat{p}(2004) = 286$, and $n\hat{q}(2004) = 582$ are all greater than 10, so both samples are large enough.

Since the conditions have been satisfied, we will model the sampling distribution of the difference in proportion with a Normal model with mean 0 and standard deviation estimated by:

$$SE_{pooled}\left(\hat{p}_{2006} - \hat{p}_{2004}\right) = \sqrt{\frac{\hat{p}_{pool}\hat{q}_{pool}}{n_{2006}} + \frac{\hat{p}_{pool}\hat{q}_{pool}}{n_{2004}}} = \sqrt{\frac{(0.369)(0.631)}{811} + \frac{(0.369)(0.631)}{868}} :$$

= 0.02356.

The observed difference between the proportions is: 0.41 – 0.33 = 0.08. We will perform a two-proportion *z*-test.

The value of *z* = 3.40 and the *P*-value = 0.0003. Since the *P*-value is low, we reject the null hypothesis. There is strong evidence that a greater proportion of teens in 2006 say their parents checked in to see what Web sites they visited than said this in 2004.

31.　Food preference

H_0: The proportion of urban residents who agree with the statement is the same as the proportion of rural residents who agree.　$\left(p_{Urban} = p_{Rural} \text{ or } p_{Urban} - p_{Rural} = 0\right)$

H_A: The proportion of urban residents who agree with the statement is different than the proportion of rural residents who agree.

$$\left(p_{Urban} \neq p_{Rural} \text{ or } p_{Urban} - p_{Rural} \neq 0\right)$$

Randomization Condition: The samples were random.

10% Condition: 646 and 154 are both less than 10% of all residents.

Independent Samples Condition: The samples were taken independently.

Success/Failure Condition: $n\hat{p}$(urban) = 417, $n\hat{q}$(urban) = 229, $n\hat{p}$(rural) = 78, and $n\hat{q}$(rural) = 76 are all greater than 10, so both samples are large enough.

Since the conditions have been satisfied, we will model the sampling distribution of the difference in proportion with a Normal model with mean 0 and standard deviation estimated by:

$$SE_{pooled}\left(\hat{p}_{Urban} - \hat{p}_{Rural}\right) = \sqrt{\frac{\hat{p}_{pool}\hat{q}_{pool}}{n_{Urban}} + \frac{\hat{p}_{pool}\hat{q}_{pool}}{n_{Rural}}} = \sqrt{\frac{(0.619)(0.381)}{646} + \frac{(0.619)(0.381)}{154}} \approx 0.0435.$$

The observed difference between the proportions is: 0.65 – 0.51 = 0.14. We will perform a two-proportion z-test.

The value of $z = 3.21$ and the *P*-value = 0.0014. Since the *P*-value is low, we reject the null hypothesis. There is strong evidence of a difference in the proportion of residents who agree with the statement. According to this sample, urban residents are more likely to agree with the statement than rural residents.

33. **Lake Ontario fish** To use the methods discussed in this chapter, the two samples must be independent. Since we have the same 44 species-size combinations at two different times, they may not be independent. If this is the case, the methods we have discussed in this chapter cannot be used to carry out a test for comparing the two proportions.

35. **Sex and reading preference**

   a) Let p_M and p_F denote the population proportions males and females who read newspapers on the subway.

   We want to test $H_0 : p_M - p_F = 0$ against $H_A : p_M - p_F \neq 0$.

   $n_M = 56$, $n_F = 50$ $\hat{p}_M = \frac{18}{56} = 0.32$, $\hat{p}_F = \frac{10}{50} = 0.20$, $\hat{p}_{pooled} = \frac{18 + 10}{56 + 50} = 0.26$

   $$SE_{pooled} = \sqrt{\hat{p}_{pooled}\hat{q}_{pooled}\left(\frac{1}{n_M} + \frac{1}{n_F}\right)} = \sqrt{0.26 \times 0.74\left(\frac{1}{56} + \frac{1}{50}\right)} = 0.085$$

   The test statistics is $z = \frac{(\hat{p}_M - \hat{p}_F) - 0}{SE_{pooled}(\hat{p}_M - \hat{p}_F)} = \frac{(0.32 - 0.20) - 0}{0.085} = 1.41$

   p-value = $2P(Z \geq 1.41) = 0.1586 > 0.05$. This means when the population proportions are equal, the observed difference is not surprising. The observed difference (or even a greater difference) can happen with probability approximately 0.16 (when the population proportions are equal). The samples observed do not provide evidence against the null hypothesis at the 5% level of significance. That is, the samples do not provide sufficient evidence of a difference in the proportion of males and females who read newspapers on the subway.

b) Let p_M and p_F denote the population proportions males and females who read novels on the subway.

We want to test $H_0 : p_M - p_F = 0$ against $H_A : p_M - p_F \neq 0$.

$n_M = 56$, $n_F = 50$ $\hat{p}_M = \dfrac{23}{56} = 0.41$, $\hat{p}_F = \dfrac{30}{50} = 0.60$, $\hat{p}_{pooled} = \dfrac{23+30}{56+50} = 0.50$

$$SE_{pooled} = \sqrt{\hat{p}_{pooled}\hat{q}_{pooled}\left(\frac{1}{n_M} + \frac{1}{n_F}\right)} = \sqrt{0.5 \times 0.5 \times \left(\frac{1}{56} + \frac{1}{50}\right)} = 0.097$$

The test statistics is $z = \dfrac{(\hat{p}_M - \hat{p}_F) - 0}{SE_{pooled}(\hat{p}_M - \hat{p}_F)} = \dfrac{(0.41 - 0.60) - 0}{0.097} = -1.96$

p-value $= 2P(Z \leq -1.96) = 0.05$. This means when the population proportions are equal, the observed difference is somewhat surprising. The observed difference (or even a greater difference) can happen with probability approximately 0.05 (when the population proportions are equal). That is, the samples provide some evidence of a difference in the proportion of males and females who read novels on the subway.

c) Yes, the probability of one or more erroneous rejections of the null hypothesis (known as the family error rate) in the series increases as the number of tests in the series increases. For a 5% level test the probability of type I error (i.e., rejecting the hypothesis when it is true) is 0.05. The probability of failing to reject the null hypothesis when it is true (i.e., the correct decision) is 0.95. If we perform 40 tests then the probability that all 40 tests will fail to reject when all of them are true is less than or equal to $0.95^{40} = 0.1285121566$, and the probability of rejecting null hypothesis in one or more of these 40 tests when all of them are true is greater than or equal to $(1 - 0.1285121566) = 0.8714878434$, which is much higher than the acceptable level (0.05).

37. Homicide victims

a) The conditional distributions are shown as percentages in the table below.

Age	Male victim	Female victim
0-11	3.24	6.79
12-29	46.76	27.78
≥ 30	50.00	65.43
Total	100	100

b) This is a two-sample test for proportions. The hypotheses are
$$H_0 : p_1 - p_2 = 0$$
$$H_a : p_1 - p_2 \neq 0$$

where p_1 and p_2 are the proportion of victims aged 30 or more in the males and females respectively. We calculate the pooled estimate of proportion
$\hat{p} = \dfrac{216 + 106}{432 + 162} = 0.5421$ and the pooled standard error

$$SE_{pooled}(\hat{p}_1 - \hat{p}_2) = \sqrt{\left(\frac{1}{n_1} + \frac{1}{n_2}\right)\hat{p}(1 - \hat{p})}$$

$$= \sqrt{\left(\frac{1}{432} + \frac{1}{162}\right)0.5421(1 - 0.5421)}$$

$$\approx 0.0459$$

The test statistic is

$$Z = \frac{\hat{p}_1 - \hat{p}_2}{SE_{pooled}(\hat{p}_1 - \hat{p}_2)}$$

$$\approx \frac{-0.1543}{0.459} \approx -0.3362$$

which gives significance probability $P < .001$ which indicates very strong evidence against the null hypothesis.

c) Using the same approach for those aged 11 or less, $\hat{p} = 0.0421$,

$SE_{pooled}(\hat{p}_1 - \hat{p}_2) \approx 0.0454$, $z = -1.9187$ and $P = 0.0550$, so there is only weak evidence against the null hypothesis of equal proportions for the under 11 age group.

d) Using the same approach for those aged 12-29, $\hat{p} = 0.4158$,

$SE_{pooled}(\hat{p}_1 - \hat{p}_2) \approx 0.0454$, $z = 4.1803$ and $P < .001$, so there is very strong evidence against the null hypothesis of equal proportions for the 12-29 age group.

e) There are proportionally fewer female than male victims in the 12-29 age group and more in the over 30 age group. An important problem with doing multiple tests is that the overall error rate, or probability of concluding at least one of the null hypotheses is incorrect, is inflated.

f) Twenty-five children under 11 years of age were murdered. The hypotheses are

$$H_0 : p = 0.5$$

$$H_a : p \neq 0.5$$

where p is the proportion who are male. The test statistic is

$$Z = \frac{\dfrac{14}{25} - 0.5}{\sqrt{\dfrac{0.5(1 - 0.5)}{25}}}$$

$$= 0.60$$

which gives significance probability $P = 0.55$. There is no evidence against the null hypothesis of no gender effect.

39. Carpal tunnel again

a)

$$M.E.\left(\hat{p}_{Surg} - \hat{p}_{Splint}\right) = 1.96\sqrt{\frac{\hat{p}_{Surg}\hat{q}_{Surg}}{n_{Surg}} + \frac{\hat{p}_{Splint}\hat{q}_{Splint}}{n_{Splint}}} =$$

$$= 1.96\sqrt{\frac{(0.5)(0.5)}{n} + \frac{(0.5)(0.5)}{n}} = 1.96\sqrt{\frac{0.5}{n}} \Rightarrow 0.05$$

$$n = 768.32$$

They would need 769 observations from each population, for a total of 1538 observations.

b) Using software: They would need to sample about 388 individuals per sample.

Part V Review: From the Data at Hand to the World at Large

1. **Herbal cancer**

   H_0: The cancer rate for those taking the herb is the same as the cancer rate for those not taking the herb $(p_{Herb} = p_{Not}$ or $p_{Herb} - p_{Not} = 0)$

   H_A: The cancer rate for those taking the herb is higher than the cancer rate for those not taking the herb $(p_{Herb} > p_{Not}$ or $p_{Herb} - p_{Not} > 0)$

3. **Birth days**

   a) If births are distributed uniformly across all days, we expect the number of births on each day to be $np = (72)(1/7) \approx 10.29$.

   b) **Randomization Condition**: The 72 births are likely to be representative of all births at the hospital with regards to day of birth.

      10% Condition: 72 births are less than 10% of the births.

      Success/Failure Condition: The expected number of births on a particular day of the week is $np = (72)(1/7) \approx 10.29$ and the expected number of births not on that particular day is $nq = (72)(6/7) \approx 61.71$. These are both greater than 10, so the sample is large enough.

      Since the conditions have been satisfied, a Normal model can be used to model the sampling distribution of the proportion of 72 births that occur on a given day of the week.

      $$\mu_{\hat{p}} = p = \tfrac{1}{7} \approx 0.1429$$

      $$\sigma(\hat{p}) = \sqrt{\frac{pq}{n}} = \sqrt{\frac{(\tfrac{1}{7})(\tfrac{6}{7})}{72}} \approx 0.04124$$

      There were seven births on Mondays, so $\hat{p} = \frac{7}{72} \approx 0.09722$. This is only about a 1.11 standard deviations below the expected proportion, so there's no evidence that this is an unusual occurrence.

   c) The 17 births on Tuesdays represent an unusual occurrence. For Tuesdays, $\hat{p} = \frac{17}{72} \approx 0.2361$, which is about 2.26 standard deviations above the expected proportion of births. There is evidence to suggest that the proportion of births on Tuesdays is higher than expected, if births are distributed uniformly across days.

   d) Some births are scheduled for the convenience of the doctor and/or the mother.

5. **Leaky gas tanks**

   a) H_0: The proportion of leaky gas tanks is 40% $(p = 0.40)$

      H_A: The proportion of leaky gas tanks is less than 40% $(p < 0.40)$

   b) **Randomization Condition**: A random sample of 27 service stations in California was taken.

      10% Condition: 27 service stations are less than 10% of all service stations in California.

Success/Failure Condition: $n_p = (27)(0.40) = 10.8$ and $np = (27)(0.60) = 16.2$ are both greater than 10, so the sample is large enough.

c) Since the conditions have been satisfied, a Normal model can be used to model the sampling distribution of the proportion, with $\mu_{\hat{p}} = p = 0.40$ and

$$\sigma(\hat{p}) = \sqrt{\frac{pq}{n}} = \sqrt{\frac{(0.40)(0.60)}{27}} \approx 0.09428 \text{ . We can perform a one-proportion z-test.}$$

The observed proportion of leaky gas tanks is $\hat{p} = \dfrac{7}{27} \approx 0.2593$.

d) Since the P-value = 0.0677 is relatively high, we fail to reject the null hypothesis. There is little evidence that the proportion of leaky gas tanks is less than 40%. The new program doesn't appear to be effective in decreasing the proportion of leaky gas tanks.

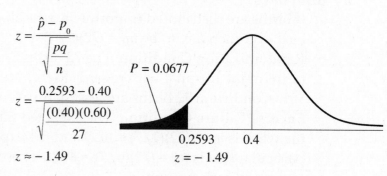

$$z = \frac{\hat{p} - p_0}{\sqrt{\dfrac{pq}{n}}}$$

$$z = \frac{0.2593 - 0.40}{\sqrt{\dfrac{(0.40)(0.60)}{27}}}$$

$z \approx -1.49$

$P = 0.0677$

$0.2593 \quad 0.4$

$z = -1.49$

e) If the program actually works, we haven't done anything *wrong*. Our methods are correct. Statistically speaking, we have committed a Type II error.

f) To decrease the probability of making this type of error, we could lower our standards of proof by raising the level of significance. This will increase the power of the test to detect a decrease in the proportion of leaky gas tanks. Another way to decrease the probability that we make a Type II error is to sample more service stations. This will decrease the variation in the sample proportion, making our results more reliable.

g) Increasing the level of significance is advantageous, since it decreases the probability of making a Type II error and increases the power of the test. However, it also increases the probability that a Type I error is made, in this case, thinking that the program is effective when it really is not effective. Increasing the sample size decreases the probability of making a Type II error and increases power, but can be costly and time-consuming.

7. Scrabble

a) The researcher believes that the true proportion of A's is within 10% of the estimated 54%, namely, between 44% and 64%.

b) A large margin of error is usually associated with a small sample, but the sample consisted of "many" hands. The margin of error is large because the standard error of the sample is large. This occurs because the true proportion of A's in a hand is close to 50%, the most difficult proportion to predict.

c) This provides no evidence that the simulation is faulty. The true proportion of A's is contained in the confidence interval. The researcher's results are consistent with 63% A's.

9. Net-newsers

a) The Pew Research Foundation believes that the true proportion of people who obtain news from the Internet is between 11% and 15%.

b) The smaller sample size in the cell sample would result in a larger standard error. This would make the margin of error larger, as well.

c)
$$\hat{p} \pm z^* \sqrt{\frac{\hat{p}\hat{q}}{n}} = (0.82) \pm 1.960 \sqrt{\frac{(0.82)(0.18)}{470}} = (78.5\%, 85.5\%)$$

We are 95% confident that between 78.5% and 85.5% of Net-newsers get news during the course of the day.

d) The sample of 470 Net-newsers is smaller than either of the earlier samples. This results in a larger margin of error.

11. Bimodal

a) The *sample's* distribution (NOT the *sampling* distribution), is expected to look more and more like the distribution of the population, in this case, bimodal.

b) The expected value of the sample's mean is expected to be μ, the population mean, regardless of sample size.

c) The variability of the sample mean, $\sigma(\bar{y})$, is $\frac{\sigma}{\sqrt{n}}$, the population standard deviation divided by the square root of the sample size, regardless of the sample size.

d) As the sample size increases, the sampling distribution model becomes closer and closer to a Normal model.

13. Twins

H_0: The proportion of preterm twin births in 1990 is the same as the proportion of preterm twin births in 2000 $(p_{1990} = p_{2000}$ or $p_{1990} - p_{2000} = 0)$

H_A: The proportion of preterm twin births in 1990 is the less than the proportion of preterm twin births in 2000 $(p_{1990} < p_{2000}$ or $p_{1990} - p_{2000} < 0)$

Randomization Condition: Assume that these births are representative of all twin births.

10% Condition: 43 and 48 are both less than 10% of all twin births.

Independent Samples Condition: The samples are from different years, so they are unlikely to be related.

Success/Failure Condition: $n\hat{p}(1990) = 20$, $n\hat{q}(1990) = 23$, $n\hat{p}(2000) = 26$, and $n\hat{q}(2000) = 22$ are all greater than 10, so both samples are large enough.

Since the conditions have been satisfied, we will perform a two-proportion z-test. We will model the sampling distribution of the difference in proportion with a

Normal model with mean 0 and standard deviation estimated by:

$$SE_{pooled}(\hat{p}_{1990} - \hat{p}_{2000}) = \sqrt{\frac{\hat{p}_{pooled}\hat{q}_{pooled}}{n_{1900}} + \frac{\hat{p}_{pooled}\hat{q}_{pooled}}{n_{2000}}} = \sqrt{\frac{\left(\frac{46}{91}\right)\left(\frac{45}{91}\right)}{43} + \frac{\left(\frac{46}{91}\right)\left(\frac{45}{91}\right)}{48}} \approx 0.1050$$

The observed difference between the proportions is: 0.4651 – 0.5417 = – 0.0766 Since the P-value = 0.2329 is high, we fail to reject the null hypothesis. There is no evidence of an increase in the proportion of preterm twin

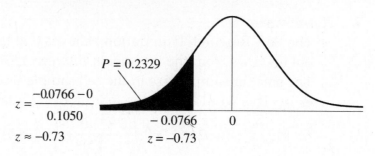

$$z = \frac{-0.0766 - 0}{0.1050}$$

$$z \approx -0.73$$

births from 1990 to 2000, at least not at this large city hospital.

15. **Eclampsia II**

   a) H_0: The proportion of pregnant women who die after developing eclampsia is the same for women taking magnesium sulfide as it is for women not taking magnesium sulfide $(p_{MS} = p_N \text{ or } p_{MS} - p_N = 0)$

   H_A: The proportion of pregnant women who die after developing eclampsia is lower for women taking magnesium sulfide than for women not taking magnesium sulfide $(p_{MS} < p_N \text{ or } p_{MS} - p_N < 0)$

   b) **Randomization Condition**: Although not specifically stated, these results are from a large-scale experiment, which was undoubtedly properly randomized. **10% Condition**: 40 and 96 are less than 10% of all pregnant women. **Independent Samples Condition**: Subjects were randomly assigned to the treatments. **Success/Failure Condition**: $n\hat{p}$(mag. sulf.) = 11, $n\hat{q}$(mag. sulf.) = 29, $n\hat{p}$(placebo) = 20, and $n\hat{q}$(placebo) = 76 are all greater than 10, so both samples are large enough. Since the conditions have been satisfied, we will perform a two-proportion z-test. We will model the sampling distribution of the difference in proportion with a Normal model with mean 0 and standard deviation estimated by:

   $$SE_{pooled}(\hat{p}_{MS} - \hat{p}_N) = \sqrt{\frac{\hat{p}_{pooled}\hat{q}_{pooled}}{n_{MS}} + \frac{\hat{p}_{pooled}\hat{q}_{pooled}}{n_N}} = \sqrt{\frac{\left(\frac{31}{136}\right)\left(\frac{105}{136}\right)}{40} + \frac{\left(\frac{31}{136}\right)\left(\frac{105}{136}\right)}{96}} \approx 0.07895.$$

   c) The observed difference between the proportions is: 0.275 – 0.2083 = 0.0667

   Since the P-value = 0.8008 is high, we fail to reject the null hypothesis. There is no evidence that the proportion of women who may die after developing

   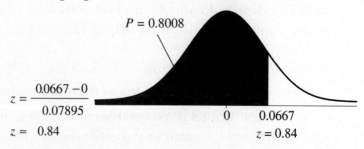

   $$z = \frac{0.0667 - 0}{0.07895}$$

   $$z \approx 0.84$$

eclampsia is lower for women taking magnesium sulfide than for women who are not taking the drug.

d) There is not sufficient evidence to conclude that magnesium sulfide is effective in preventing death when eclampsia develops.

e) If magnesium sulfide is effective in preventing death when eclampsia develops, then we have made a Type II error.

f) To increase the power of the test to detect a decrease in death rate due to magnesium sulfide, we could increase the sample size or increase the level of significance.

g) Increasing the sample size lowers variation in the sampling distribution, but may be costly. The sample size is already quite large. Increasing the level of significance increases power by increasing the likelihood of rejecting the null hypothesis, but increases the chance of making a Type I error, namely thinking that magnesium sulfide is effective when it is not.

17. No appetite for privatization

a) The actual margin of error (ME) is

$$ME = 1.96\sqrt{\frac{\hat{p}(1-\hat{p})}{1000}} \approx 0.0201, \text{ or about } 2\%.$$

b) We cannot do this without knowing the sample size for Quebec. If we are told Quebec's population is 23% of Canada's, we could base a calculation on an approximate sample size for Quebec of $0.23(1000) = 230$, which gives

$$ME = 1.96\sqrt{\frac{0.16(1-0.16)}{230}} \approx 0.0474, \text{ or about } 4.74\%.$$

c) The margin of error is inversely proportional to the square root of the sample size. Increasing the sample size by a factor f, reduces the MOE for both all of Canada and just Quebec by $\frac{1}{\sqrt{f}}$, so we want $\frac{1}{\sqrt{f}} = 1/4$ or $f = 16$. The sample size for Canada would need to increase from 1000 to $1000 \times 16 = 16000$.

19. Teen deaths 2005

a) H_0: The percentage of fatal accidents involving teenage girls is 12.6%, the same as the overall percentage of fatal accidents involving teens ($p = 0.126$)
H_A: The percentage of fatal accidents involving teenage girls is lower than 12.6%, the overall percentage of fatal accidents involving teens ($p < 0.126$)
Randomization Condition: Assume that the 388 accidents observed are representative of all accidents.
10% Condition: The sample of 388 accidents is less than 10% of all accidents.
Success/Failure Condition: $np = (388)(0.126) = 48.888$ and $nq = (388)(0.874) = 339.112$ are both greater than 10, so the sample is large enough.
The conditions have been satisfied, so a Normal model can be used to model the sampling distribution of the proportion, with $\mu_{\hat{p}} = p = 0.126$ and

$$\sigma(\hat{p}) = \sqrt{\frac{pq}{n}} = \sqrt{\frac{(0.126)(0.874)}{388}} \approx 0.01685$$

We can perform a one-proportion z-test. The observed proportion of fatal accidents involving teen girls is $\hat{p} = \frac{44}{388} \approx 0.1134$.

$$z = \frac{\hat{p} - p_0}{\sqrt{\frac{pq}{n}}} = \frac{0.1134 - 0.126}{\sqrt{\frac{(0.126)(0.874)}{388}}} = -0.748$$

Since the P-value = 0.227 is high, we fail to reject the null hypothesis. There is little evidence that the proportion of fatal accidents involving teen girls is less than the overall proportion of fatal accidents involving teens.

b) If the proportion of fatal accidents involving teenage girls is really 12.6%, we expect to see the observed proportion, 11.34%, or lower in about 22.7% of samples of size 388 simply due to sampling variation.

21. Largemouth bass

a) One would expect many small fish, with a few large fish.

b) We cannot determine the probability that a largemouth bass caught from the lake weighs over 3 pounds because we don't know the exact shape of the distribution. We know that it is NOT Normal.

c) It would be quite risky to attempt to determine whether or not the mean weight of five fish was over 3 pounds. With a skewed distribution, a sample of size 5 is not large enough for the Central Limit Theorem to guarantee that a Normal model is appropriate to describe the distribution of the mean.

d) A sample of 60 randomly selected fish is large enough for the Central Limit Theorem to guarantee that a Normal model is appropriate to describe the sampling distribution of the mean, as long as 60 fish is less than 10% of the population of all the fish in the lake.

The mean weight is $m = 3.5$ pounds, with standard deviation $s = 2.2$ pounds.

Since the sample size is sufficiently large, we can model the sampling distribution of the mean weight of 60 fish with a Normal model, with

$m_{\bar{y}} = 3.5$ pounds and standard deviation $s(\bar{y}) = \frac{2.2}{\sqrt{60}} \approx 0.284$ pounds.

According to the Normal model, the probability that 60 randomly selected fish average more than 3 pounds is approximately 0.961.

$z = \dfrac{3 - 3.5}{\dfrac{2.2}{\sqrt{60}}}$

$z \approx -1.76$

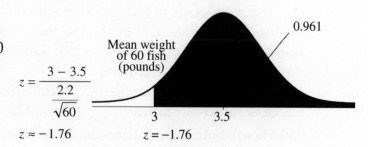

$z = -1.76$

23. **Language**
    a) **Randomization Condition:** 60 people were selected at random.
       10% Condition: The 60 people represent less than 10% of all people.
       Success/Failure Condition: $np = (60)(0.80) = 48$ and $nq = (60)(0.20) = 12$ are both greater than 10.
       Therefore, the sampling distribution model for the proportion of 60 randomly selected people who have left-brain language control is Normal, with

       $$m_{\hat{p}} = p = 0.80 \text{ and standard deviation }_{s\ (\hat{p})} = \sqrt{\frac{pq}{n}} = \sqrt{\frac{(0.80)(0.20)}{60}} \approx 0.0516 .$$

    b) According to the Normal model, the probability that over 75% of these 60 people have left-brain language control is approximately 0.834.

    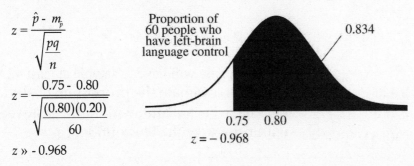

    $$z = \frac{\hat{p} - m_{\hat{p}}}{\sqrt{\frac{pq}{n}}}$$

    $$z = \frac{0.75 - 0.80}{\sqrt{\frac{(0.80)(0.20)}{60}}}$$

    $$z \approx -0.968$$

    c) If the sample had consisted of 100 people, the probability would have been higher. A larger sample results in a smaller standard deviation for the sample proportion.

    d) Answers may vary. Let's consider three standard deviations below the expected proportion to be "almost certain." It would take a sample of (exactly!) 576 people to make sure that 75% would be 3 standard deviations below the expected percentage of people with left-brain language control.

       Using round numbers for n instead of z, about 500 people in the sample would make the probability of choosing a sample with at least 75% of the people having left-brain language control is a whopping 0.997. It all depends on what "almost certain" means to you.

       $$z = \frac{\hat{p} - m_{\hat{p}}}{\sqrt{\frac{pq}{n}}}$$

       $$-3 = \frac{0.75 - 0.80}{\sqrt{\frac{(0.80)(0.20)}{n}}}$$

       $$n = \frac{(-3)^2 (0.80)(0.20)}{(0.75 - 0.80)^2} = 576$$

25. **Crohn's disease**
    a) **Independence Assumption**: It is reasonable to think that the patients would respond to infliximab independently of each other.
       Randomization Condition: Assume that the 573 patients are representative of all Crohn's disease sufferers.

10% Condition: 573 patients are less than 10% of all sufferers of Crohn's disease.
Success/Failure Condition: $n\hat{p}$ = 335 and $n\hat{q}$ = 238 are both greater than 10.

Since the conditions are met, we can use a one-proportion z-interval to estimate the percentage of Crohn's disease sufferers who respond positively to infliximab.

$$\hat{p} \pm z^* \sqrt{\frac{\hat{p}\hat{q}}{n}} = \left(\frac{335}{573}\right) \pm 1.960 \sqrt{\frac{\left(\frac{335}{573}\right)\left(\frac{238}{573}\right)}{573}} = (54.4\%, 62.5\%)$$

b) We are 95% confident that between 54.4% and 62.5% of Crohn's disease sufferers would respond positively to infliximab.

c) 95% of random samples of size 573 will produce intervals that contain the true proportion of Crohn's disease sufferers who respond positively to infliximab.

27. Alcohol abuse

$$ME = z^* \sqrt{\frac{\hat{p}\hat{q}}{n}}$$

$$0.04 = 1.645 \sqrt{\frac{(0.5)(0.5)}{n}}$$

$$n = \frac{(1.645)^2 (0.5)(0.5)}{(0.04)^2}$$

$$n \approx 423$$

The university will have to sample at least 423 students to estimate the proportion of students who have been drunk within the past week to within ± 4%, with 90% confidence.

29. Preemies

a) **Randomization Condition**: Assume that these kids are representative of all kids.
10% Condition: 242 and 233 are less than 10% of all kids.
Independent Samples Condition: The groups are independent.
Success/Failure Condition: $n\hat{p}$ (preemies) = (242)(0.74) = 179, $n\hat{q}$ (preemies) = (242)(0.26) = 63, $n\hat{p}$ (normal weight) = (233)(0.83) = 193, and $n\hat{q}$ (normal weight) = 40 are all greater than 10, so the samples are both large enough.
Since the conditions have been satisfied, we will find a two-proportion z-interval.

$$\left(\hat{p}_N - \hat{p}_P\right) \pm z^* \sqrt{\frac{\hat{p}_N \hat{q}_N}{n_N} + \frac{\hat{p}_P \hat{q}_P}{n_P}}$$

$$= (0.83 - 0.74) \pm 1.960 \sqrt{\frac{(0.83)(0.17)}{233} + \frac{(0.74)(0.26)}{242}} = (0.017, 0.163)$$

We are 95% confident that between 1.7% and 16.3% more normal birth-weight children graduated from high school than children who were born premature.

b) Since the interval for the difference in percentage of high school graduates is above 0, there is evidence normal birth-weight children graduate from high school at a greater rate than premature children.

c) If preemies do not have a lower high school graduation rate than normal birthweight children, then we made a Type I error. We rejected the null hypothesis of "no difference" when we shouldn't have.

31. Fried PCs

a) H_0: The computer is undamaged
   H_A: The computer is damaged

b) The biggest advantage is that all of the damaged computers will be detected, since, historically, damaged computers never pass all the tests. The disadvantage is that only 80% of undamaged computers pass all the tests. The engineers will be classifying 20% of the undamaged computers as damaged.

c) In this example, a Type I error is rejecting an undamaged computer. To allow this to happen only 5% of the time, the engineers would reject any computer that failed three or more tests, since 95% of the undamaged computers fail two or fewer tests.

d) The power of the test in part c) is 20%, since only 20% of the damaged machines fail 3 or more tests.

e) By declaring computers "damaged" if they fail two or more tests, the engineers will be rejecting only 7% of undamaged computers. From 5% to 7% is an increase of 2% in α. Since 90% of the damaged computers fail two or more tests, the power of the test is now 90%, a substantial increase.

33. Name recognition

a) The company wants evidence that the athlete's name is recognized more often than 25%.

b) Type I error means that fewer than 25% of people will recognize the athlete's name, yet the company offers the athlete an endorsement contract anyway. In this case, the company is employing an athlete that doesn't fulfill their advertising needs.
   Type II error means that more than 25% of people will recognize the athlete's name, but the company doesn't offer the contract to the athlete. In this case, the company is letting go of an athlete that meets their advertising needs.

c) If the company uses a 10% level of significance, the company will hire more athletes that don't have high enough name recognition for their needs. The risk of committing a Type I error is higher.
   At the same level of significance, the company is less likely to lose out on athletes with high name recognition. They will commit fewer Type II errors.

35. NIMBY

Randomization Condition: Not only was the sample random, but Gallup randomly divided the respondents into groups.
10% Condition: 502 and 501 are less than 10% of all adults.
Independent Samples Condition: The groups are independent.

Success/Failure Condition: The number of respondents in favour, and opposed, in both groups are all greater than 10, so the samples are both large enough.

Since the conditions have been satisfied, we will find a two-proportion z-interval.

$$(\hat{p}_1 - \hat{p}_2) \pm z^* \sqrt{\frac{\hat{p}_1 \hat{q}_1}{n_1} + \frac{\hat{p}_2 \hat{q}_2}{n_2}}$$

$$= (0.53 - 0.40) \pm 1.960 \sqrt{\frac{(0.53)(0.47)}{502} + \frac{(0.40)(0.60)}{501}} = (0.07, 0.19)$$

We are 95% confident that the proportion of U.S. adults who favour nuclear energy is between 7 and 19 percentage points higher than the proportion that would accept a nuclear plant near their area.

37. **Home ice**
   a) The hypotheses are

$$H_0 : p_1 - p_2 = 0$$
$$H_a : p_1 - p_2 \neq 0$$

   where p_1 and p_2 are the home winning proportions for the Vancouver Canucks and Calgary Flames respectively. The pooled estimate of proportion is $\hat{p} = \frac{24 + 27}{41 + 41} \approx 0.6220$ and the pooled standard error

$$SE_{pooled}(\hat{p}_1 - \hat{p}_2) = \sqrt{\left(\frac{1}{n_1} + \frac{1}{n_2}\right)\hat{p}(1 - \hat{p})}$$

$$= \sqrt{\left(\frac{1}{41} + \frac{1}{41}\right)0.6220(1 - 0.6220)}$$

$$\approx 0.107094$$

   The test statistic, using $\hat{p} = 0.5854$ and $\hat{p}_2 = 0.6585$ is

$$Z = \frac{\hat{p}_1 - \hat{p}_2}{SE_{pooled}(\hat{p}_1 - \hat{p}_2)}$$

$$\approx \frac{-0.0732}{0.107094} \approx -0.6835$$

   which gives significance probability $P < .49$ which means that there is no evidence against the null hypothesis that the two rates are equal.
   b) For the Canucks, $\hat{p} = \frac{24 + 21}{41 + 41} = 0.5488$, $SE_{pooled}(\hat{p}_1 - \hat{p}_2) \approx 0.1099$, $\hat{p}_1 = 0.5487$, $\hat{p}_2 = 0.5854$, $z = -0.6658$ and $P = 0.51$, so there is no evidence of a home-away difference in winning proportion. For the Flames, $\hat{p} = \frac{27 + 19}{41 + 41} = 0.561$, $SE_{pooled}(\hat{p}_1 - \hat{p}_2) \approx 0.1096$, $\hat{p}_1 = 0.5610$, $\hat{p}_2 = 0.6585$, $z = 1.7802$, and

$P = 0.075$, so there is weak evidence against the null hypothesis of no difference in winning proportions for home and away games for the Flames.

c) When you look at all the teams in the NHL, you expect some large differences in home and away winning percentages. If the winning proportion is the same for home and away games, then the number won at home is binomial with probability .5 and n equal to the total number of games won. For the Islanders, $n = 26$ and $P(X \geq 17) = .0843$, so it is not all that unusual to win such a large proportion at home.

d) Publication bias results when insignificant results are not reported. Because the number of insignificant studies not reported is large, many of the reported significant results could be Type I errors. When researchers select their hypotheses on the basis of observed results, significance is also overstated, i.e., results appear more significant than they actually are.

39. CADUMS survey

a) Based on a random sample of 1443, the margin of error should be

$$\text{ME} = 1.96\sqrt{\frac{0.529(1 - 0.529)}{1443}}$$
$$= 0.0258$$

or 2.6%. The published ME is larger at 4.7%. This is not a random sample, but a stratified sample where random samples are taken within each province. In addition, the results are post-stratified by age, to focus on the 15–24 age group. The published ME takes these factors into account.

b) A 99% confidence interval requires adjustment of the confidence level from 0.95 to 0.99 in the margin of error, using $\hat{p} \pm 2.58\ \text{ME}\ /\ 1.96$. Here $\hat{p} = 0.529$, and $\text{ME} = 0.047$, so the confidence interval is $0.529 \pm 2.58(0.047)/1.96$ or $(0.4671, 0.5909)$. An interval constructed in this way will contain the true proportion 0.99 most of the time.

1. **t-models, part I**
   a) 1.74
   b) 2.37
   c) 0.0524
   d) 0.0889

3. **t-models, part III** As the number of degrees of freedom increases, the centres of t-models do not change. The spread of t-models decreases as the number of degrees of freedom increases, and the shape of the distribution becomes closer to Normal.

5. **Cattle**
   a) Not correct. A confidence interval is for the mean weight gain of the population of all cows. It says nothing about individual cows. This interpretation also appears to imply that there is something special about the interval that was generated, when this interval is actually one of many that could have been generated, depending on the cows that were chosen for the sample.
   b) Not correct. A confidence interval is for the mean weight gain of the population of all cows, not individual cows.
   c) Not correct. We don't need a confidence interval about the average weight gain for cows in this study. We are certain that the mean weight gain of the cows in this study is 56 pounds. Confidence intervals are for the mean weight gain of the population of all cows.
   d) Not correct. This statement implies that the average weight gain varies. It doesn't. We just don't know what it is, and we are trying to find it. The average weight gain is either between 45 and 67 pounds, or it isn't.
   e) Not correct. This statement implies that there is something special about our interval, when this interval is actually one of many that could have been generated, depending on the cows that were chosen for the sample. The correct interpretation is that 95% of samples of this size will produce an interval that will contain the mean weight gain of the population of all cows.

7. **Meal plan**
   a) Not correct. The confidence interval is not about the individual students in the population.
   b) Not correct. The confidence interval is not about individual students in the sample. In fact, we know exactly what these students spent, so there is no need to estimate.
   c) Not correct. We know that the mean cost for students in this sample was $1467.
   d) Not correct. A confidence interval is not about other sample means.
   e) This is the correct interpretation of a confidence interval. It estimates a population parameter.

9. **Pulse rates**
   a) We are 95% confident the interval 70.9 to 74.5 beats per minute contains the true mean heart rate.
   b) The width of the interval is about 74.5 – 70.9 = 3.6 beats per minute. The margin of error is half of that, about 1.8 beats per minute.
   c) The margin of error would have been larger. More confidence requires a larger critical value of *t*, which increases the margin of error.

11. **CEO compensation** We should be hesitant to trust this confidence interval, since the conditions for inference are not met. The distribution is highly skewed and there is an outlier.

13. **Normal temperature**
    a) **Randomization Condition**: The adults were randomly selected.
       Nearly Normal Condition: The sample of 52 adults is large, and the histogram shows no serious skewness, outliers, or multiple modes.
       The people in the sample had a mean temperature of 36.83°C and a standard deviation in temperature of 0.38. Since the conditions are satisfied, the sampling distribution of the mean can be modelled by a Student's *t* model, with 52 – 1 = 51 degrees of freedom. We will use a one-sample *t*-interval with 98% confidence for the mean body temperature.
       (By hand, use $t_{50}^* \approx 2.403$ from the table.)
    b) $\bar{y} \pm t_{n-1}^* \left(\dfrac{s}{\sqrt{n}} \right) = 36.83 \pm t_{51}^* \left(\dfrac{0.38}{\sqrt{52}} \right) \approx (36.70, 36.96)$
    c) We are 98% confident that the interval 36.7°C to 36.96°C contains the true mean body temperature for adults.
    d) 98% of all random samples of size 52 will produce intervals that contain the true mean body temperature of adults.
    e) Since the interval is completely below the body temperature of 37°C, there is strong evidence that the true mean body temperature of adults is lower than 37°C.

15. **Normal temperatures, part II**
    a) The 90% confidence interval would be narrower than the 98% confidence interval. We can be more precise with our interval when we are less confident.
    b) The 98% confidence interval has a greater chance of containing the true mean body temperature of adults than the 90% confidence interval, but the 98% confidence interval is less precise (wider) than the 90% confidence interval.
    c) The 98% confidence interval would be narrower if the sample size were increased from 52 people to 500 people. The smaller standard error would result in a smaller margin of error.
    d) Our sample of 52 people gave us a 98% confidence interval with a margin of error of (36.96 – 36.7)/2 =0.13°C. To get a margin of error of 0.05, less than half of

that, we need a sample over four times as large. It should be safe to use $t^*_{100} \approx 2.364$ from the table, since the sample will need to be larger than 101. Or we could use $z^* \approx 2.326$, since we expect the sample to be large. We need a sample of about 313 people to estimate the mean body temperature of adults to within 0.05°C.

$$ME = t^*_{n-1}\left(\frac{s}{\sqrt{n}}\right)$$

$$0.05 = 2.326\left(\frac{0.38}{\sqrt{n}}\right)$$

$$n = \frac{(2.326)^2(0.38)^2}{(0.05)^2}$$

$$n \approx 313$$

17. Speed of light

a) $\bar{y} \pm t^*_{n-1}\left(\frac{s}{\sqrt{n}}\right) = 756.22 \pm t^*_{22}\left(\frac{107.12}{\sqrt{23}}\right) \approx (709.9, 802.5)$

b) We are 95% confident that the interval 299,709.9 to 299,802.5 km/sec contains the speed of light.

c) We have assumed that the measurements are independent of each other and that the distribution of the population of all possible measurements is Normal. The assumption of independence seems reasonable, but it might be a good idea to look at a display of the measurements made by Michelson to verify that the Nearly Normal Condition is satisfied.

19. Departures 2011

a) **Randomization Condition:** Since there is no time trend, the monthly on-time departure rates should be independent. This is not a random sample, but should be representative.

Nearly Normal Condition: The histogram looks unimodal, and slightly skewed to the left. Since the sample size is 201, this should not be of concern.

b) The on-time departure rates in the sample had a mean of 80.752% and a standard deviation of 4.594%. Since the conditions have been satisfied, construct a one-sample t-interval, with 201 – 1 = 200 degrees of freedom, at 90% confidence.

$$\bar{y} \pm t^*_{n-1}\left(\frac{s}{\sqrt{n}}\right) = 80.752 \pm t^*_{200}\left(\frac{4.594}{\sqrt{201}}\right) \approx (80.22, 81.29)$$

c) We are 90% confident that the interval from 80.22% to 81.29% contains the true mean monthly percentage of on-time flight departures.

d) If the number of flights differs each month, we should construct a separate interval for each month. Otherwise, we will not have the correct t* value, and our interval with be too wide or too narrow for that month.

21. Farmed salmon, second look The 95% confidence interval lies entirely above the 0.08 ppm limit. This is evidence that mirex contamination is too high, and consistent with rejecting the null hypothesis. We used an upper-tail test, so the P-value should therefore be smaller than $\frac{1}{2}(1 - 0.95) = 0.025$, and it was.

23. **Pizza** If in fact the mean cholesterol of pizza eaters does not indicate a health risk, then only 7 out of every 100 samples would be expected to have mean cholesterol as high or higher than the mean cholesterol observed in the sample.

25. **TV safety**
    a) The inspectors are performing an upper-tail test. They need to prove that the stands will support 500 pounds (or more) easily.
    b) The inspectors commit a Type I error if they certify the stands as safe when they are not.
    c) The inspectors commit a Type II error if they decide the stands are not safe when they are.

27. **TV safety revisited**
    a) The value of α should be decreased. This means a smaller chance of declaring the stands safe under the null hypothesis that they are not safe.
    b) The power of the test is the probability of correctly detecting that the stands can safely hold over 500 pounds.
    c) The company could redesign the stands so that their strength is more consistent, as measured by the standard deviation. Redesigning the manufacturing process is likely to be quite costly.

       The company could increase the number of stands tested. This costs them both time to perform the test and money to pay the quality control personnel.

       The company could increase α, effectively lowering their standards for what is required to certify the stands "safe." This is a big risk, since there is a greater chance of Type I error, namely allowing unsafe stands to be sold.

       The company could make the stands stronger, increasing the mean amount of weight that the stands can safely hold. This type of redesign is expensive.

29. **Marriage**
    a) H_0: The mean age at which Canadian women first marry is 22.6 years ($\mu = 22.6$).
       H_A: The mean age at which Canadian women first marry is greater than 22.6 years ($\mu > 22.6$)
    b) **Randomization Condition:** The 40 women were selected randomly.
       Nearly Normal Condition: The population of ages of women at first marriage is likely to be skewed to the right. It is much more likely that there are women who marry for the first time at an older age than at a very young age. We should examine the distribution of the sample to check for serious skewness and outliers, but with a large sample of 40 women, it should be safe to proceed.
    c) Since the conditions for inference are satisfied, we can model the sampling distribution of the mean age of women at first marriage with $N\left(22.6, \dfrac{\sigma}{\sqrt{n}}\right)$. Since we do not know σ, the standard deviation of the population, $\sigma(\bar{y})$ will be

estimated by $SE(\bar{y}) = \frac{s}{\sqrt{n}}$, and we will use a Student's t model, with $40 - 1 = 39$ degrees of freedom.

d) The mean age at first marriage in the sample was 27.2 years, with a standard deviation in age of 5.3 years. Use a one-sample t-test, modelling the sampling distribution of $\bar{y}$ with a t-distribution.

The P-value is < 0.00001.

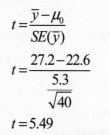

$$t = \frac{\bar{y} - \mu_0}{SE(\bar{y})}$$

$$t = \frac{27.2 - 22.6}{\frac{5.3}{\sqrt{40}}}$$

$$t = 5.49$$

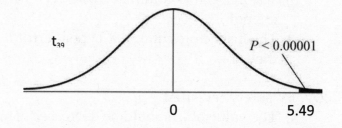

e) If the mean age at first marriage is still 22.6 years, there is a near-zero chance of getting a sample mean of 27.2 years or older simply from natural sampling variation.

f) Since the P-value is low, we reject the null hypothesis. We have very strong evidence to suggest that the mean age of women at first marriage has increased from 22.6 years, the mean in 1960.

31. Ruffles

a) **Randomization Condition**: The six bags were not selected at random, but it is reasonable to think that these bags are representative of all bags of chips.

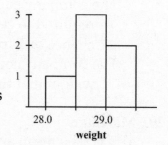

Nearly Normal Condition: The histogram of the weights of chips in the sample is nearly normal.

b) $\bar{y} \approx 28.78$ grams, $s \approx 0.40$ grams

c) Since the conditions for inference have been satisfied, use a one-sample t-interval, with $6 - 1 = 5$ degrees of freedom, at 95% confidence.

$$\bar{y} \pm t^*_{n-1}\left(\frac{s}{\sqrt{n}}\right) = 28.78 \pm t^*_s\left(\frac{0.40}{\sqrt{6}}\right) \approx (28.36, 29.21)$$

d) We are 95% confident that the mean weight of the contents of Ruffles bags is between 28.36 and 29.21 grams.

e) Since the interval is above the stated weight of 28.3 grams, there is evidence that the company is filling the bags to more than the stated weight, on average.

33. Popcorn

a) Hopps made a Type I error. He mistakenly rejected the null hypothesis that the proportion of unpopped kernels was 10% (or higher).

b) H_0: The mean proportion of unpopped kernels is 10% ($\mu = 10$)

H_A: The mean proportion of unpopped kernels is lower than 10% $(\mu < 10)$

Randomization Condition: The eight bags were randomly selected.

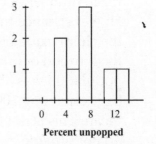

Nearly Normal Condition: The histogram of the percentage of unpopped kernels is unimodal and roughly symmetric.

Percent unpopped

The bags in the sample had a mean percentage of unpopped kernels of 6.775% and a standard deviation in percentage of unpopped kernels of 3.637%. Since the conditions for inference are satisfied, we can model the sampling distribution of the mean percentage of unpopped kernels with a Student's t model, with 8 – 1 = 7 degrees of freedom. We will perform a one-sample t-test.

Since the P-value = 0.0203 is low, we reject the null hypothesis. There is evidence to suggest the mean percentage of unpopped kernels is less than 10% at this setting.

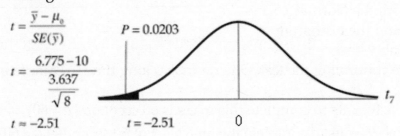

$$t = \frac{\bar{y} - \mu_0}{SE(\bar{y})}$$

$$t = \frac{6.775 - 10}{\frac{3.637}{\sqrt{8}}}$$

$P = 0.0203$

$t \approx -2.51$ $t = -2.51$ 0

35. **Chips Ahoy**
    a) **Randomization Condition**: The bags of chips were randomly selected.
    Nearly Normal Condition: The Normal probability plot is reasonably straight, and the histogram of the number of chips per bag is unimodal and symmetric.

    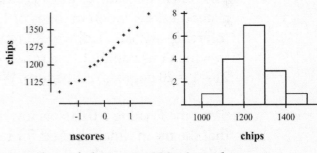

    nscores   chips

    b) The bags in the sample had with a mean number of chips or 1238.19, and a standard deviation of 94.282 chips. Since the conditions for inference have been satisfied, use a one-sample t-interval, with 16 – 1 = 15 degrees of freedom, at 95% confidence.

    $$\bar{y} \pm t^*_{n-1}\left(\frac{s}{\sqrt{n}}\right) = 1238.19 \pm t^*_{15}\left(\frac{94.282}{\sqrt{16}}\right) \approx (1187.9, 1288.4)$$

    We are 95% confident that the mean number of chips in an 18-ounce bag of Chips Ahoy cookies is between 1187.9 and 1288.4.

    c) H_0: The mean number of chips per bag is 1000 $(\mu = 1000)$

    H_A: The mean number of chips per bag is greater than 1000 $(\mu > 1000)$

Since the confidence interval is well above 1000, there is strong evidence that the mean number of chips per bag is well above 1000.

However, since the "1000 Chip Challenge" is about individual bags, not means, the claim made by Nabisco may not be true. If the mean was around 1188 chips, the low end of our confidence interval, and the standard deviation of the population was about 94 chips, our best estimate obtained from our sample, a bag containing 1000 chips would be about 2 standard deviations below the mean. This is not likely to happen, but not an outrageous occurrence. These data do not provide evidence that the "1000 Chip Challenge" is true.

37. Maze

a) **Independence Assumption**: It is reasonable to think that the rats' times will be independent, as long as the times are for different rats. **Nearly Normal Condition**: There is an outlier in both the Normal probability plot and the histogram that should probably be 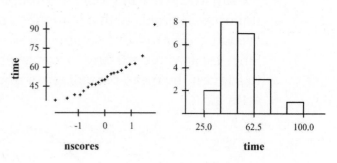 eliminated before continuing the test. One rat took a long time to complete the maze.

b) H_0: The mean time for rats to complete this maze is 60 seconds ($\mu = 60$)

H_A: The mean time for rats to complete this maze is not 60 seconds ($\mu \neq 60$)

The rats in the sample finished the maze with a mean time of 52.21 seconds and a standard deviation in times of 13.5646 seconds. Since the conditions for inference are satisfied, we can model the sampling distribution of the mean time in which rats complete the maze with a Student's t model, with $21 - 1 = 20$ degrees of freedom. We will perform a one-sample t-test.

Since the P-value = 0.0160 is low, we reject the null hypothesis. There is evidence that the mean time required for rats to finish the maze is not 60 seconds. Our evidence suggests that the mean time is actually less than 60 seconds.

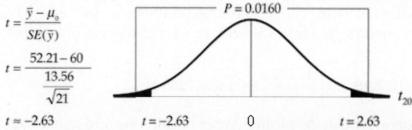

$$t = \frac{\bar{y} - \mu_0}{SE(\bar{y})}$$

$$t = \frac{52.21 - 60}{\frac{13.56}{\sqrt{21}}}$$

$$t \approx -2.63$$

c) Without the outlier, the rats in the sample finished the maze with a mean time of 50.13 seconds and standard deviation in times of 9.90 seconds. Since the conditions for inference are satisfied, we can model the sampling distribution of the mean time in which rats complete the maze with a Student's t model, with

20 – 1 = 19 degrees of freedom. We will perform a one-sample *t*-test.
This test results in a value of $t = -4.46$, and a two-sided P-value = 0.0003. Since the P-value is low, we reject the null hypothesis. There is evidence that the mean time required for rats to finish the maze is not 60 seconds. Our evidence suggests that the mean time is actually less than 60 seconds.

d) According to both tests, there is evidence that the mean time required for rats to complete the maze is different than 60 seconds. The maze does not meet the "one-minute average" requirement. It should be noted that the test without the outlier is the appropriate test. The one slow rat made the mean time required seem much higher than it probably was.

e) Let *p* be the proportion of rats finishing the maze in less than 60 seconds. Since there are no times equal to 60 seconds, we can use all of the data.
H_0: the median is 60 seconds, or $p = 0.5$
H_a: the median is not 60 seconds, or $p \neq 0.5$
$$\hat{p} = 16/21 = 0.7619 \qquad\qquad \sigma(\hat{p}) = \sqrt{\frac{pq}{n}} = \sqrt{(0.5*0.5/21)} = 0.10911$$

$z = (0.7619 - 0.5)/0.10911 = 2.40$ P-value = 0.0164 (two-sided)
We have good evidence to reject the null hypothesis that the median time for completion is 60 seconds, just as in part d).

39. Driving distance 2011

a) $\bar{y} \pm t^*_{n-1}\left(\dfrac{s}{\sqrt{n}}\right) = 291.09 \pm t^*_{185}\left(\dfrac{8.343}{\sqrt{186}}\right) \approx (289.9, 292.3)$

b) These data are not a random sample of golfers. The top professionals are not representative of all golfers and were not selected at random. We might consider the 2011 data to represent the population of all professional golfers, past, present, and future.

c) The data are means for each golfer, so they are less variable than if we looked at separate drives.

d) Using several drives for each golfer would violate the Independence Assumption.

41. Worst of times

a) For these values on percentage changes, we have the following summary statistics:
Sample size $(n) = 20$

Sample mean $\left(\bar{y}\right) = 108.99$

Sample standard deviation $(s) = 71.871$

The confidence interval is given by $\bar{y} \pm t^* \dfrac{s}{\sqrt{n}}$.

Substituting values, the 95% confidence interval for the population mean increase is $108.99 \pm 2.093 \times \dfrac{71.87}{\sqrt{20}} = (75.35, 142.63)$.

b)

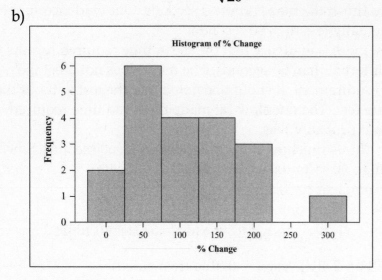

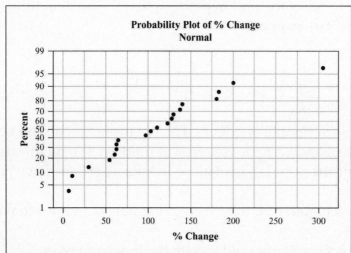

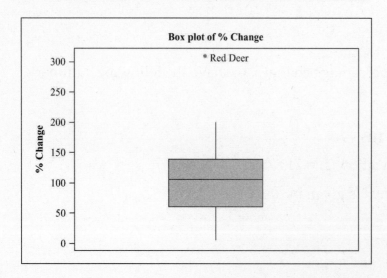

The plots suggest that the distribution of the data is close enough to Normal but there is one outlier (Red Deer). When there are outliers in the data set, the sample mean and the standard deviation are not good summary statistics of the data set, and so the 95% confidence level quoted in a) may not be trustworthy. The summary statistics and the confidence intervals calculated from the data set with and without the outlier are shown below. The mean is approximately 10 percentage points higher with the outlier and the standard deviation is about 15 percentage points higher.

	n	Mean	StDev	SE Mean	95% CI
With the outlier	20	108.990	71.871	16.071	(75.353 142.627)
Without the outlier	19	98.7105	56.7602	13.0217	(71.3530-126.0681)

c) No, because our interval is for an average change per urban area, not an average per capita (which would only agree if all urban areas were the same size)

43. Simulations

a) The simulated sample from a normal distribution mean 50 and standard deviation 10 is shown below (Note: This is a random sample and so the answers can vary.):

38.9915 40.6694 42.2333 56.6027 53.8246 41.7169 56.4824

58.9814 66.2086 40.9988 58.9337 56.9360 47.1902 44.5696

50.3561 40.4574 55.8165 66.7234 56.1157 67.8447

Sample mean = 52.08
Sample standard deviation = 9.46
Standard error of the mean = 2.12
90% confidence interval is $52.08 \pm 1.729 \times 2.12 = (48.42, 55.74)$
This interval includes the mean (i.e., 50).

b) The results for 100 samples are given below. 91 of the intervals contain 50. We expect 90% percent of intervals to contain 50, if we repeated these simulations many times. If X = number of intervals containing 50, then X has a binomial distribution with $n = 100$ and $p = 0.95$.

Sample	Lower limit	Upper limit	Is mu in CI? (1 = Yes 0 = No)
1	46.8287	54.7402	1
2	45.2213	50.4444	1
3	43.1479	50.1435	1
4	45.0871	53.9720	1
5	43.8966	50.6573	1

Sample	Lower limit	Upper limit	Is mu in CI? (1 = Yes 0 = No)
6	46.7444	54.0840	1
7	45.5834	53.9410	1
8	47.4073	54.8735	1
9	45.8292	51.7334	1
10	47.7203	52.7239	1
11	44.8570	53.0214	1
12	50.1789	56.9975	0
13	47.0817	54.2550	1
14	47.7295	56.4808	1
15	48.6491	57.0253	1
16	47.0764	53.8848	1
17	42.3624	50.1233	1
18	49.8208	56.0376	1
19	46.3226	55.8385	1
20	45.5904	52.5858	1
21	47.6078	54.3333	1
22	41.0293	48.9811	0
23	41.9557	49.0181	0
24	47.1299	54.2891	1
25	47.1213	53.8481	1
26	48.1212	55.1473	1
27	46.5877	52.9893	1
28	44.0114	52.1599	1
29	48.5973	55.3574	1
30	45.6141	51.3604	1
31	49.5235	56.3526	1
32	46.1029	55.1506	1
33	41.7496	49.9462	0
34	49.1933	54.8805	1
35	46.3342	53.0526	1
36	47.0429	55.5142	1
37	45.6500	51.9249	1
38	43.4494	50.2173	1
39	45.1662	53.9872	1

Sample	Lower limit	Upper limit	Is mu in CI? (1 = Yes 0 = No)
40	48.1884	55.7717	1
41	43.6584	51.9277	1
42	43.1976	52.0532	1
43	43.3492	51.0945	1
44	45.1286	54.7027	1
45	44.1183	53.6761	1
46	45.3881	53.5367	1
47	48.4881	54.5354	1
48	46.0861	54.1598	1
49	47.8200	55.6915	1
50	46.7662	54.7076	1
51	50.0724	56.3251	0
52	44.3637	52.0437	1
53	48.8991	54.1511	1
54	45.0726	53.4418	1
55	47.5560	54.2997	1
56	44.1319	51.4971	1
57	41.2206	49.3686	0
58	45.6447	50.6546	1
59	43.7943	50.4571	1
60	48.0205	54.4572	1
61	45.9123	52.5493	1
62	47.4707	53.4410	1
63	40.5179	51.3647	1
64	43.2859	50.7090	1
65	43.5471	50.1373	1
66	47.5367	55.4627	1
67	44.7113	50.9052	1
68	48.8710	54.7114	1
69	42.6696	52.8379	1
70	45.6380	51.4373	1
71	49.0097	54.1843	1
72	46.3687	53.5640	1
73	44.6466	53.6845	1

Sample	Lower limit	Upper limit	Is mu in CI? (1 = Yes 0 = No)
74	46.3326	55.3837	1
75	45.4664	53.3785	1
76	43.7267	52.8088	1
77	45.5380	53.0625	1
78	45.1098	52.2788	1
79	47.9032	55.1251	1
80	46.7072	54.1474	1
81	47.7600	55.2346	1
82	47.2041	53.6313	1
83	47.5978	54.1414	1
84	46.1906	54.4134	1
85	47.4681	53.8763	1
86	46.6361	53.4053	1
87	42.2656	51.0899	1
88	48.1179	54.9510	1
89	45.2856	52.8882	1
90	45.4159	55.3781	1
91	51.2429	58.3487	0
92	46.3859	52.8781	1
93	43.3430	49.5763	0
94	44.8461	52.5728	1
95	47.7771	54.5224	1
96	48.2824	55.6703	1
97	44.5222	51.4175	1
98	44.0169	49.9961	0
99	45.0104	54.5010	1
100	48.4244	55.7409	1

45. Still more simulations

a) The simulated sample from the exponential distribution mean 1 is shown below (Note: This is a random sample and so the answers can vary.):

0.23950   2.97284   2.23162   2.59763   0.08291   0.45010   0.26260

2.00925   0.06010   4.15170   0.20565   0.24763   0.63370   0.99105

1.16028

The histogram of this sample is shown below. It is right skewed.

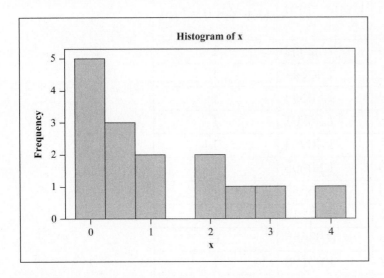

Sample mean = 1.21977
Sample standard deviation = 1.27320
Standard error of the mean = 0.32874
90% confidence interval is $1.21977 \pm 1.761 \times 0.32874 = (0.6408, 1.7987)$
This interval includes the mean (i.e., 1.0).

b) The results for 100 samples are given below. 82 of these intervals contain the mean 1.0. The formula for the confidence interval we learned in this chapter (and used in this question) is not appropriate when the distribution of the sample mean is not Normal. The sample mean of a sample of size 15 from an exponential distribution (a skewed distribution) will not have a Normal distribution and so this interval will not have the requested confidence level (i.e., 90%). If we change the sample size to 100, then the distribution of the sample mean will become approximately Normal and so approximately 90% of the CIs will contain 1.0.

Sample	Lower limit	Upper limit	Is 1.0 in CI? (1 = Yes 0 = No)
1	0.41321	1.08837	1
2	0.59322	1.13652	1
3	0.53257	1.01576	1
4	0.67703	1.55177	1
5	0.37853	0.89310	0
6	0.57111	1.71702	1
7	0.38795	1.17508	1

Sample	Lower limit	Upper limit	Is 1.0 in CI? (1 = Yes 0 = No)
8	0.59990	1.52553	1
9	0.61096	1.49044	1
10	0.82439	1.98845	1
11	0.65996	1.50281	1
12	0.55939	1.40904	1
13	0.64906	1.80605	1
14	0.71240	1.46553	1
15	0.68941	2.06949	1
16	0.42698	0.93655	0
17	0.75072	1.63158	1
18	0.51172	1.37070	1
19	0.70406	2.15330	1
20	0.74053	1.75792	1
21	0.50909	1.10211	1
22	0.47399	0.92792	0
23	0.42207	1.58466	1
24	0.44710	2.81677	1
25	0.65776	1.18138	1
26	0.55948	1.29414	1
27	0.31993	0.98749	0
28	0.64789	1.36053	1
29	0.63524	2.04012	1
30	0.46635	1.71979	1
31	0.60636	1.71701	1
32	0.35535	1.04123	1
33	0.51970	1.14547	1
34	0.55030	1.21178	1
35	0.55910	1.81254	1
36	0.62463	1.70598	1
37	0.94693	2.01598	1
38	0.42573	0.92784	0
39	0.75008	1.57588	1
40	0.49416	1.04601	1
41	0.87979	1.71881	1

Sample	Lower limit	Upper limit	Is 1.0 in CI? (1 = Yes 0 = No)
42	0.43638	0.96242	0
43	0.38834	0.90343	0
44	0.63330	2.10864	1
45	0.57412	1.62677	1
46	0.45400	1.24766	1
47	0.63798	1.51278	1
48	0.27058	1.23743	1
49	0.65048	1.39562	1
50	0.53115	1.14980	1
51	0.44540	1.00229	1
52	0.61006	1.21031	1
53	0.58005	1.13960	1
54	0.86680	1.69323	1
55	0.46848	0.89850	0
56	0.65990	2.23803	1
57	0.80994	1.92091	1
58	0.47791	1.13617	1
59	0.73114	1.32412	1
60	1.02448	2.64439	0
61	0.25996	0.85328	0
62	0.62518	1.41445	1
63	0.61772	1.47174	1
64	0.35537	0.97258	0
65	0.72694	1.71476	1
66	0.39975	0.95079	0
67	0.38694	1.00107	1
68	0.51612	1.78867	1
69	0.65677	1.34919	1
70	0.40058	1.13681	1
71	0.31929	0.92408	0
72	0.89482	2.23752	1
73	0.37230	0.93398	0
74	0.50438	1.22648	1
75	0.58450	1.49718	1

Sample	Lower limit	Upper limit	Is 1.0 in CI? (1 = Yes 0 = No)
76	0.59250	1.33373	1
77	0.59785	1.31385	1
78	0.91520	1.83678	1
79	0.38645	1.01354	1
80	0.26898	0.62632	0
81	0.54904	1.75814	1
82	0.73623	1.81129	1
83	0.64125	1.63378	1
84	0.81907	1.58136	1
85	0.51564	1.09377	1
86	0.46291	1.41097	1
87	0.35770	0.96339	0
88	0.55769	1.21368	1
89	0.96044	1.83062	1
90	0.57312	1.39409	1
91	0.76557	1.47910	1
92	0.48797	1.17056	1
93	0.76376	1.44517	1
94	0.80645	1.65747	1
95	0.87193	1.91061	1
96	0.46853	1.96917	1
97	0.58745	1.26259	1
98	0.31282	0.81307	0
99	0.27260	0.94509	0
100	0.64076	1.79878	1

47. Even more simulations

a) The simulated sample from the Normal distribution mean 50 and standard deviation 5 is shown below (Note: This is a random sample and so the answers can vary.):

46.7670 54.9289 48.1935 48.5884 59.6762

For this sample mean = 51.6308, Standard deviation = 5.4848, Standard error of the mean = 2.4529. The value of the t-statistic = 0.66, p-value = 0.543 > 0.10 and so we do not reject the null hypothesis. Passed a good batch. Since the null hypothesis is true, the error of rejecting is the possible error here. That is a Type I error.

b) The results for the 100 samples are shown below. We expect to reject 10% of the tests. Note that this is the Type I error and the probability of a Type I error is the level of the test. The distribution of X is binomial with ($n = 100$, $p = 0.10$).

Sample	Sample mean	stdev	*t*-statistic	*p*-value	H₀ rejected? (1 = yes 0 = no)
1	46.2748	3.80742	–2.18781	0.093922	1
2	48.0603	4.49512	–0.96491	0.389232	0
3	52.0605	4.29720	1.07220	0.344010	0
4	50.5021	7.56920	0.14833	0.889260	0
5	52.8143	4.42250	1.42294	0.227836	0
6	52.6813	6.65392	0.90106	0.418502	0
7	51.3238	7.08891	0.41757	0.697703	0
8	51.8063	5.60505	0.72061	0.511014	0
9	49.4705	3.20815	–0.36907	0.730782	0
10	48.8749	9.25011	–0.27197	0.799107	0
11	49.2397	3.54700	–0.47932	0.656741	0
12	48.9067	4.65915	–0.52471	0.627518	0
13	46.4743	5.01645	–1.57158	0.191146	0
14	47.3446	4.07167	–1.45829	0.218511	0
15	55.4176	5.73580	2.11200	0.102260	0
16	52.0481	3.79141	1.20793	0.293606	0
17	48.8231	4.50261	–0.58448	0.590276	0
18	48.2878	1.95543	–1.95792	0.121845	0
19	47.2371	6.64394	–0.92987	0.405072	0
20	48.3731	4.07371	–0.89300	0.422325	0
21	46.1506	4.56140	–1.88704	0.132202	0
22	46.5372	1.20608	–6.42002	0.003026	1
23	50.3264	5.74471	0.12704	0.905038	0
24	49.9674	5.67110	–0.01286	0.990355	0
25	54.5470	5.89554	1.72459	0.159690	0
26	48.0832	2.01620	–2.12579	0.100683	0
27	52.4429	6.84136	0.79844	0.469334	0
28	48.3628	5.51689	–0.66360	0.543240	0
29	46.5241	3.88520	–2.00049	0.116051	0

Sample	Sample mean	stdev	*t*-statistic	*p*-value	H$_0$ rejected? (1 = yes 0 = no)
30	48.1883	6.12731	–0.66115	0.544653	0
31	50.0410	3.28960	0.02787	0.979098	0
32	48.6584	3.32977	–0.90096	0.418550	0
33	49.7359	4.24247	–0.13919	0.896025	0
34	52.1436	4.70633	1.01849	0.366038	0
35	47.9616	3.09919	–1.47069	0.215331	0
36	54.7241	4.49776	2.34857	0.078638	1
37	51.0487	5.65881	0.41440	0.699841	0
38	48.5205	2.26601	–1.45996	0.218079	0
39	50.4033	6.88545	0.13097	0.902122	0
40	49.0995	5.54758	–0.36297	0.734993	0
41	50.0813	7.09393	0.02562	0.980788	0
42	53.6849	2.43121	3.38910	0.027553	1
43	51.0525	4.38593	0.53657	0.620017	0
44	51.0827	6.08720	0.39770	0.711160	0
45	49.4778	5.00597	–0.23326	0.827013	0
46	48.4272	5.83624	–0.60261	0.579264	0
47	44.8766	6.06139	–1.89004	0.131744	0
48	51.5127	4.53481	0.74591	0.497172	0
49	51.4978	2.75521	1.21557	0.290984	0
50	51.4714	4.14206	0.79434	0.471465	0
51	50.8118	8.87236	0.20460	0.847875	0
52	50.4946	7.10700	0.15563	0.883865	0
53	50.3808	6.37916	0.13347	0.900269	0
54	52.8106	6.29231	0.99880	0.374416	0
55	49.2606	7.79924	–0.21200	0.842474	0
56	55.0104	4.84249	2.31361	0.081708	1
57	47.8081	3.14241	–1.55971	0.193840	0
58	51.5332	3.08816	1.11014	0.329179	0
59	50.4982	3.38917	0.32871	0.758861	0
60	45.6191	3.25169	–3.01258	0.039450	1
61	47.7067	2.74388	–1.86885	0.135011	0
62	49.0524	1.40988	–1.50289	0.207291	0
63	47.9821	1.84027	–2.45193	0.070299	1

Sample	Sample mean	stdev	*t*-statistic	*p*-value	H₀ rejected? (1 = yes 0 = no)
64	47.2050	6.16037	–1.01453	0.367708	0
65	52.6395	6.08581	0.96981	0.387060	0
66	48.9599	3.54128	–0.65674	0.547213	0
67	53.1695	6.63599	1.06799	0.345693	0
68	46.8135	2.26222	–3.14966	0.034525	1
69	45.3079	7.41778	–1.41441	0.230147	0
70	51.4109	5.79544	0.54437	0.615112	0
71	52.8343	2.85959	2.21631	0.090985	1
72	47.9641	2.91773	–1.56028	0.193711	0
73	50.1445	4.03327	0.08010	0.940002	0
74	49.2659	4.77267	–0.34392	0.748228	0
75	52.4377	3.05791	1.78258	0.149236	0
76	49.4968	5.32183	–0.21144	0.842883	0
77	49.7751	3.82581	–0.13147	0.901754	0
78	48.7420	3.11140	–0.90410	0.417071	0
79	48.7134	6.64032	–0.43325	0.687170	0
80	47.6129	8.30410	–0.64277	0.555364	0
81	49.3152	2.40850	–0.63579	0.559472	0
82	49.7653	5.08627	–0.10319	0.922780	0
83	51.1871	2.43146	1.09166	0.336328	0
84	48.2109	6.80774	–0.58765	0.588338	0
85	50.3646	6.02874	0.13525	0.898949	0
86	54.8934	7.05274	1.55145	0.195738	0
87	49.9285	6.76787	–0.02362	0.982288	0
88	50.9467	7.78841	0.27181	0.799219	0
89	50.0038	4.77965	0.00178	0.998665	0
90	50.1442	2.83570	0.11373	0.914932	0
91	48.5917	4.56316	–0.69009	0.528090	0
92	50.2928	5.45898	0.11993	0.910324	0
93	48.3497	6.72995	–0.54831	0.612641	0
94	47.0805	4.56754	–1.42925	0.226144	0
95	48.5884	3.88943	–0.81153	0.462589	0
96	50.7051	2.32611	0.67783	0.535063	0
97	49.6041	4.86090	–0.18212	0.864346	0

Sample	Sample mean	stdev	t-statistic	p-value	H₀ rejected? (1 = yes 0 = no)
98	50.1793	4.36562	0.09182	0.931254	0
99	52.7694	6.71625	0.92202	0.408696	0
100	51.6308	5.48483	0.66485	0.542515	0

Summary

Sample size:	5.00000
True mean:	50.0000
Standard deviation:	5.00000
Number of samples:	100.000
Level of Sig:	0.100000
Proportion rejected:	0.0900000

49. **Calculating power**

a) $z^* = 1.645$ for $\alpha = 0.05$ and so the criterion is to reject the null hypothesis if $\dfrac{\bar{y} - 0.08}{s / \sqrt{150}} > 1.645$.

b) Substituting $s = 0.05$ and solving this for $\bar{y}$, we get:

$$\bar{y} > 0.08 + 1.645 \times \frac{0.05}{\sqrt{150}} = 0.08671568438.$$ That is, our criterion now is to reject the null hypothesis if $\bar{y} > 0.08671568438$.

c) If the mean $\mu = 0.09$. The probability of rejecting the null hypothesis is:

$$P\left(Z > \frac{0.08671568438 - 0.09}{0.05 / \sqrt{150}}\right) = P(Z > -0.80) = 1 - 0.2119 = 0.7881.$$

i.e., the power of the test when $\mu = 0.09$ is 0.7881.

d) Using software (MINITAB), the power at sample sizes 150, 75, and 38, with a difference 0.01 (i.e., 0.09 – 0.08) and $\alpha = 0.05$, is 0.786, 0.528 and 0.332 respectively. Note that the power decreases as the sample size decreases.

Power and Sample Size

1-Sample t-Test

Testing mean = null (versus > null)

Calculating power for mean = null + difference

Alpha = 0.05

Assumed standard deviation = 0.05

Difference	Sample Size	Power
0.01	150	0.786254
0.01	75	0.528434
0.01	38	0.331957

Chapter 21: Comparing Means

1. **Hot dogs** Yes, the 95% confidence interval would contain 0. The high P-value means that we lack evidence of a difference, so 0 is a possible value for $\mu_{Meat} - \mu_{Beef}$.

3. **Hot dogs and fat**
   a) Plausible values for $\mu_{Meat} - \mu_{Beef}$ are all negative, so the mean fat content is probably higher for beef hot dogs.
   b) The fact that the confidence interval does not contain 0 indicates that the difference is significant.
   c) The corresponding alpha level is 10%.

5. **Hot dogs, 2nd helping**
   a) False. The confidence interval is about means, not about individual hot dogs.
   b) False. The confidence interval is about means, not about individual hot dogs.
   c) True.
   d) False. Confidence intervals based on other samples will also try to estimate the true difference in population means. There's no reason to expect other samples to conform to this result.
   e) True.

7. **Learning math**
   a) The margin of error of this confidence interval is (11.427 – 5.573)/2 = 2.927 points.
   b) The margin of error for a 98% confidence interval would have been larger. The critical value of t^* is larger for higher confidence levels. We need a wider interval to increase the likelihood that we catch the true mean difference in test scores within our interval. In other words, greater confidence comes at the expense of precision.
   c) We are 95% confident that the mean score for the CPMP math students will be between 5.573 and 11.427 points higher on this assessment than the mean score of the traditional students.
   d) Since the entire interval is above 0, there is strong evidence that students who learn with CPMP will have higher mean scores is algebra than those in traditional programs.

9. **CPMP, again**
   a) H_0: The mean score of CPMP students is the same as the mean score of traditional students. $\left(\mu_C = \mu_T \text{ or } \mu_C - \mu_T = 0\right)$

   H_A: The mean score of CPMP students is different from the mean score of traditional students. $\left(\mu_C \neq \mu_T \text{ or } \mu_C - \mu_T \neq 0\right)$

b) **Independent Groups Assumption**: Scores of students from different classes should be independent.

Randomization Condition: Although not specifically stated, classes in this experiment were probably randomly assigned to either CPMP or traditional curricula.

10% Condition: 312 and 265 are less than 10% of all students.

Nearly Normal Condition: We don't have the actual data, so we can't check the distribution of the sample. However, the samples are large. The Central Limit Theorem allows us to proceed.

Since the conditions are satisfied, we can use a two-sample *t*-test with 583 degrees of freedom (from the computer).

c) If the mean scores for the CPMP and traditional students are really equal, there is less than a 1 in 10 000 chance of seeing a difference as large or larger than the observed difference just from natural sampling variation.

d) Since the P-value < 0.0001, reject the null hypothesis. There is strong evidence that the CPMP students have a different mean score than the traditional students. The evidence suggests that the CPMP students have a higher mean score.

11. Commuting

a) **Independent Groups Assumption**: Since the choice of route was determined at random, the commuting times for Route A are independent of the commuting times for Route B.

Randomization Condition: The man randomly determined which route he would travel on each day.

Nearly Normal Condition: The histograms of travel times for the routes are roughly symmetric and show no outliers. (Given)

Since the conditions are satisfied, it is appropriate to model the sampling distribution of the difference in means with a Student's *t*-model, with 33.1 degrees of freedom (from the approximation formula). We will construct a two-sample *t*-interval, with 95% confidence.

$$(\bar{y}_B - \bar{y}_A) \pm t^*_{df} \sqrt{\frac{s_B^2}{n_B} + \frac{s_A^2}{n_A}} = (43 - 40) \pm t^*_{33.1} \sqrt{\frac{2^2}{20} + \frac{3^2}{20}} \approx (1.36, 4.64)$$

We are 95% confident that Route B has a mean commuting time between 1.36 and 4.64 minutes longer than the mean commuting time of Route A.

b) Since 5 minutes is beyond the high end of the interval, there is no evidence that Route B is an average of 5 minutes longer than Route A. It appears that the old-timer may be exaggerating the average difference in commuting time.

13. Cereal

Independent Groups Assumption: The percentage of sugar in the children's cereals is unrelated to the percentage of sugar in adults' cereals.

Randomization Condition: It is reasonable to assume that the cereals are representative of all children's cereals and adults' cereals in regard to sugar content.

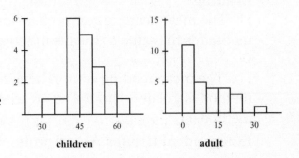

children   adult

Nearly Normal Condition: The histogram of adult cereal sugar content is skewed to the right, but the sample sizes are of reasonable size. The Central Limit Theorem allows us to proceed.

Since the conditions are satisfied, it is appropriate to model the sampling distribution of the difference in means with a Student's *t*-model, with 42 degrees of freedom (from the approximation formula). We will construct a two-sample *t*-interval, with 95% confidence.

$$(\bar{y}_C - \bar{y}_A) \pm t^*_{df}\sqrt{\frac{s_C^2}{n_C} + \frac{s_A^2}{n_A}} = (46.8 - 10.1536) \pm t^*_{42}\sqrt{\frac{6.41838^2}{19} + \frac{7.61239^2}{28}} \approx (32.49, 40.80)$$

We are 95% confident that children's cereals have a mean sugar content that is between 32.49% and 40.80% higher than the mean sugar content of adults' cereals.

15. Hurricanes

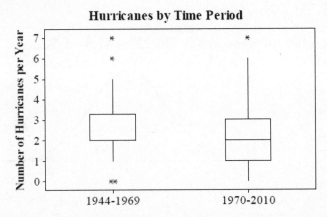

These data present some concerns. First, the sample is not from a randomized experiment—we have to assume that the actual number of hurricanes in a given year is a random sample of the hurricanes that might occur under similar weather conditions. Also, the data for 1944–1969 are not symmetric and have four outliers. Nevertheless, the data from both groups show the same degree of moderate right skewness, and their sample sizes are large enough (n > 20) and similar enough in shape to trust the robustness of the t-test.

17. Reading

H_0: The mean reading comprehension score of students who learn by the new method is the same as the mean score of students who learn by traditional methods. ($\mu_N = \mu_T$ or $\mu_N - \mu_T = 0$)

H_A: The mean reading comprehension score of students who learn by the new method is greater than the mean score of students who learn by traditional methods. ($\mu_N > \mu_T$ or $\mu_N - \mu_T > 0$)

Independent Groups Assumption: Student scores in one group should not have an impact on the scores of students in the other group.

Randomization Condition: Students were randomly assigned to classes.

Nearly Normal Condition: The histograms of the scores are unimodal and symmetric.

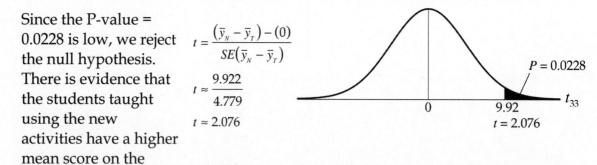

Since the conditions are satisfied, it is appropriate to model the sampling distribution of the difference in means with a Student's *t*-model, with 33 degrees of freedom (from the approximation formula). We will perform a two-sample *t*-test. We know:

$$\bar{y}_N = 51.7222 \qquad \bar{y}_T = 41.8$$
$$s_N = 11.7062 \qquad s_T = 17.4495$$
$$n_N = 18 \qquad n_T = 20$$

The sampling distribution model has mean 0, with standard error:

$$SE(\bar{y}_N - \bar{y}_T) = \sqrt{\frac{11.7062^2}{18} + \frac{17.4495^2}{20}} \approx 4.779 \ .$$

The observed difference between the mean scores is $51.7222 - 41.8 \approx 9.922$.

Since the P-value = 0.0228 is low, we reject the null hypothesis. There is evidence that the students taught using the new activities have a higher mean score on the reading comprehension test than the students taught using traditional methods.

$$t = \frac{(\bar{y}_N - \bar{y}_T) - (0)}{SE(\bar{y}_N - \bar{y}_T)}$$

$$t \approx \frac{9.922}{4.779}$$

$$t \approx 2.076$$

$P = 0.0228$

$t = 2.076$

19. Baseball

a) The boxplots of the average number of home runs hit at the ballparks in the two leagues are shown below. Both distributions appear at least roughly symmetric, with roughly the same centre, around 1 home run. The distribution of average number of home runs hit appears a bit more spread out for the American League.

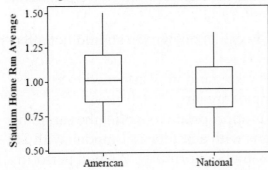

Average Number of Home Runs per Stadium

b) $\bar{y} \pm t^*_{n-1}\left(\dfrac{s}{\sqrt{n}}\right) = 1.027857 \pm t^*_{13}\left(\dfrac{0.22027}{\sqrt{14}}\right) = (0.90, 1.16)$

    We are 95% confident that the mean number of home runs hit per game in American League stadiums is between 0.90 and 1.16.

c) The average of 1.354 home runs hit per game in Coors Field is not unusual. It is the highest average in the National League, but by no means an outlier.

d) If you attempt to use two confidence intervals to assess a difference in means, you are actually adding standard deviations. But it's the variances that add, not the standard deviations. The two-sample difference of means procedure takes this into account.

e) $(\bar{y}_A - \bar{y}_N) \pm t^*_{d_f}\sqrt{\dfrac{s_A^2}{n_A} + \dfrac{s_N^2}{n_N}} = (1.027857 - 0975375) \pm t^*_{26,77}\sqrt{\dfrac{0.22027^2}{14} + \dfrac{0.20399^2}{16}}$

    $= (-0.107, 0.212)$

f) We are 95% confident that the mean number of home runs in American League stadiums is between 0.107 home runs fewer and 0.212 home runs more than the mean number of home runs in National League stadiums.

g) Since the interval contains 0, there is no evidence of a difference in the mean number of home runs hit per game in the stadiums of the two leagues.

21. Job satisfaction A two-sample *t*-procedure is not appropriate for these data because the two groups are not independent. They are before and after satisfaction scores for the same workers. Workers that have high levels of job satisfaction before the exercise program is implemented may tend to have higher levels of job satisfaction than other workers after the exercise program as well.

23. Sex and violence

a) Since the P-value = 0.136 is high, we fail to reject the null hypothesis. There is no evidence of a difference in the mean number of brands recalled by viewers of sexual content and viewers of violent content.

b) H_0: The mean number of brands recalled is the same for viewers of sexual content and viewers of neutral content. $(\mu_S = \mu_N \text{ or } \mu_S - \mu_N = 0)$

H$_A$: The mean number of brands recalled is different for viewers of sexual content and viewers of neutral content. ($\mu_s \neq \mu_N$ or $\mu_s - \mu_N \neq 0$)

Independent Groups Assumption: Recall of one group should not affect recall of another.

Randomization Condition: Subjects were randomly assigned to groups.

Nearly Normal Condition: The samples are large.

Since the conditions are satisfied, it is appropriate to model the sampling distribution of the difference in means with a Student's *t*-model, with 214 degrees of freedom (from the approximation formula). We will perform a two-sample *t*-test.

The sampling distribution model has mean 0, with standard error:

$$SE\ (\bar{y}_S - \bar{y}_N) = \sqrt{\frac{1.76^2}{108} + \frac{1.77^2}{108}} \approx 0.24 \ .$$

The observed difference between the mean scores is 1.71 – 3.17 = – 1.46.

Since the P-value = 5.5×10^{-9} is low, we reject the null hypothesis. There is strong evidence that the mean number of brand names recalled is different for viewers of sexual content and viewers of neutral content. The evidence suggests that viewers of neutral ads remember more brand names on average than viewers of sexual content.

$$t = \frac{(\bar{y}_S - \bar{y}_N) - (0)}{SE\ (\bar{y}_S - \bar{y}_N)}$$

$$t \approx \frac{-1.46}{0.24}$$

$$t \approx -6.08$$

c) Using a pooled t-test,

$$S_p = \sqrt{\frac{(n_V - 1)s_V^2 + (n_S - 1)S_S^2}{n_V + n_S - 2}} = \sqrt{\frac{107\left(1.87^2\right) + 107\left(1.76^2\right)}{214}} = 1.8158$$

The observed test statistic is

$$t = \frac{(\bar{y}_V - \bar{y}_S) - 0}{S_p \sqrt{\frac{1}{n_V} + \frac{1}{n_S}}} = \frac{0.37}{1.8158\sqrt{\frac{1}{108} + \frac{1}{108}}} = 1.4974$$

The p-value is 0.136, so our conclusion is the same as part a). The pooled variance t-test makes sense in this case because this is an experiment and the standard deviations are similar between groups.

25. Sex and violence II

a) H$_0$: The mean number of brands recalled is the same for viewers of violent content and viewers of neutral content. ($\mu_V = \mu_N$ or $\mu_V - \mu_N = 0$)

H$_A$: The mean number of brands recalled is different for viewers of violent content and viewers of neutral content. ($\mu_V \neq \mu_N$ or $\mu_V - \mu_N \neq 0$)

Independent Groups Assumption: Recall of one group should not affect recall of another.

Randomization Condition: Subjects were randomly assigned to groups.

Nearly Normal Condition: The samples are large.

Since the conditions are satisfied, it is appropriate to model the sampling distribution of the difference in means with a Student's *t*-model, with 201.96 degrees of freedom (from the approximation formula). We will perform a two-sample *t*-test.

The sampling distribution model has mean 0, with standard error:

$$SE\left(\bar{y}_V - \bar{y}_N\right) = \sqrt{\frac{1.61^2}{101} + \frac{1.62^2}{103}} \approx 0.226 \cdot$$

The observed difference between the mean scores is $3.02 - 4.65 = -1.63$.

$$t = \frac{\left(\bar{y}_V - \bar{y}_N\right) - (0)}{SE\left(\bar{y}_V - \bar{y}_N\right)}$$

$$t \approx \frac{-1.63}{0.226}$$

$$t \approx -7.21$$

Since the P-value $= 1.1 \times 10^{-11}$ is low, we reject the null hypothesis. There is strong evidence that the mean number of brand names recalled is different for viewers of violent content and viewers of neutral content. The evidence suggests that viewers of neutral ads remember more brand names on average than viewers of violent content.

b) $\left(\bar{y}_N - \bar{y}_S\right) \pm t_{df}^* \sqrt{\dfrac{s_N^2}{n_N} + \dfrac{s_S^2}{n_S}} = (4.65 - 2.72) \pm t_{204.8}^* \sqrt{\dfrac{1.62^2}{103} + \dfrac{1.85^2}{106}} \approx (1.456, 2.404)$

We are 95% confident that the mean number of brand names recalled 24 hours later is between 1.46 and 2.40 higher for viewers of shows with neutral content than for viewers of shows with sexual content.

27. Hungry?

H_0: The mean number of ounces of ice cream people scoop is the same for large and small bowls. $\left(\mu_{big} = \mu_{small} \text{ or } \mu_{big} - \mu_{small} = 0\right)$

H_A: The mean number of ounces of ice cream people scoop is the different for large and small bowls. $\left(\mu_{big} \neq \mu_{small} \text{ or } \mu_{big} - \mu_{small} \neq 0\right)$

Independent Groups Assumption: The amount of ice cream scooped by individuals should be independent.

Randomization Condition: Subjects were randomly assigned to groups.

Nearly Normal Condition: Assume that this condition is met.

Since the conditions are satisfied, it is appropriate to model the sampling distribution of the difference in means with a Student's *t*-model, with 34 degrees of freedom (from the approximation formula). We will perform a two-sample *t*-test. The sampling distribution model has mean 0, with standard error:

$$SE(\bar{y}_{big} - \bar{y}_{small}) = \sqrt{\frac{2.91^2}{22} + \frac{1.84^2}{26}} \approx 0.7177 \text{ oz.}$$

The observed difference between the mean amounts is 6.58 – 5.07 = 1.51 oz.

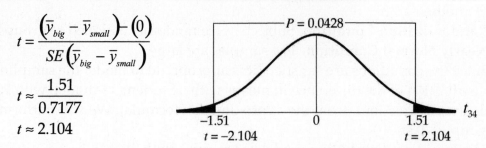

$$t = \frac{\left(\bar{y}_{big} - \bar{y}_{small}\right) - (0)}{SE\left(\bar{y}_{big} - \bar{y}_{small}\right)}$$

$$t \approx \frac{1.51}{0.7177}$$

$$t \approx 2.104$$

Since the P-value of 0.0428 is low, we reject the null hypothesis. There is strong evidence that the mean amount of ice cream people put into a bowl is related to the size of the bowl. People tend to put more ice cream into the large bowl, on average, than the small bowl.

29. **Battle of the sexes** The boxplots for the two are shown below. The data for males has an extreme outlier. The *t*-tests use the means and the standard deviations are non-resistant to outliers, and so ideally we should remove them before using *t*-tests.

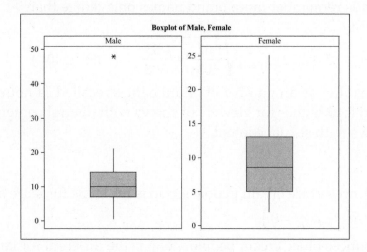

After removing the extreme outlier, the boxplots and some summary statistics are shown below. We see that the spreads in the boxplots (and also the standard deviations in the summary statistics) are very similar and so we can use the pooled *t*-test.

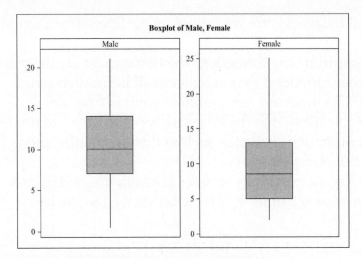

Descriptive Statistics: Male, Female

Variable	N	Mean	SE Mean	StDev
Male	45	10.322	0.820	5.498
Female	47	9.309	0.777	5.326

To test whether the mean distance is shorter for women than that for men, we test the null hypothesis $H_0 : \mu_M - \mu_F = 0$ against $H_A : \mu_M - \mu_F > 0$.

$$s_{pooled} = \sqrt{\frac{(n_M - 1)s_M^2 + (n_F - 1)s_F^2}{(n_M - 1) + (n_F - 1)}} = 5.4107$$

$$SE_{pooled}(\bar{Y}_M - \bar{Y}_F) = s_{pooled}\sqrt{\frac{1}{n_M} + \frac{1}{n_F}} = 5.4107\sqrt{\frac{1}{45} + \frac{1}{47}} = 1.1285$$

$$t = \frac{\bar{Y}_M - \bar{Y}_F - 0}{SE_{pooled}(\bar{Y}_M - \bar{Y}_F)} = \frac{10.32 - 9.31 - 0}{1.1285} = 0.90$$ and $p - Value = P(t_{90} > 0.90) > 0.10$ (the

exact value from software is 0.186). Even if the population means were equal, there is a high chance (about 19 in 100) for $\bar{Y}_M - \bar{Y}_F$ to be greater than or equal to 1.013 (i.e. 10.322 – 9.309) and so the data do not provide evidence against the hull hypothesis. We do not reject the null hypothesis. That is, the data do not provide sufficient evidence to conclude that women are better at parallel parking.

31. Crossing Ontario
a)

$$(\bar{y}_M - \bar{y}_W) \pm t_{df}^*\sqrt{\frac{s_M^2}{n_M} + \frac{s_W^2}{n_W}} = (1196.75 - 1271.59) \pm t_{37.67}^*\sqrt{\frac{304.369^2}{20} + \frac{261.111^2}{22}} \approx (-252.89, 103.21)$$

We are 95% confident that the interval –252.89 to 103.20 minutes (–74.84 ± 178.05 minutes) contains the true difference in mean crossing times between men and women. Because the interval includes zero, we cannot be confident that there is any difference at all.

b) **Independent Groups Assumption**: The times from the two groups are likely to be independent of one another, provided that these were all individual swims.
Randomization Condition: The times are not a random sample from any identifiable population, but it is likely that the times are representative of times from swimmers who might attempt a challenge such as this. Hopefully, these times were recorded from different swimmers.
Nearly Normal Condition: The distributions of times are both unimodal, with no outliers. The distribution of men's times is somewhat skewed to the left.

c) Using a pooled t-test,

$$s_p = \sqrt{\frac{(n_F - 1)s_F^2 + (n_M - 1)s_M^2}{n_F + n_M - 2}} - \sqrt{\frac{21(261.111^2) + 19(304.369^2)}{40}} = 282.4857$$

The 95% confidence interval is

$$\left(\bar{y}_M - \bar{y}_F\right) \pm t_{40}^* s_p \sqrt{\frac{1}{n_F} + \frac{1}{n_M}} = -74.84 \pm t_{40}^* 282.4857 \sqrt{\frac{1}{22} + \frac{1}{20}} = (-251.2, 101.6)$$

which is nearly the same confidence interval as part a). The pooled t-test might be appropriate because the standard deviations of the two groups are similar.

33. Running heats

H_0: The mean time to finish is the same for heats 2 and 5.
$(\mu_2 = \mu_5 \text{ or } \mu_2 - \mu_5 = 0)$
H_A: The mean time to finish is not the same for heats 2 and 5.
$(\mu_2 \neq \mu_5 \text{ or } \mu_2 - \mu_5 \neq 0)$
Independent Groups Assumption: The two heats were independent.
Randomization Condition: Runners were randomly assigned.
Nearly Normal Condition: The boxplot shows an outlier in the distribution of times in heat 2. We will perform the test twice, once with the outlier and once without.

Since the conditions are satisfied, it is appropriate to model the sampling distribution of the difference in means with a Student's *t*-model, with 10.82 degrees of freedom (from the approximation formula). We will perform a two-sample *t*-test. The sampling distribution model has mean 0, with standard error:

$$SE(\bar{y}_2 - \bar{y}_5) = \sqrt{\frac{1.69319^2}{7} + \frac{1.20055^2}{7}} \approx 0.7845 .$$

$$t = \frac{(\bar{y}_2 - \bar{y}_5) - (0)}{SE(\bar{y}_2 - \bar{y}_5)}$$

The observed difference between mean times is 52.3557 – 52.3286 = 0.0271.

$$t \approx \frac{0.0271}{0.7845}$$

Since the P-value = 0.97 is high, we fail to reject the null hypothesis. $t \approx 0.035$

There is no evidence that the mean time to finish differs between the two heats.

Without the outlier, it is appropriate to model the sampling distribution of the difference in means with a Student's t-model, with 8.83 degrees of freedom (from the approximation formula). We will perform a two-sample t-test.

The sampling distribution model has mean 0, with standard error:

$$SE(\bar{y}_2 - \bar{y}_5) = \sqrt{\frac{0.56955^2}{6} + \frac{1.20055^2}{7}} \approx 0.5099 \ .$$

The observed difference between mean times is 51.7467 – 52.3286 = –0.5819.

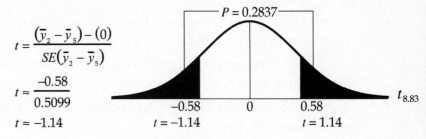

Since the P-value = 0.2837 is high, we fail to reject the null hypothesis. There is no evidence that the mean time to finish differs between the two heats.

$$t = \frac{(\bar{y}_2 - \bar{y}_5) - (0)}{SE(\bar{y}_2 - \bar{y}_5)}$$

$$t \approx \frac{-0.58}{0.5099}$$

$$t \approx -1.14$$

35. Tees

H_0: The mean ball velocity is the same for regular and Stinger tees.
$(\mu_S = \mu_R$ or $\mu_S - \mu_R = 0)$

H_A: The mean ball velocity is higher for the Stinger tees. $(\mu_S > \mu_R$ or $\mu_S - \mu_R > 0)$

Assuming the conditions are satisfied, it is appropriate to model the sampling distribution of the difference in means with a Student's t-model, with 7.03 degrees of freedom (from the approximation formula). We will perform a two-sample t-test.

The sampling distribution model has mean 0, with standard error:

$$SE(\bar{y}_S - \bar{y}_R) = \sqrt{\frac{.41^2}{6} + \frac{.89^2}{6}} \approx 0.4000 \ .$$

The observed difference between the mean velocities is 128.83 – 127 = 1.83.
$$t = \frac{(\bar{y}_S - \bar{y}_R) - (0)}{SE(\bar{y}_S - \bar{y}_R)}$$

Since the P-value = 0.0013, we reject the null hypothesis. There is strong evidence that the mean ball velocity for Stinger tees is higher than the mean velocity for regular tees.
$$t \approx \frac{1.83}{0.4000}$$
$$t \approx 4.57$$

37. Statistics journals

Independent Groups Assumption: These were articles submitted for publication in two different journals.

Randomization Condition: It is necessary to assume that these articles are representative of all articles of this type with respect to publication delay.

Nearly Normal Condition: The samples are large, so as long as there are no outliers in the data, it is okay to proceed.

Since the conditions are satisfied, it is appropriate to model the sampling distribution of the difference in means with a Student's t-model, with 338.30 degrees of freedom (from the approximation formula). We will construct a two-sample t-interval, with 90% confidence.

$$(\bar{y}_{Ap} - \bar{y}_{Am}) \pm t^*_{df}\sqrt{\frac{s^2_{Ap}}{n_{Ap}} + \frac{s^2_{Am}}{n_{Am}}} = (31 - 21) \pm t^*_{338.30}\sqrt{\frac{12^2}{209} + \frac{8^2}{288}} \approx (8.43, 11.57)$$

We are 90% confident that the mean number of months by which publication is delayed is between 8.43 and 11.57 months greater for *Applied Statistics* than it is for *The American Statistician*.

39. Rap

a) H_0: The mean memory test score is the same for those who listen to rap as it is for those who listen to no music. ($\mu_R = \mu_N$ or $\mu_R - \mu_N = 0$)

H_A: The mean memory test score is lower for those who listen to rap than it is for those who listen to no music. ($\mu_R < \mu_N$ or $\mu_R - \mu_N < 0$)

Independent Groups Assumption: The groups are not related in regards to memory score.

Randomization Condition: Subjects were randomly assigned to groups.

Nearly Normal Condition: We don't have the actual data. We will assume that the distributions of the populations of memory test scores are Normal.

Since the conditions are satisfied, it is appropriate to model the sampling distribution of the difference in means with a Student's *t*-model, with 20.00 degrees of freedom (from the approximation formula). We will perform a two-sample *t*-test.

The sampling distribution model has mean 0, with standard error:

$$SE(\bar{y}_R - \bar{y}_N) = \sqrt{\frac{3.99^2}{29} + \frac{4.73^2}{13}} \approx 1.5066$$

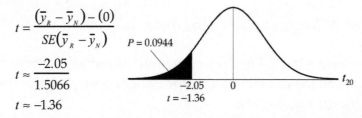

$$t = \frac{(\bar{y}_R - \bar{y}_N) - (0)}{SE(\bar{y}_R - \bar{y}_N)}$$

$$t \approx \frac{-2.05}{1.5066}$$

$$t \approx -1.36$$

$P = 0.0944$

The observed difference between the mean number of objects remembered is $10.72 - 12.77 = -2.05$.

Since the P-value = 0.0944 is high, we fail to reject the null hypothesis. There is little evidence that the mean number of objects remembered by those who listen to rap is lower than the mean number of objects remembered by those who listen to no music.

b) We did not conclude that there was a difference in the number of items remembered.

41. Tee it up again The margin of error of the CI is given by:

$$ME = t^*_{n-1}\sqrt{\frac{s_1^2}{n_1} + \frac{s_2^2}{n_2}}$$

Using $z^* = 1.96$ for t^*_{n-1} and setting each standard deviation to 3.0 and using equal sample sizes,

$$ME = 1.96\sqrt{\frac{9}{n} + \frac{9}{n}} = 1.96\sqrt{\frac{18}{n}} = 1$$

and $\sqrt{n} = \frac{1.96 \times \sqrt{18}}{1} = 8.32$. $n = 69$. This is large enough.

43. Fore!
The required sample size is 27 for each sample. The software output may vary depending on the software used. The MINITAB output is as follows:
Power and Sample Size
2-Sample t Test
Testing mean 1 = mean 2 (versus not =)
Calculating power for mean 1 = mean 2 + difference
Alpha = 0.05
Assumed standard deviation = 3

Difference	Sample Target Size	Power	Actual Power
3	27	0.95	0.950077

The sample size is for each group.

1. **More eggs?**
   a) Randomly assign 50 hens to each of the two kinds of feed. Compare the mean egg production of the two groups at the end of one month.

   100 hens — Random — Group 1 — 50 hens - regular feed / Group 2 — 50 hens - additive — Compare egg production

   b)

   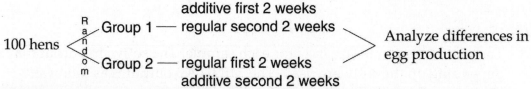

   100 hens — Random — Group 1 — additive first 2 weeks, regular second 2 weeks / Group 2 — regular first 2 weeks, additive second 2 weeks — Analyze differences in egg production

   Randomly divide the 100 hens into two groups of 50 hens each. Feed the hens in the first group the regular feed for two weeks, then switch to the additive for two weeks. Feed the hens in the second group the additive for two weeks, and then switch to the regular feed for two weeks. Subtract each hen's "regular" egg production from her "additive" egg production, and analyze the mean difference in egg production.
   c) The matched pairs design in part b) is the stronger design. Hens vary in their egg production regardless of feed. This design controls for that variability by matching the hens with themselves.

3. **Sex sells?**
   a) Randomly assign half of the volunteers to watch ads with sexual images, and assign the other half to watch ads without sexual images. Record the number of items remembered. Then have each group watch the other type of ad. Record the number of items recalled. Examine the difference in the number of items remembered for each person.

   Volunteers — Random — Group 1 — Ads with sexual images first / Group 2 — Ads with no sexual images first — Analyze mean difference in number of items remembered

   b) Randomly assign half of the volunteers to watch ads with sexual images, and assign the other half to watch ads without sexual images. Record the number of items remembered. Compare the mean number of products remembered by each group.

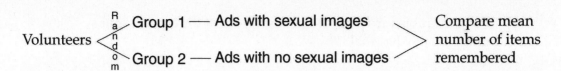

5. **Diabetes and weight loss**
   a) The paired *t*-Test is appropriate here. The weight measurement before and after surgical treatment is paired by subject.
   H_0: There is no difference in weight due to the surgical treatment.
   Since the P-value is very low, we conclude that the mean weight is lower after surgical treatment than before.
   The test statistic is t-distributed with 29 degrees of freedom.
   b) The two-sample *t*-Test is appropriate here. The subjects are not paired across the control and experimental groups.
   H_0: There is no difference in weight loss between the two treatment groups.
   Since the P-value is very low, we conclude that the mean weight loss of the group receiving surgical treatment is higher than the group receiving conventional treatment.
   The test statistic is t-distributed. (The number of degrees of freedom depends on our assumptions on the variances of the two populations.)

7. **Friday the 13th, I**
   a) The paired *t*-Test is appropriate, since we have pairs of Fridays in five different months. Data from adjacent Fridays within a month may be more similar than randomly chosen Fridays.
   b) Since the P-value = 0.0212, there is evidence that the mean number of cars on the M25 motorway on Friday the 13th is less than the mean number of cars on the previous Friday.
   c) We don't know if these Friday pairs were selected at random. Obviously, if these are the Fridays with the largest differences, this will affect our conclusion. The Nearly Normal Condition appears to be met by the differences, but the sample size of five pairs is small.

9. **Online insurance I** Adding variances requires that the variables be independent. These price quotes are for the same cars, so they are paired. Drivers quoted high insurance premiums by the local company will be likely to get a high rate from the online company, too.

11. **Online insurance II**
   a) The histogram would help you decide whether the online company offers cheaper insurance. We are concerned with the difference in price, not the distribution of each set of prices.
   b) Insurance cost is based on risk, so drivers are likely to see similar quotes from each company, making the differences relatively smaller.

c) The price quotes are paired. The histogram of differences looks approximately Normal.

13. Online insurance III

H_0: The mean difference between online and local insurance rates is zero. $(\mu_{Local-Online} = 0)$

H_A: The mean difference between online and local insurance rates is greater than zero. $(\mu_{Local-Online} > 0)$

Since the conditions are satisfied (in a previous exercise), the sampling distribution of the difference can be modelled with a Student's *t*-model with $10 - 1 = 9$ degrees of freedom.

$$t = \frac{\bar{d} - 0}{\frac{s}{\sqrt{n}}}$$

$$t = \frac{45.9 - 0}{\frac{175.663}{\sqrt{10}}}$$

$$t \approx 0.83$$

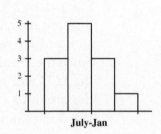

We will use a paired *t-Test*, with $\bar{d} = 45.9$.

Since the P-value = 0.215 is high, we fail to reject the null hypothesis. There is no evidence that online insurance premiums are lower on average.

15. Temperatures

Paired Data Assumption: The data are paired by city.
Randomization Condition: These cities might not be representative of all European cities, so be cautious in generalizing the results.

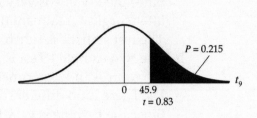

Normal Population Assumption: The histogram of differences between January and July mean temperature is roughly unimodal and symmetric.

Since the conditions are satisfied, the sampling distribution of the difference can be modelled with a Student's *t*-model with $12 - 1 = 11$ degrees of freedom. We will find a paired *t*-interval, with 90% confidence.

$$\bar{d} \pm t^*_{n-1}\left(\frac{s_d}{\sqrt{n}}\right) = 36.8333 \pm t^*_{11}\left(\frac{8.66375}{\sqrt{12}}\right) \approx (32.3, 41.3)$$

We are 90% confident that the average high temperature in European cities in July is an average of between 18.0° to 22.9° higher than in January.

17. **Push-ups**
    Independent Groups Assumption: The group
    of boys is independent of the group of girls.
    Randomization Condition: Assume that
    students are assigned to gym classes at random.

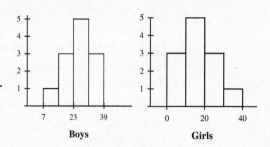
Boys   Girls

Nearly Normal Condition: The histograms of
the number of push-ups from each group are
roughly unimodal and symmetric.
Since the conditions are satisfied, it is appropriate to model the sampling
distribution of the difference in means with a Student's *t*-model, with 21 degrees
of freedom (from the approximation formula). We will construct a two-sample *t*-
interval, with 90% confidence.

$$(\bar{y}_B - \bar{y}_G) \pm t^*_{df}\sqrt{\frac{s_B^2}{n_B} + \frac{s_G^2}{n_G}} = (23.8333 - 16.5000) \pm t^*_{21}\sqrt{\frac{7.20900^2}{12} + \frac{8.93919^2}{12}} \approx (1.6, 13.0)$$

We are 90% confident that, at Gossett High, the mean number of push-ups that
boys can do is between 1.6 and 13.0 more than the mean number for the girls.

19. **Job satisfaction**
    a) Use a paired *t*-Test.
       Paired Data Assumption: The data are before and after
       job satisfaction rating for the same workers.
       Randomization Condition: The workers were randomly
       selected to participate.

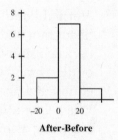
After-Before

       Nearly Normal Condition: The histogram of differences
       between before and after job satisfaction ratings is roughly
       unimodal and symmetric.
    b) H$_0$: The mean difference in before and after job satisfaction scores is zero, and
       the exercise program is not effective at improving job satisfaction. $(\mu_d = 0)$
       H$_A$: The mean difference in before and after job satisfaction scores is greater
       than zero, and the exercise program is effective at improving job satisfaction.
       $(\mu_d > 0)$
       Since the conditions are satisfied, the sampling distribution of the difference
       can be modelled with a Student's *t*-model with 10 – 1 = 9 degrees of freedom.
       We will use a paired *t*-Test, with $\bar{d} = 8.5$ (after-before).

       Since the P-value = 0.0029 is low, we
       reject the null hypothesis. There is
       evidence that the mean job
       satisfaction rating has increased since
       the implementation of the exercise
       program.

$$t = \frac{\bar{d} - 0}{\frac{s_d}{\sqrt{n}}}$$

$$t = \frac{8.5 - 0}{\frac{7.47217}{\sqrt{10}}}$$

$$t \approx 3.60$$

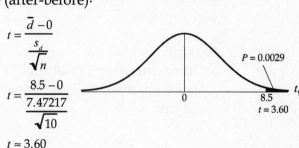
P = 0.0029
t_9
8.5
t = 3.60

c) Since we have rejected the null hypothesis, we would be making a Type I error if we are wrong.

d) H_0: The median difference in before and after job satisfaction scores is zero.

H_A: The median difference in before and after job satisfaction scores is greater than zero.

Taking differences in Job Satisfaction Index as After – Before, we get the following: –, +, +, –, +, +, +, +, +, + (no ties). The p-value for 8 +'s and 2's is $P(X = 0) + P(X = 1) + P(X = 2) = 0.0547$, where X is the number of –'s. Thus, we would not reject the null hypothesis, unlike in part b).

21. Yogourt

H_0: The mean difference in calories between servings of strawberry and vanilla yogourt is zero. $(\mu_d = 0)$

H_A: The mean difference in calories between servings of strawberry and vanilla yogourt is different from zero. $(\mu_d \neq 0)$

Paired Data Assumption: The yogourt is paired by brand.

Randomization Condition: Assume that these brands are representative of all brands.

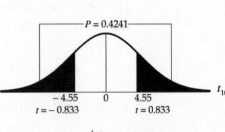

S-V   S-V

With outlier   **Without outlier**

Normal Population Assumption: The histogram of differences in calorie content between strawberry and vanilla shows an outlier, brand Great Value. When the outlier is eliminated, the histogram of differences is roughly unimodal and symmetric. When Great Value yogourt is removed, the conditions are satisfied. The sampling distribution of the difference can be modelled with a Student's t-model with

$11 - 1 = 10$ degrees of freedom.

We will use a paired t-Test, with $\bar{d} \approx 4.54545$.

Since the P-value = 0.4241 is high, we fail to reject the null hypothesis. There is no evidence of a mean difference in calorie content between strawberry yogourt and vanilla yogourt.

$$t = \frac{\bar{d} - 0}{\frac{s_d}{\sqrt{n}}}$$

$$t = \frac{4.54545 - 0}{\frac{18.0907}{\sqrt{11}}}$$

$$t \approx 0.833$$

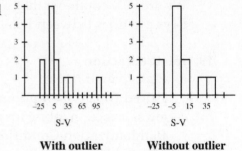

23. Braking

a) **Randomization Condition**: These stops are probably representative of all such stops for this type of car, but not for all cars.

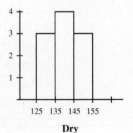

Dry

Nearly Normal Condition: A histogram of the stopping distances is roughly unimodal and symmetric.

The stops in the sample had a mean stopping distance of 139.4 feet, and a standard deviation of 8.09938 feet. Since the conditions have been satisfied, construct a one-sample *t*-interval, with 10 – 1 = 9 degrees of freedom, at 95% confidence.

$$\bar{y} \pm t^*_{n-1}\left(\frac{s}{\sqrt{n}}\right) = 139.4 \pm t^*_9\left(\frac{8.09938}{\sqrt{10}}\right) \approx (133.6, 145.2)$$

We are 95% confident that the mean dry pavement stopping distance for this type of car is between 133.6 and 145.2 feet.

b) **Independent Groups Assumption**: The wet pavement stops and dry pavement stops were made under different conditions and not paired in any way.
Randomization Condition: These stops are probably representative of all such stops for this type of car, but not for all cars.

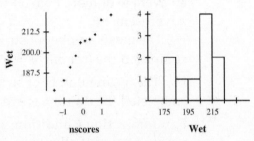

Nearly Normal Condition: The histogram of dry pavement stopping distances is roughly unimodal and symmetric (from part a), but the histogram of wet pavement stopping distances is a bit skewed. Since the Normal probability plot looks fairly straight, we will proceed.

Since the conditions are satisfied, it is appropriate to model the sampling distribution of the difference in means with a Student's *t*-model, with 13.8 degrees of freedom (from the approximation formula). We will construct a two-sample *t*-interval, with 95% confidence.

$$\left(\bar{y}_W - \bar{y}_D\right) \pm t^*_{df}\sqrt{\frac{s_W^2}{n_W} + \frac{s_D^2}{n_D}} = (202.4 - 139.4) \pm t^*_{13.8}\sqrt{\frac{15.07168^2}{10} + \frac{8.09938^2}{10}} \approx (51.4, 74.6)$$

We are 95% confident that the mean stopping distance on wet pavement is between 51.4 and 74.6 feet longer than the mean stopping distance on dry pavement.

25. **Math and gender 2009**
    a) **Randomization Condition**: These countries are not random, having been selected from writers of the PISA, but they are probably representative of all countries.
    Nearly Normal Condition: A histogram shows a roughly normal distribution with some left skew.
    If we consider each country as a data point, we can model this situation with a paired *t*-Test since males and females both come from the same country.

$$\bar{d} \pm t^{*}_{n-1}\left(\frac{s_d}{\sqrt{n}}\right) = 10.1875 \pm t^{*}_{15}\,(9.894/\sqrt{16}) = (5.851, 14.524)$$

b) Since the confidence interval does not include zero, we have strong evidence that there is a non-zero difference between genders in math scores.
Since the confidence interval includes 10, we cannot conclude that the mean difference is at least 10 points.

c) It is an estimate of average difference in MATH scores, not scores.

27. Strike three

a) Since 60% of 50 pitches is 30 pitches, the Little Leaguers would have to throw an average of more than 30 strikes to give support to the claim made by the advertisements.

H_0: The mean number of strikes thrown by Little Leaguers who have completed the training is 30. ($\mu_A = 30$)

H_A: The mean number of strikes thrown by Little Leaguers who have completed the training is greater than 30. ($\mu_A > 30$)

Randomization Condition: Assume that these players are representative of all Little League pitchers.

Nearly Normal Condition: The histogram of the number of strikes thrown after the training is roughly unimodal and symmetric.

After

The pitchers in the sample threw a mean of 33.15 strikes, with a standard deviation of 2.32322 strikes. Since the conditions for inference are satisfied, we can model the sampling distribution of the mean number of strikes thrown with a Student's *t* model, with 20 – 1 = 19 degrees of freedom. We will perform a one-sample *t*-Test.

$$t = \frac{\bar{y}_A - \mu_0}{\dfrac{s_A}{\sqrt{n_A}}}$$

Since the P-value = 3.92×10^{-6} is very low, we reject the null hypothesis. There is strong evidence that the mean number of strikes that Little Leaguers can throw after the training is more than 30. (This test says nothing about the effectiveness of the training; just that Little Leaguers can throw more than 60% strikes on average after completing the training. This might not be an improvement.)

$$t = \frac{33.15 - 30}{\dfrac{2.32322}{\sqrt{20}}}$$

$$t = 6.06$$

b) H_0: The mean difference in number of strikes thrown before and after the training is zero. $(\mu_d = 0)$

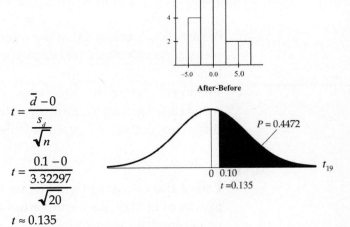

After-Before

H_A: The mean difference in number of strikes thrown before and after the training is greater than zero. $(\mu_d > 0)$

Paired Data Assumption: The data are paired by pitcher.

$$t = \frac{\bar{d} - 0}{\frac{s_d}{\sqrt{n}}}$$

$$t = \frac{0.1 - 0}{\frac{3.32297}{\sqrt{20}}}$$

$$t \approx 0.135$$

$P = 0.4472$

t_{19}

$0 \quad 0.10$

$t = 0.135$

Randomization Condition: Assume that these players are representative of all Little League pitchers.

Normal Population Assumption: The histogram of differences is roughly unimodal and symmetric.

Since the conditions are satisfied, the sampling distribution of the difference can be modelled with a Student's t-model with $20 - 1 = 19$ degrees of freedom.

We will use a paired t-*Test*, with $\bar{d} = 0.1$.

Since the P-value = 0.4472 is high, we fail to reject the null hypothesis. There is no evidence of a mean difference in number of strikes thrown before and after the training. The training does not appear to be effective.

c) H_0: The median difference in number of strikes thrown before and after training is zero.

H_A: The median difference in number of strikes thrown before and after training is greater than zero.

Taking differences in number of strikes thrown as After – Before, we get the following: $+,+,+,-,-,-,0,+,+,+,0,+,-,-,-,-,+,-,-,-$ (two ties). The

p-value for 8 +'s and 10 –'s is $P(X = 0) + P(X = 1) + \cdots + P(X = 10) = 0.760$,

where X is the number of –'s. Thus, we would not reject the null hypothesis, just like in part b).

29. Wheelchair marathon 2010

a) The data are certainly paired. Even if the individual times show a trend of improving speed over time, the differences may well be independent of each other. They are subject to random year-to-year fluctuations, and we may believe that these data are representative of similar races. We don't have any information with which to check the Nearly Normal Condition.

b) $\bar{d}\pm t^*_{n-1}\left(\dfrac{S_d}{\sqrt{n}}\right)=-3.57\pm t^*_{33}\left(\dfrac{35.083}{\sqrt{34}}\right)=(-15.81,8.67)$

We are 95% confident that the interval –15.51 to 8.67 minutes contains the true mean time difference for women's wheelchair times and men's running times.

c) The interval contains zero, so we would not reject a null hypothesis of no mean difference at a significance level of 0.05. We are unable to discern a difference between the female wheelchair times and the male running times.

31. **BST**
a) **Paired Data Assumption**: We are testing the same cows, before and after injections of BST. **Randomization Condition**: These cows are likely to be representative of all Ayrshires.

   Normal Population Assumption: We don't have the list of individual differences, so we can't look at a histogram. The sample is large, so we may proceed.
   Since the conditions are satisfied, the sampling distribution of the difference can be modelled with a Student's *t*-model with 60 – 1 = 59 degrees of freedom. We will find a paired *t*-interval, with 95% confidence.

b) $\bar{d}\pm t^*_{n-1}\left(\dfrac{S_d}{\sqrt{n}}\right)=14\pm t^*_{59}\left(\dfrac{5.2}{\sqrt{60}}\right)\approx(12.66,15.34)$

c) We are 95% confident that the mean increase in daily milk production for Ayrshire cows after BST injection is between 12.66 and 15.34 pounds.

d) 25% of 47 pounds is 11.75 pounds. According to the interval generated in part b), the average increase in milk production is more than this, so the farmer can justify the extra expense for BST.

33. **Sample size & margin of error** Width of the 95% confidence interval is 500, so the margin of error (*ME*) = 250.

$ME=t^*_{n-1}\dfrac{s}{\sqrt{n}}$

If the sample size is large enough the value of t^*_{n-1} will be close to $z^*=1.96$.

Substituting these values we have $\sqrt{n}=1.96\times\dfrac{1139.6}{250}=8.934464$ and *n* = 80. This sample size is large enough for the use of z^* in place of t^*_{n-1}.

35. **Sample size and power** Using MINITAB, to attain a power of 0.90 for the test of hypothesis when there is an actual annual mean mileage reduction of 500 miles, we need a sample of size 57.
Power and Sample Size
1-Sample *t-Test*

Testing mean = null (versus not = null)
Calculating power for mean = null + difference
Alpha = 0.05
Assumed standard deviation = 1139.6

Sample Target
Difference	Size	Power	Actual Power
500	57	0.9	0.902376

The probability that the test will conclude that the mean difference is non-zero when the true mean difference is 500 is 0.90.

1. **Which test?**
   a) Chi-square test of independence. We have one sample and two variables. We want to see if the variable *account type* is independent of the variable *trade type*.
   b) Some other statistics test. The variable *account size* is quantitative, not counts.
   c) Chi-square test of homogeneity. We have two samples (residential and nonresidential students), and one variable, *courses*. We want to see if the distribution of *courses* is the same for the two groups.

3. **Dice**
   a) If the die were fair, you'd expect each face to show 10 times.
   b) Use a chi-square test for goodness-of-fit. We are comparing the distribution of a single variable (outcome of a die roll) to an expected distribution.
   c) H_0: The die is fair. (All faces have the same probability of coming up.)
      H_A: The die is not fair. (Some faces are more or less likely to come up than others.)
   d) **Counted Data Condition:** We are counting the number of times each face comes up.
      Randomization Condition: Die rolls are random and independent of each other.
      Expected Cell Frequency Condition: We expect each face to come up 10 times, and 10 is greater than 5.
   e) Under these conditions, the sampling distribution of the test statistic is χ^2 on $6 - 1 = 5$ degrees of freedom. We will use a chi-square goodness-of-fit test.
   f)

Face	Observed	Expected	Residual = $(Obs - Exp)$	$(Obs - Exp)^2$	Component = $\dfrac{(Obs - Exp)^2}{Exp}$
1	11	10	1	1	0.1
2	7	10	–3	9	0.9
3	9	10	–1	1	0.1
4	15	10	5	25	2.5
5	12	10	2	4	0.4
6	6	10	–4	16	1.6

   $$\sum = 5.6$$

   g) Since the *P*-value = 0.3471 is high, we fail to reject the null hypothesis. There is no evidence that the die is unfair.

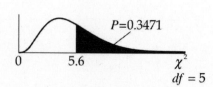

5. **Nuts**
   a) The weights of the nuts are quantitative. Chi-square goodness-of-fit requires counts.
   b) In order to use a chi-square test, you could count the number of each type of nut. However, it's not clear whether the company's claim was a percentage by number or a percentage by weight.

7. **Blood type**

   H_0: The distribution of blood types of residents on Bongainville Island represents the distribution of blood types of the general population.

   H_A: The distribution of blood types of residents on Bongainville Island does not represent the distribution of blood types of the general population.

   Counted Data Condition: The data are counts.

   Randomization Condition: These blood types are representative of all blood types.

   Expected Cell Frequency Condition: One cell has expected count less than 5, the chi-square procedures are not appropriate. Cells would have to be combined in order to proceed.

Blood Type	Observed	Expected
A	74	48.72
B	12	11.6
AB	11	3.48
O	19	52.2

   It seems reasonable to combine Blood type B with blood type AB.

   Once the cells are combined, all expected counts are greater than 5.

   Under these conditions, the sampling distribution of the test statistic is χ^2 on 3 – 1 =

Blood Type	Observed	Expected
A	74	48.72
B & AB	12 + 11 = 23	11.6 + 3.48 = 15.08
O	19	52.2

   2 degrees of freedom. We will use a chi-square goodness-of-fit test.

   $$\chi^2 = \sum_{\text{all cells}} \frac{(\text{Obs} - \text{Exp})^2}{\text{Exp}} = \frac{(74 - 48.72)^2}{48.72} + \frac{(23 - 15.08)^2}{15.08} + \frac{(19 - 52.2)^2}{52.2} \approx 38.39$$

   With $\chi^2 \approx 38.39$, on 3 degrees of freedom, the *P*-value = 0.0000 is small, so we reject the null hypothesis. There is strong evidence that the distribution of blood types of residents on Bongainville Island does not represent the distribution of blood types of the general population.

9. **Fruit flies**
   a) H_0: The ratio of traits in this type of fruit fly is 9:3:3:1, as genetic theory predicts.

   H_A: The ratio of traits in this type of fruit fly is not 9:3:3:1.

   Counted Data Condition: The data are counts.

   Randomization Condition: Assume that these flies are representative of all fruit flies of this type.

trait	Observed	Expected
YN	59	56.25
YS	20	18.75
EN	11	18.75
ES	10	6.25

Expected Cell Frequency Condition: The expected counts are all greater than 5.

Under these conditions, the sampling distribution of the test statistic is χ^2 on $4 - 1 = 3$ degrees of freedom. We will use a chi-square goodness-of-fit test.

$$\chi^2 = \sum_{all\ cells} \frac{(Obs - Exp)^2}{Exp} = \frac{(59 - 56.25)^2}{56.25} + \frac{(20 - 18.75)^2}{18.75} + \frac{(11 - 18.75)^2}{18.75} + \frac{(10 - 6.25)^2}{6.25} \approx 5.671$$

With $\chi^2 \approx 5.671$, on 3 degrees of freedom, the P-value = 0.1288 is high, so we fail to reject the null hypothesis. There is no evidence that the ratio of traits is different than the theoretical ratio predicted by the genetic model. The observed results are consistent with the genetic model.

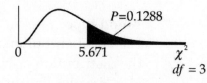

trait	Observed	Expected
YN	118	112.5
YS	40	37.5
EN	22	37.5
ES	20	12.5

b) With $\chi^2 \approx 11.342$, on 3 degrees of freedom, the P-value = 0.0100 is low, so we reject the null hypothesis. There is strong evidence that the ratio of traits is different than the theoretical ratio predicted by the genetic model. Specifically, there is evidence that the normal wing length occurs less frequently than expected and the short wing length occurs more frequently than expected.

c) At first, this seems like a contradiction. We have two samples with exactly the same ratio of traits. The smaller of the two provides no evidence of a difference, yet the larger one provides strong evidence of a difference. This is explained by the sample size. In general, large samples decrease the proportion of variation from the true ratio. Because of the relatively small sample in the first test, we are unwilling to say that there is a difference. There just isn't enough evidence. But the larger sample allows us to be more certain about the difference.

11. **Hurricane frequencies**

a) We would expect $96/16 = 6$ hurricanes per time period.

b) We are comparing the distribution of the number of hurricanes, a single variable, to a theoretical distribution. A Chi Square test for goodness-of-fit is appropriate.

c) H_0: The number of large hurricanes remains constant over decades.
H_A: The number of large hurricanes has changed.

d) There are 16 time periods, so there are $16 - 1 = 15$ degrees of freedom.

e) $P(\chi^2_{df=15} > 12.67) \approx 0.63$

f) The very high P-value means that these data offer no evidence that the number of large hurricanes has changed.

g) If the final period is only 6 years rather than 10, and already 7 hurricanes have been observed, perhaps this decade will have an unusually large number of such hurricanes.

13. Childbirth, part 1

a) There are two variables, breastfeeding and having an epidural, from a single group of births. We will perform a chi square test for independence.

b) H_0: Breastfeeding success is independent of having an epidural.
H_A: There is an association between breastfeeding success and having an epidural.

15. Childbirth, part 2

a) The table has 2 rows and 2 columns, so there is $(2-1) \times (2-1) = 1$ degree of freedom.

b) We expect $\dfrac{474}{1178} \approx 40.2\%$ of all babies to not be breastfeeding after six months, so we expect that 40.2% of the 396 epidural babies, or 159.34, to not be breastfeeding after six months.

c) Breastfeeding behaviour should be independent for these babies. They are fewer than 10% of all babies, and we assume they are representative of all babies. We have counts, and all the expected cells are at least 5.

17. Childbirth, part 3

a) $\dfrac{\left(Obs - Exp\right)^2}{Exp} = \dfrac{\left(190 - 159.34\right)^2}{159.34} = 5.90$

b) $P(\chi^2_{df=1} > 14.87) < 0.0001$

c) The *P*-value is very low, so reject the null hypothesis. There's strong evidence of an association between having an epidural and subsequent success in breastfeeding.

19. Childbirth, part 4

a) $c = \dfrac{Obs - Exp}{\sqrt{Exp}} = \dfrac{190 - 159.34}{\sqrt{159.34}} = 2.43$

b) It appears that babies whose mothers had epidurals during childbirth are much more likely to not be breastfeeding six months later.

21. Childbirth, part 5
These factors would not have been mutually exclusive. There would be yes or no responses for every baby for each.

23. *Titanic*

a) $P(\text{crew}) = \dfrac{885}{2201} \approx 0.402$

b) $P(\text{third and alive}) = \dfrac{178}{2201} \approx 0.081$

c) $P(\text{alive} \mid \text{first}) = \dfrac{P(\text{alive and first})}{P(\text{first})} = \dfrac{{}^{202}\!/_{2201}}{{}^{325}\!/_{2201}} = \dfrac{202}{325} \approx 0.622$

d) The overall chance of survival is $\dfrac{710}{2201} \approx 0.323$, so we would expect about 32.3% of the crew, or about 285.48 members of the crew, to survive.

e) H_0: Survival was independent of status on the ship.
    H_A: Survival depended on status on the ship.

f) The table has 2 rows and 4 columns, so there are $(2-1) \times (4-1) = 3$ degrees of freedom.

g) With $\chi^2 \approx 187.8$, on 3 degrees of freedom, the P-value is essentially 0, so we reject the null hypothesis. There is strong evidence that survival depended on status. First-class passengers were more likely to survive than any other class or crew.

25. Titanic again

First class passengers were most likely to survive, while third class passengers and crew were under-represented among the survivors.

27. Birth order and study choice

a) This is a chi-square test for independence. There is one sample of students, and two variables, *Birth Order* and *Discipline*. We want to know if *Discipline* choice is independent of *Birth Order*.

b) H_0: Discipline choice is independent of birth order.
    H_A: There is an association between discipline choice and birth order.

c) **Counted Data Condition:** The data are counts.
    Randomization Condition: This is not a random sample of students, but there is no reason to think that this group of students isn't representative, at least of students in a Statistics class.
    Expected Cell Frequency Condition: The expected counts are low for both the Social Science and Professional disciplines for both third and fourth or higher birth order. We'll keep an eye on these when we calculate the standardized residuals.

d) There are 4 rows and 4 columns, for $(4 - 1)(4 - 1) = 9$ degrees of freedom.

e) With a P-value this low, we reject the null hypothesis. There is some evidence of an association between birth order and discipline choice.

f) Unfortunately, 3 of the 4 largest standardized residuals are in cells with expected counts less than 5. We should be very wary of drawing conclusions from this test.

29. Cranberry juice

a) This is an experiment. Volunteers were assigned to treatment groups, each of which drank a different beverage.

b) We are concerned with the proportion of urinary tract infections among three different groups. We will use a chi-square test for homogeneity.

c) H_0: The proportion of urinary tract infection is the same for each group.
   H_A: The proportion of urinary tract infection is different among the groups.

d) **Counted Data Condition:** The data are counts.
   Randomization Condition: Although not specifically stated, we will assume that the women were randomly assigned to treatments.
   Expected Cell Frequency Condition: The expected counts are all greater than 5.

	Cranberry (Obs / Exp)	Lactobacillus (Obs / Exp)	Control (Obs / Exp)
Infection	8 / 15.333	20 / 15.333	18 / 15.333
No infection	42 / 34.667	30 / 34.667	32 / 34.667

e) The table has 2 rows and 3 columns, so there are $(2-1) \times (3-1) = 2$ degrees of freedom.

f) $\chi^2 = \sum\limits_{all\ cells} \dfrac{(Obs - Exp)^2}{Exp} \approx 7.776$

   P-value ≈ 0.020.

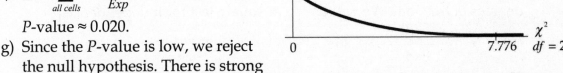

g) Since the P-value is low, we reject the null hypothesis. There is strong evidence of difference in the proportion of urinary tract infections for cranberry juice drinkers, lactobacillus drinkers, and women that drink neither of the two beverages.

h) A table of the standardized residuals is below, calculated by using $c = \dfrac{Obs - Exp}{\sqrt{Exp}}$.

	Cranberry	Lactobacillus	Control
Infection	−1.87276	1.191759	0.681005
No infection	1.245505	−0.79259	−0.45291

There is evidence that women who drink cranberry juice are less likely to develop urinary tract infections, and women who drank lactobacillus are more likely to develop urinary tract infections.

31. Computer-assisted ordering

a) We have one group, categorized according to two variables, *computer assisted ordering* and *delivery time*, so we will perform a chi-square test for independence.

b) H_0: delivery time is independent from computer-assisted ordering.

H_A: There is an association between delivery time and computer-assisted ordering.

c) **Counted Data Condition:** The data are counts.

Randomization Condition: Although not specifically stated, we will assume that the sampling of 40 firms was conducted randomly.

	Below Industry Average (Obs / Exp)	Above Industry Average (Obs / Exp)	Above Industry Average (Obs / Exp)
Not computer-assisted	4 / 8.4	12 / 9.6	8 / 6
Computer-assisted	10 / 5.6	4 / 6.4	2 / 4

	Below Industry Average (Obs / Exp)	Above Industry Average (Obs / Exp)
Not computer-assisted	4 / 8.4	20 / 15.6
Computer-assisted	10 / 5.6	6 / 10.4

Expected Cell Frequency Condition: One cell has expected count less than 5, the chi-square procedures are not appropriate. Cells would have to be combined in order to proceed. It seems reasonable to combine last two columns.

Once the cells are combined, all expected counts are greater than 5.

d) Under these conditions, the sampling distribution of the test statistic is χ^2 on 1 degrees of freedom. We will use a chi-square test for independence.

$$\chi^2 = \sum_{\text{all cells}} \frac{(\text{Obs} - \text{Exp})^2}{\text{Exp}} \approx 8.86$$

The *P*-value ≈ 0.003

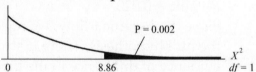

e) Since the *P*-value ≈ 0.003 is small, we reject the null hypothesis. There is evidence of an association between delivery time and computer-assisted ordering.

33. Grades

a) We have two groups, students of Professor Alpha and students of Professor Beta, and we are concerned with the distribution of one variable, *grade*. We will perform a chi-square test for homogeneity.

b) H_0: The distribution of grades is the same for the two professors.

H_A: The distribution of grades is different for the two professors.

c) The expected counts are organized in the table below:

	Prof. Alpha	Prof. Beta
A	6.667	5.333
B	12.778	10.222
C	12.222	9.778
D	6.111	4.889
F	2.222	1.778

Since three cells have expected counts less than 5, the chi-square procedures are not appropriate. Cells would have to be combined in order to proceed. (We will do this in another exercise.)

35. **Grades again**
    a) **Counted Data Condition:** The data are counts. **Randomization Condition:** Assume that these students are representative of all students that have ever taken courses from the professors. **Expected Cell Frequency Condition:** The expected counts are all greater than 5.

	Prof. Alpha (Obs / Exp)	Prof. Beta (Obs / Exp)
A	3 / 6.667	9 / 5.333
B	11 / 12.778	12 / 10.222
C	14 / 12.222	8 / 9.778
Below C	12 / 8.333	3 / 6.667

    b) Under these conditions, the sampling distribution of the test statistic is χ^2 on 3 degrees of freedom, instead of 4 degrees of freedom before the change in the table. We will use a chi-square test for homogeneity.

    c) With $\chi^2 = \sum\limits_{all\ cells} \dfrac{(Obs - Exp)^2}{Exp} \approx 9.306$, the P-value ≈ 0.0255.

    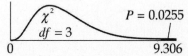

    Since P-value = 0.0255 is low, we reject the null hypothesis. There is evidence that the grade distributions for the two professors are different. Professor Alpha gives fewer as and more grades below C than Professor Beta.

37. **Eating in front of the TV**
    a) Displays may vary. Stacked bar charts or comparative pie charts are appropriate.
    b) H_0: Response to the statement is independent of age.
    H_A: There is an association between the response to the statement and age.
    Counted Data Condition: The data are counts.
    Randomization Condition: These data are from a random sample.
    Expected Cell Frequency Condition: All of the expected counts are much greater than 5.
    Under these conditions, the sampling distribution of the test statistic is χ^2 on 8 degrees of freedom. We will use a chi-square test for independence.

    With $\chi^2 = \sum\limits_{all\ cells} \dfrac{(Obs - Exp)^2}{Exp} \approx 190.96$, the P-value < 0.001.

    Since the P-value is so small, reject the null hypothesis. There is strong evidence to suggest that response is not independent of age.

39. **Racial steering**

    H_0: There is no association between race and section of the complex in which people live.

    H_A: There is an association between race and section of the complex in which people live.

    Counted Data Condition: The data are counts.

    Randomization Condition: Assume that the recently rented apartments are representative of all apartments in the complex.

    Expected Cell Frequency Condition: The expected counts are all greater than 5.

	White (Obs / Exp)	Black (Obs / Exp)
Section A	87 / 76.179	8 / 18.821
Section B	83 / 93.821	34 / 23.179

    Under these conditions, the sampling distribution of the test statistic is χ^2 on 1 degree of freedom. We will use a chi-square test for independence.

    $$\chi^2 = \sum_{all\ cells} \frac{(Obs - Exp)^2}{Exp} \approx \frac{(87 - 76.179)^2}{76.179} + \frac{(8 - 18.821)^2}{18.821} + \frac{(83 - 93.821)^2}{93.821} + \frac{(34 - 23.179)^2}{23.179}$$

    $$\approx 1.5371 + 6.2215 + 1.2481 + 5.0517$$

    $$\approx 14.058$$

    With $\chi^2 \approx 14.058$, on 1 degree of freedom, the P-value ≈ 0.0002.

    Since the P-value ≈ 0.0002 is low, we reject the null hypothesis. There is strong evidence of an association between race and the section of the

    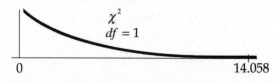

    apartment complex in which people live. An examination of the components shows us that white renters are much more likely to rent in Section A (component = 6.2215), and black renters are much more likely to rent in Section B (component = 5.0517).

41. **Steering revisited**

    a) H_0: The proportion of white renters who live in Section A is the same as the proportion of black renters who live in Section A.
    $(p_W = p_B \text{ or } p_W - p_B = 0)$

    H_A: The proportion of white renters who live in Section A is different than the proportion of black renters who live in Section A.
    $(p_W \neq p_B \text{ or } p_W - p_B \neq 0)$

    Independence assumption: Assume that people rent independent of the section.
    Independent samples condition: The groups are not associated.
    Success/Failure condition: $n\hat{p}$ (white) = 87, $n\hat{q}$ (white) = 83, $n\hat{p}$ (black) = 8, and $n\hat{q}$ (black) = 34. These are not all greater than 10, since the number of black renters in Section A is only 8, but it is close to 10, and the others are large. It should be safe to proceed.

Since the conditions have been satisfied, we will model the sampling distribution of the difference in proportion with a Normal model with mean 0 and standard deviation estimated by

$$SE_{pooled}(\hat{p}_W - \hat{p}_B) = \sqrt{\frac{\hat{p}_{pooled}\hat{q}_{pooled}}{n_W} + \frac{\hat{p}_{pooled}\hat{q}_{pooled}}{n_B}} = \sqrt{\frac{\left(\frac{95}{212}\right)\left(\frac{117}{212}\right)}{170} + \frac{\left(\frac{95}{212}\right)\left(\frac{117}{212}\right)}{42}} \approx 0.0856915 \, .$$

The observed difference between the proportions is: 0.5117647 – 0.1904762 = 0.3212885.

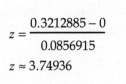

$$z = \frac{0.3212885 - 0}{0.0856915}$$

$$z \approx 3.74936$$

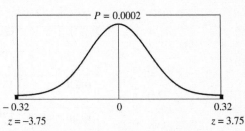

(You have to use a *ridiculous* number of decimal places to get this to come out "right." This is to done to illustrate the point of the question. DO NOT DO THIS! Use technology.)

Since the *P*-value = 0.0002 is low, we reject the null hypothesis. There is strong evidence of a difference in the proportion of white renters and black renters living in Section A. The evidence suggests that the proportion of all white residents living in Section A is much higher than the proportion of all black residents living in Section A.

The value of *z* for this test was approximately 3.74936. $z^2 \approx (3.74936)^2 \approx 14.058$, the same as the value for χ^2 in Exercise 37.

b) The resulting *P*-values were both approximately 0.0002. The two tests are equivalent.

43. Race and education

H_0: Race is independent of education level.

H_A: There is an association between race and education level.

Counted Data Condition: The data are counts.

Randomization Condition: Assume that the sample was taken randomly.

Expected Cell Frequency Condition: The expected counts are all greater than 5.

Under these conditions, the sampling distribution of the test statistic is χ^2 on 9 degrees of freedom. We will use a chi-square test for independence.

With $\chi^2 = \sum_{all\ cells} \frac{(Obs - Exp)^2}{Exp} \approx 2815.968$, on 9 degrees of freedom, the *P*-value is essentially 0.

Since the *P*-value is so low, we reject the null hypothesis. There is strong evidence of an association between race and education level. Hispanic people are more likely to not have a high school diploma, and fewer white people fail to graduate from high school than expected.

45. Race and education, part 2

H_0: Race is independent of education level.

H_A: There is an association between race and education level.

Counted Data Condition: The data are counts.

Randomization Condition: Assume that the sample was taken randomly.

	HS Diploma (Obs / Exp)	College Grad (Obs / Exp)	Adv. Degree (Obs / Exp)
Black	1598 / 1605.1	549 / 538.01	117 / 120.93
Hispanic	1269 / 1261.9	412 / 422.99	99 / 95.074

Expected Cell Frequency Condition: The expected counts are all greater than 5. Under these conditions, the sampling distribution of the test statistic is χ^2 on 2 degrees of freedom. We will use a chi-square test for independence.

With $\chi^2 = \sum\limits_{all\ cells} \dfrac{(Obs - Exp)^2}{Exp} \approx 0.870$, on 2 degrees of freedom, the P-value ≈ 0.6471.

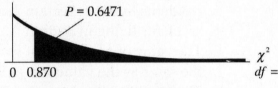

Since the P-value is high, we fail to reject the null hypothesis. There is no evidence of an association between race and educational performance. Black people and Hispanic people appear to have similar distributions of educational opportunity.

47. Ranking universities

H_0: The top 100 universities have the same distribution by region as the top 500 universities.

H_A: The top 100 universities have a different distribution by region than the top 500 universities.

Counted Data Condition: The number of universities in each region is reported.

Randomization Condition: We aren't attempting to generalize our findings, so a random sample is not required.

Expected Cell Frequency Condition: The expected counts are all much greater than 5.

Under these conditions, the sampling distribution of the test statistic is χ^2 on 3 – 1 = 2 degrees of freedom. We will use a chi-square goodness-of-fit test.

Region	Observed	Expected
Americas	55	39.8406
Europe	37	41.6335
Other	8	18.5259

$$\chi^2 = \sum_{all\ cells} \frac{(Obs - Exp)^2}{Exp} = \frac{(55 - 39.8406)^2}{39.8406} + \frac{(37 - 41.6335)^2}{41.6335} + \frac{(8 - 18.5259)^2}{18.5259}$$

$$= 5.7681 + 0.5157 + 5.9805 \approx 12.26$$

With $\chi^2 \approx 12.26$, on 2 degrees of freedom, the *P*-value is 0.002, so we reject the null hypothesis. There is strong evidence that the top 100 universities have a different distribution by region than the top 500 universities. The top 100 universities seem to be more heavily concentrated in the Americas, and sparser in the Asia/Africa/Pacific region.

Part VI Review: Learning About the World

1. **Crawling**

   a) H₀: The mean age at which babies begin to crawl is the same whether the babies were born in January or July. $\left(\mu_{Jan} = \mu_{July} \text{ or } \mu_{Jan} - \mu_{July} = 0\right)$

   H_A: There is a difference in the mean age at which babies begin to crawl, depending on whether the babies were born in January or July.
   $\left(\mu_{Jan} \neq \mu_{July} \text{ or } \mu_{Jan} - \mu_{July} \neq 0\right)$

   Independent Groups Assumption: The groups of January and July babies are independent.

   Randomization Condition: Although not specifically stated, we will assume that the babies are representative of all babies.

   Nearly Normal Condition: We don't have the actual data, so we can't check the distribution of the sample. However, the samples are fairly large, so the robustness of the t procedure allows us to proceed.

   Since the conditions are satisfied, it is appropriate to model the sampling distribution of the difference in means with a Student's t-model, with 43.68 degrees of freedom (from the approximation formula).

   We will perform a two-sample t-test. The standard error of the difference of the means is: $SE(\bar{y}_{Jan} - \bar{y}_{July}) = \sqrt{\dfrac{7.08^2}{32} + \dfrac{6.91^2}{21}} \approx 1.9596$.

   The observed difference between the mean ages is 29.84 – 33.64 = –3.8 weeks.

   Since the P-value = 0.0590 is fairly low, we reject the null hypothesis. There is some evidence that mean age at which babies crawl is different for January and June babies. January babies appear to crawl a bit earlier than July babies, on average. Since the evidence is

   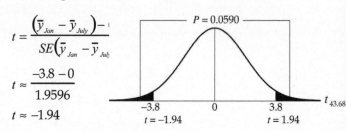

   not strong, we might want to do some more research into this claim.

   b) H₀: The mean age at which babies begin to crawl is the same whether the babies were born in April or October. $\left(\mu_{Apr} = \mu_{Oct} \text{ or } \mu_{Apr} - \mu_{Oct} = 0\right)$

   H_A: There is a difference in the mean age at which babies begin to crawl, depending on whether the babies were born in April or October.
   $\left(\mu_{Apr} \neq \mu_{Oct} \text{ or } \mu_{Apr} - \mu_{Oct} \neq 0\right)$

   The conditions (with minor variations) were checked in part a).

   Since the conditions are satisfied, it is appropriate to model the sampling distribution of the difference in means with a Student's t-model, with 59.40 degrees of freedom (from the approximation formula).

We will perform a two-sample *t*-test. The sampling distribution model has mean 0, with standard error: $SE(\bar{y}_{Apr} - \bar{y}_{Oct}) = \sqrt{\dfrac{6.21^2}{26} + \dfrac{7.29^2}{44}} \approx 1.6404$.

The observed difference between the mean ages is 31.84 – 33.35 = –1.51 weeks.

Since the P-value = 0.3610 is high, we fail to reject the null hypothesis. There is no evidence that mean age at which babies crawl is different for April and October babies.

$$t = \frac{(\bar{y}_{Apr} - \bar{y}_{Oct}) - (0)}{SE(\bar{y}_{Apr} - \bar{y}_{Oct})}$$

$$t \approx \frac{-1.51 - 0}{1.6404}$$

$$t \approx -0.92$$

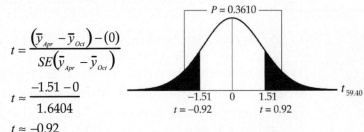

c) These results are not consistent with the researcher's claim. We have slight evidence in one test and no evidence in the other. The researcher will have to do better than this to convince us!

3. **BC birth weights**
   a) For first nations, the confidence interval is
      $\bar{y} \pm 1.96 s / \sqrt{n} = 3645 \pm 1.96(466) / \sqrt{20578}$ or (3638.6, 3651.4). For Immigrants of Chinese origin, the interval is (3386.7, 3399.3). Note that with such a large sample size, we may use normal approximation and use $z^* = 1.96$.
   b) To determine significance, we can simply check whether the value from the BC population, 3558, is in either confidence interval. It isn't, so both groups differ significantly from the BC population.
   c) We will perform a two-sample *z*-test since sample sizes are large and CLT applies. The sampling distribution model has mean 0, with standard error:

   $$SE(\bar{y}_{FN} - \bar{y}_{Chi}) = \sqrt{\frac{466^2}{20578} + \frac{391^2}{14586}} = 4.586.$$

   The observed difference between the mean ages is 3645 – 3393 = 252 grams, and the statistic 252/4.586 = 54.9. With such a large statistic, the P-value is small and we reject the null hypothesis that the mean difference in birth weights between First Nation and Immigrants of Chinese origin is 0. Using normal approximation, the confidence interval is $(\bar{y}_{FN} - \bar{y}_{Chi}) \pm 1.96 \times SE(\bar{y}_{FN} - \bar{y}_{Chi})$ or (243.01, 260.99).

5. **Pottery**
   Independent Groups Assumption: The pottery samples are from two different sites.
   Randomization Condition: It is reasonable to think that the pottery samples are representative of all pottery at that site with respect to aluminum oxide content.
   Nearly Normal Condition: The histograms of

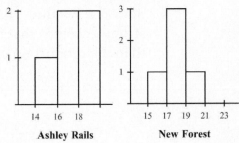

aluminum oxide content are roughly unimodal and symmetric.

Since the conditions are satisfied, it is appropriate to model the sampling distribution of the difference in means with a Student's t-model, with 7.96 degrees of freedom (from the approximation formula). We will use df = 7 for conservative reason. We will construct a two-sample t-interval, with 95% confidence.

$$(\bar{y}_{AR} - \bar{y}_{NF}) \pm t^*_{df}\sqrt{\frac{s^2_{AR}}{n_{AR}} + \frac{s^2_{NF}}{n_{NF}}} = (17.32 - 18.18) \pm t^*_{7}\sqrt{\frac{1.65892^2}{5} + \frac{1.77539^2}{5}} \approx (-3.37, 1.65)$$

We are 95% confident that the difference in the mean percentage of aluminum oxide content of the pottery at the two sites is between –3.37% and 1.65%. Since 0 is in the interval, there is no evidence that the aluminum oxide content at the two sites is different. It would be reasonable for the archaeologists to think that the same ancient people inhabited the sites.

7. **Gehrig**

a) H_0: The proportion of ALS patients who were athletes is the same as the proportion of patients with other disorders who were athletes.

$$\left(p_{ALS} = p_{Other} \text{ or } p_{ALS} - p_{Other} = 0\right)$$

H_A: The proportion of ALS patients who were athletes is greater than the proportion of patients with other disorders who were athletes.

$$\left(p_{ALS} > p_{Other} \text{ or } p_{ALS} - p_{Other} > 0\right)$$

Randomization Condition: This is NOT a random sample. We must assume that these patients are representative of all patients with neurological disorders.

10% Condition: 280 and 151 are both less than 10% of all patients with disorders.

Independent Samples Condition: The groups are independent.

Success/Failure Condition: $n\hat{p}$ (ALS) = (280)(0.38) = 106, $n\hat{q}$ (ALS) = (280)(0.62) = 174, $n\hat{p}$ (Other) = (151)(0.26) = 39, and $n\hat{q}$ (Other) = (151)(0.74) = 112 are all greater than 10, so the samples are both large enough.

Since the conditions have been satisfied, we will model the sampling distribution of the difference in proportion with a Normal model with mean 0 and standard deviation estimated by:

$$SE_{pooled}\left(\hat{p}_{ALS} - \hat{p}_{Other}\right) = \sqrt{\frac{(0.338)(0.662)}{280} + \frac{(0.338)(0.662)}{151}} \approx 0.0478$$

where $\hat{p}_{pooled} = \dfrac{0.38 \times 280 + 0.26 \times 151}{280 + 151} = 0.338$.

The observed difference between the proportions is 0.38 – 0.26 = 0.12.

Since the P-value = 0.0060 is very low, we reject the null hypothesis. There is strong evidence that the proportion of ALS patients who are athletes is greater than the proportion of patients with other disorders who are athletes.

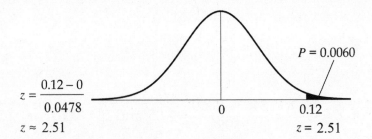

$$z = \frac{0.12 - 0}{0.0478}$$

$z \approx 2.51$

b) This was a retrospective observational study. In order to make the inference, we must assume that the patients studied are representative of all patients with neurological disorders.

9. **Genetics**

H_0: The proportions of traits are as specified by the ratio 1:3:3:9.
H_A: The proportions of traits are not as specified.
Counted Data Condition: The data are counts.
Randomization Condition: Assume that these students are representative of all people.
Expected Cell Frequency Condition: The expected counts (shown in the table) are all greater than 5.

Trait	Observed	Expected	Residual = $(Obs - Exp)$	$(Obs - Exp)^2$	Component = $\dfrac{(Obs - Exp)^2}{Exp}$
Attached, noncurling	10	7.625	2.375	5.6406	0.73975
Attached, curling	22	22.875	– 0.875	0.7656	0.03347
Free, noncurling	31	22.875	8.125	66.0156	2.8859
Free, curling	59	68.625	– 9.625	92.6406	1.35

$\sum \approx 5.01$

Under these conditions, the sampling distribution of the test statistic is χ^2 on 4 – 1 = 3 degrees of freedom. We will use a chi-square goodness-of-fit test.
$\chi^2 = 5.01$. Since the P-value = 0.1711 is high, we fail to reject the null hypothesis.
There is no evidence that the proportions of traits are anything other than 1:3:3:9.

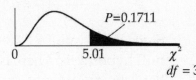

11. Babies

We test the hypotheses

$H_0 : \mu = 3360$

$H_a : \mu \neq 3360$

which gives a test statistic of $Z = \dfrac{\bar{x} - \mu}{\dfrac{s}{\sqrt{n}}} = \dfrac{3463 - 3360}{\dfrac{595}{\sqrt{112}}} = 1.85$ and $P = 0.067$, or $0.05 < P$

< 0.10 using normal approximation. There is only weak evidence against the null hypothesis that the means are the same. (We can use either z or t_{111} since the sample is very large and small difference between the two models)

13. Feeding fish

a) If there is no difference in the average fish sizes, the chance of observing a difference this large, or larger, just by natural sampling variation is 0.1%.

b) There is evidence that largemouth bass that are fed a natural diet are larger. The researchers would advise people who raise largemouth bass to feed them a natural diet.

c) If the advice is incorrect, the researchers have committed a Type I error.

15. Twins

H_0: There is no association between duration of pregnancy and level of prenatal care.

H_A: There is an association between duration of pregnancy and level of prenatal care.

	Preterm (induced or Caesarean) (Obs/Exp)	Preterm (without procedures) (Obs/Exp)	Term or postterm (Obs/Exp)
Intensive	18 /16.676	15 / 15.579	28 / 28.745
Adequate	46 / 42.101	43 / 39.331	65 / 72.568
Inadequate	12 / 17.223	13 / 16.090	38 / 29.687

Counted Data Condition: The data are counts.

Randomization Condition: Assume that these pregnancies are representative of all twin births.

Expected Cell Frequency Condition: The expected counts are all greater than 5. Under these conditions, the sampling distribution of the test statistic is χ^2 on 4 degrees of freedom. We will use a chi-square test for independence.

$\chi^2 = \displaystyle\sum_{all\ cells} \dfrac{(Obs - Exp)^2}{Exp} \approx 6.14$, and the P-value ≈ 0.1887.

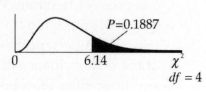

Since the P-value ≈ 0.1887 is high, we fail to reject the null hypothesis. There is no evidence of an association between duration of pregnancy and level of prenatal care in twin births.

17. **Age**
   a) **Independent Groups Assumption**: The group of patients with and without cardiac disease are not related in any way.
   Randomization Condition: Assume that these patients are representative of all people.

   Normal Population Assumption: We don't have the actual data, so we will assume that the population of ages of patients is Normal.
   Since the conditions are satisfied, it is appropriate to model the sampling distribution of the difference in means with a Student's *t*-model, with 670 degrees of freedom (from the approximation formula). We will construct a two-sample *t*-interval, with 95% confidence.

   $$(\bar{y}_{Card} - \bar{y}_{None}) \pm t^*_{df}\sqrt{\frac{s^2_{Card}}{n_{Card}} + \frac{s^2_{None}}{n_{None}}} = (74.0 - 69.8) \pm t^*_{670}\sqrt{\frac{7.9^2}{450} + \frac{8.7^2}{2397}} \approx (3.39, 5.01)$$

   We are 95% confident that the mean age of patients with cardiac disease is between 3.39 and 5.01 years higher than the mean age of patients without cardiac disease.
   b) Older patients are at greater risk for a variety of health problems. If an older patient does not survive a heart attack, the researchers will not know to what extent depression was involved, because there will be a variety of other possible variables influencing the death rate. Additionally, older patients may be more (or less) likely to be depressed than younger ones.

19. **Computer use**
   a) It is unlikely that an equal number of boys and girls were contacted strictly by chance. It is likely that this was a stratified random sample, stratified by gender.
   b) **Randomization Condition**: The teens were selected at random.
   10% Condition: 620 boys and 620 girls are both less than 10% of all teens.
   Independent Groups Assumption: The groups of boys and girls are not paired or otherwise related in any way.
   Success/Failure Condition: $n\hat{p}$ (boys) = (620)(0.77) = 477, $n\hat{q}$ (boys) = (620)(0.23) = 143, $n\hat{p}$ (girls) = (620)(0.65) = 403, and $n\hat{q}$ (girls) = (620)(0.35) = 217 are all greater than 10, so the samples are both large enough.
   Since the conditions have been satisfied, we will find a two-proportion *z*-interval.

   $$(\hat{p}_B - \hat{p}_G) \pm z^*\sqrt{\frac{\hat{p}_B\hat{p}_B}{n_B} + \frac{\hat{p}_G\hat{q}_G}{n_G}} = (0.77 - 0.65) \pm 1.960\sqrt{\frac{(0.77)(0.23)}{620} + \frac{(0.65)(0.35)}{620}} =$$

   $(0.070, 0.170)$
   We are 95% confident that the proportion of computer gamers is between 7.0% and 17.0% higher for boys than for girls.

c) Since the interval lies entirely above 0, there is evidence that a greater percentage of boys play computer games than girls.

21. **Forest fires**

a) There are $n = 100$ fires, with mean $\bar{y} = 6056$ and standard deviation $s = 10811.98$. The 90% confidence interval is $6065 \pm 1.65(10811.98) / \sqrt{100} =$ or $(4260.35, 7850.77)$.

b) The interval does contain the actual value of 6398. If we took 200 independent random samples from the population and counted the number of intervals which contained the actual mean 6398, this number would follow a binomial distribution with index $n = 200$ and probability $p = 0.10$.

c) The distribution is extremely skewed, as shown in the histogram below. However, the sample size is quite large, so we expect the Central Limit Theorem to apply. The confidence interval should be valid.

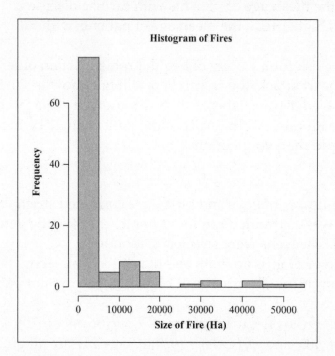

d) The median is an appropriate measure of centre in this case. The sign test could be used to assess whether the true median has a particular value. A confidence interval would contain all values for the median for which the result of the sign test is not significant.

e) A transformation is appropriate here. The histogram for the logarithm transformation is shown below, and the distribution is much more symmetric than before.

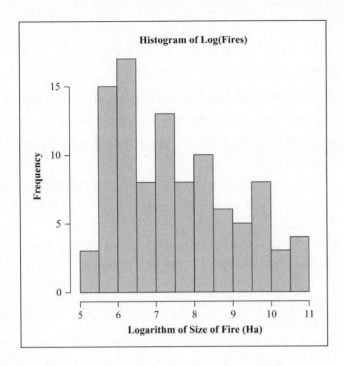

Histogram of Log(Fires)

23. Meals

H_0: The university student's mean daily food expense is \$10. ($\mu = 10$)

H_A: The university student's mean daily food expense is greater than \$10. ($\mu > 10$)

Randomization Condition: Assume that these days are representative of all days.

Nearly Normal Condition: The histogram of daily expenses is fairly unimodal and symmetric. It is reasonable to think that this sample came from a Normal population.

The expenses in the sample had a mean of 11.4243 dollars and a standard deviation of 8.05794 dollars. Since the conditions for inference are satisfied, we can model the sampling distribution of the mean daily expense with a Student's t model, with $14 - 1 = 13$

degrees of freedom: $t_{13}\left(10, \dfrac{8.05794}{\sqrt{14}}\right)$.

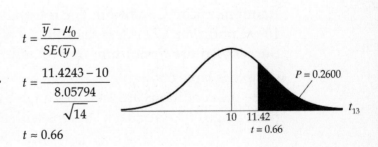

We will perform a one-sample t-test. Since the P-value = 0.2600 is high, we fail to reject the null hypothesis. There is no evidence that the student's average spending is more than \$10 per day.

$$t = \frac{\overline{y} - \mu_0}{SE(\overline{y})}$$

$$t = \frac{11.4243 - 10}{\dfrac{8.05794}{\sqrt{14}}}$$

$$t \approx 0.66$$

25. Teach for America

H_0: The mean score of students with certified teachers is the same as the mean score of students with uncertified teachers. $(m_C = m_U \text{ or } m_C - m_U = 0)$

H_A: The mean score of students with certified teachers is greater than the mean score of students with uncertified teachers. $(m_C > m_U \text{ or } m_C - m_U > 0)$

Independent Groups Assumption: The certified and uncertified teachers are independent groups.

Randomization Condition: Assume the students studied were representative of all students.

Nearly Normal Condition: We don't have the actual data, so we can't look at the graphical displays, but the sample sizes are large, so we can proceed.

Since the conditions are satisfied, it is appropriate to model the sampling distribution of the difference in means with a Student's t-model, with 86 degrees of freedom (from the approximation formula).

We will perform a two-sample t-test. The sampling distribution model has mean 0, with standard error: $SE(\bar{y}_C - \bar{y}_U) = \sqrt{\dfrac{9.31^2}{44} + \dfrac{9.43^2}{44}} \approx 1.9977$.

The observed difference between the mean scores is 35.62 – 32.48 = 3.14.

Since the P-value = 0.0598 is fairly high, we fail to reject the null hypothesis. There is little evidence that students with certified teachers had mean scores higher than students with uncertified teachers.

$$t = \frac{(\bar{y}_c - \bar{y}_U) - (0)}{SE(\bar{y}_c - \bar{y}_U)}$$

$$t \approx \frac{3.14}{1.9977}$$

$$t \approx 1.57$$

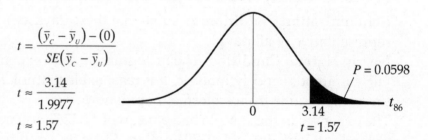

However, since the P-value is not extremely high, further investigation is recommended.

27. Streams

Randomization Condition: The researchers randomly selected 172 streams.

10% Condition: 172 is less than 10% of all streams.

Success/Failure Condition: $n\hat{p} = 69$ and $n\hat{q} = 103$ are both greater than 10, so the sample is large enough.

Since the conditions are met, we can use a one-proportion z-interval to estimate the percentage of Adirondack streams with a shale substrate.

$$\hat{p} \pm z^* \sqrt{\frac{\hat{p}\hat{q}}{n}} = \left(\frac{69}{172}\right) \pm 1.960 \sqrt{\frac{\left(\frac{69}{172}\right)\left(\frac{103}{172}\right)}{172}} = (32.8\%, 47.4\%)$$

We are 95% confident that between 32.8% and 47.4% of Adirondack streams have a shale substrate.

29. **PISA reading scale**

   a) The standard deviation of 97 does not match the stated standard error because the standard deviation calculation assumes that the test scores are independent. Since the 22 000 subjects were sampled from 1000 different schools, there is a dependence between the subjects based on what school they went to. We will assume that the standard error presented (2.4) accounts for this dependence in its calculation.

   The confidence interval for the Canadian population mean reading score is as follows (with big samples, just use z* instead of t*):

   $$\bar{x} \pm z^* SE$$
   $$= 527 \pm (1.96)(2.4)$$
   $$= 527 \pm 4.704$$
   $$= (522.296, 531.704)$$

   We therefore conclude that the 95% confidence interval for reading scores is (522.296, 531.704). The observations are assumed to be independent and approximately normally distributed, with the previously stated assumption about the standard error.

   b) As before, the problem with applying the Pythagorean rule is that it assumes that the variances of the two groups (males and females) are independent, and because of the dependence on school this is not true. We will once again assume the stated standard error of the difference between the means is calculated to account for this difference.

   $$\bar{d} \pm z^* SE = 32 \pm 1.96 \times 2.3 = (27.49, 36.51)$$

   Therefore, males are expected to score (27.49, 36.51) points less than females. The male and female groups are independent, and due to the very large sample sizes, we can relax the assumption that the data are approximately normal.

31. **Online testing**

   a)　H_0: The mean difference in the scores between Test A and Test B is zero. $(\mu_d = 0)$.

   H_A: The mean difference in the scores is not zero. $(\mu_d \neq 0)$

   Paired Data Assumption: The data are paired by student.
   Randomization Condition: The volunteers were randomized with respect to the order in which they took the tests and which form they took in each environment.
   Nearly Normal Condition: The distribution of differences is roughly unimodal and symmetric.

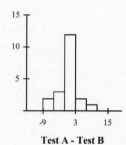

Since the conditions are satisfied, the sampling distribution of the difference can be modelled with a Student's *t*-model with 20 – 1 = 19 degrees of freedom,

$$t_{19}\left(\frac{3.52584}{\sqrt{20}}\right).$$

We will use a paired *t*-test, (Test A – Test B) with $\bar{d} = 0.3$.

Since the P-value = 0.7078 is very high, we fail to reject the null hypothesis. There is no evidence of that the mean difference in score is different from zero. It is reasonable to think that Test A and Test B are equivalent in terms of difficulty.

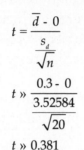

$$t = \frac{\bar{d} - 0}{\dfrac{s_d}{\sqrt{n}}}$$

$$t \approx \frac{0.3 - 0}{\dfrac{3.52584}{\sqrt{20}}}$$

$$t \approx 0.381$$

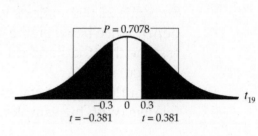

b) H_0: The mean difference between paper scores and online scores is zero $(\mu_d = 0)$.

   H_A: The mean difference between paper scores and online scores is not zero $(\mu_d \neq 0)$.

   Paired Data Assumption: The data are paired.
   Randomization Condition: The volunteers were randomized with respect to the order in which they took the tests and which form they took in each environment.

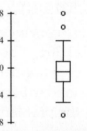

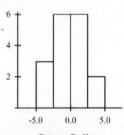

   Nearly Normal Condition: The boxplot of the distribution of differences is shows three outliers: students 3, 10, and 17. With these outliers removed, the histogram is roughly unimodal and symmetric. Since the conditions are satisfied, the sampling distribution of the difference can be modelled with a Student's *t*-model with 17 – 1 = 16 degrees of freedom,

$$t_{16}\left(0, \frac{2.26222}{\sqrt{17}}\right).$$

We will use a paired *t*-test, (Paper – Online) with $\bar{d} = -0.647059$.

Since the P-value = 0.2555 is high, we fail to reject the null hypothesis. There is no evidence that the mean difference in score is different from zero. It is reasonable to think that paper and online tests are equivalent in difficulty.

$$t = \frac{\bar{d} - 0}{\dfrac{s_d}{\sqrt{n}}}$$

$$t \approx \frac{-0.647059 - 0}{\dfrac{2.26222}{\sqrt{17}}}$$

$$t \approx -1.179$$

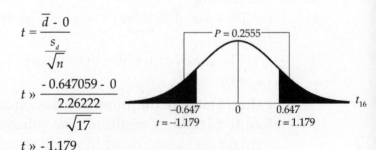

33. Irises

   a) Parallel boxplots of the distributions of petal lengths for the
      two species of flower are shown at the right. No units are
      specified, but millimetres seems like a reasonable guess.

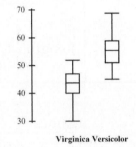

   b) The petals of *versicolor* are generally longer than the petals of
      virginica. Both distributions have about the same range, and
      both distributions are fairly symmetric.

   c) **Independent Groups Assumption**: The two species of
      flowers are independent.
      Randomization Condition: It is reasonable to assume that these flowers are
      representative of their species..
      Nearly Normal Condition: The boxplots show distributions of petal lengths that
      are reasonably symmetric with no outliers. Additionally, the samples are large.
      Since the conditions are satisfied, it is appropriate to model the sampling
      distribution of the difference in means with a Student's *t*-model, with 97.92
      degrees of freedom (from the approximation formula). We will construct a two-
      sample *t*-interval, with 95% confidence.

$$(\bar{y}_{Ver} - \bar{y}_{Vir}) \pm t^*_{df} \sqrt{\frac{s^2_{Ver}}{n_{Ver}} + \frac{s^2_{Vir}}{n_{Vir}}} = (55.52 - 43.22) \pm t^*_{97.92} \sqrt{\frac{5.519^2}{50} + \frac{5.362^2}{50}} \approx (10.14, 14.46)$$

   d) We are 95% confident the mean petal length of *versicolor* irises is between 10.14
      and 14.46 millimetres longer than the mean petal length of *virginica* irises.

   e) Since the interval is completely above 0, there is strong evidence that the mean
      petal length of *versicolor* irises is greater than the mean petal length of *virginica*
      irises.

35. World Series

   H_0: World Series teams are evenly matched. (The lengths of the World Series in
   years since 1922 are consistent with what probability would dictate for evenly
   matched teams.)

   H_A: World Series teams are not evenly matched. (The lengths of the World Series in
   years since 1922 are not consistent with what probability would dictate for evenly
   matched teams.)

   Counted Data Condition: We are counting the number of games in each World
   Series.
   Randomization Condition: These series are representative of all possible match-
   ups.
   Expected Cell Frequency Condition: All expected counts are greater than 5.
   Under these conditions, the sampling distribution of the test statistic is X^2 on 4 – 1
   = 3 degrees of freedom. We will use a chi-square goodness-of-fit test.

Length	Observed	Expected	Residual $=$ $(Obs - Exp)$	$(Obs - Exp)^2$	Component $=$ $\dfrac{(Obs - Exp)^2}{Exp}$
4 games	19	11.125	7.875	62.016	5.574
5 games	18	22.25	-4.25	18.063	0.812
6 games	19	27.8125	-8.8125	77.660	2.792
7 games	33	27.8125	5.1875	26.910	0.968

$$X^2 = 10.146$$

With degrees of freedom as 3, this gives a P-value of 0.0174. Thus, we reject the null hypothesis and there is some evidence that teams in the World Series are not evenly matched. If they were, we would expect to see fewer four-game series.

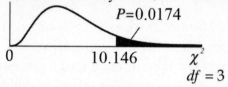

P=0.0174

0   10.146 χ^2

$df = 3$

37. **Lay's**

a) H_0: The mean weight of bags of Lay's is 35.4 grams. $(\mu = 35.4)$

H_A: The mean weight of bags of Lay's is less than 35.4 grams. $(\mu < 35.4)$

b) **Randomization Condition**: It is reasonable to think that the six bags are representative of all bags of Lay's.
Nearly Normal Condition: The histogram of bag weights shows one unusually heavy bag. Although not technically an outlier, it probably should be excluded for the purposes of the test. (We will leave it in for the preliminary test, then remove it and test again.)

c) The bags in the sample had a mean weight of 35.5333 grams and a standard deviation in weight of 0.450185 grams. Since the conditions for inference are satisfied, we can model the sampling distribution of the mean weight of bags of Lays with a Student's t model, with $6 - 1 = 5$ degrees of freedom,

$$t_5\left(35.4, \frac{0.450185}{\sqrt{6}}\right).$$

We will perform a one-sample t-test.
Since the P-value = 0.7497 is high, we fail to reject the null hypothesis. There is no evidence to suggest that the mean weight of bags of Lays is less than 35.4 grams.

$$t = \frac{\bar{y} - \mu_0}{SE(\bar{y})}$$

$$t = \frac{35.5333 - 35.4}{\frac{0.450185}{\sqrt{6}}}$$

$$t = 0.726$$

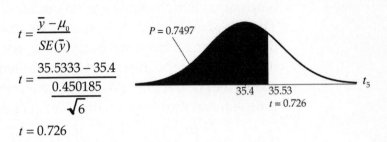

d) With the one unusually high value removed, the mean weight of the 5 remaining bags is 35.36 grams, with a standard deviation in weight of 0.167332 grams. Since the conditions for inference are satisfied, we can model the sampling distribution of the mean weight of bags of Lay's with a Student's t model, with $5 - 1 = 4$ degrees of freedom, $t_4\left(35.4, \frac{0.167332}{\sqrt{5}}\right).$

We will perform a one-sample t-test.
Since the P-value = 0.3107 is high, we fail to reject the null hypothesis. There is no evidence to suggest that the mean weight of bags of Lay's is less than 35.4 grams.

$$t = \frac{\bar{y} - \mu_0}{SE(\bar{y})}$$

$$t = \frac{35.36 - 35.4}{\frac{0.167332}{\sqrt{5}}}$$

$$t = -0.53$$

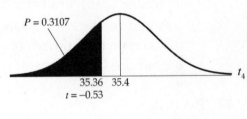

e) Neither test provides evidence that the mean weight of bags of Lay's is less than 35.4 grams. It is reasonable to believe that the mean weight of the bags is the same as the stated weight. However, the sample sizes are very small, and the tests have very little power to detect lower mean weights. It would be a good idea to weigh more bags.

39. And it means?

a) The margin of error is $\dfrac{(2391 - 1644)}{2} = \$373.50$.

b) The insurance agent is 95% confident that the mean loss claimed by clients after home burglaries is between $1644 and $2391.

c) 95% of all random samples of this size will produce intervals that contain the true mean loss claimed.

41. Hamsters

a) **Randomization Condition**: Assume these litters are representative of all litters.

Nearly Normal Condition: We don't have the actual data, so we can't look at a graphical display. However, since the sample size is large, the Central Limit Theorem guarantees that the distribution of averages will be approximately Normal, as long as there are no outliers.

The litters in the sample had a mean size of 7.72 baby hamsters and a standard deviation of 2.5 baby hamsters. Since the conditions are satisfied, the sampling distribution of the mean can be modelled by a Student's t- model, with $47 - 1 = 46$ degrees of freedom. We will use a one-sample t-interval with 90% confidence for the mean number of baby hamsters per litter.

$$\bar{y} \pm t^*_{n-1}\left(\frac{s}{\sqrt{n}}\right) = 7.72 \pm \left(\frac{2.5}{\sqrt{47}}\right) \approx (7.11, 8.33)$$

We are 90% confident that the mean number of baby hamsters per litter is between 7.11 and 8.33.

b) A 98% confidence interval would have a larger margin of error. Higher levels of confidence come at the price of less precision in the estimate.

$$ME = t^*_{24}\left(\frac{s}{\sqrt{n}}\right)$$

c) A quick estimate using z gives us a sample size of about 25 litters. Using this estimate, $t^*_{24} = 2.064$ at 95% confidence. We need a sample of about 27 litters in order to estimate the number of baby hamsters per litter to within one baby hamster.

$$1 = 2.064\left(\frac{2.5}{\sqrt{n}}\right)$$

$$n = \frac{(2.064)^2(2.5)^2}{(1)^2}$$

$$n \approx 27$$

43. Recruiting

a) **Randomization Condition**: Assume that these students are representative of all admitted graduate students.

10% Condition: Two groups of 500 students are both less than 10% of all students.

Independent Groups Assumption: The groups are from two different years.

Success/Failure Condition: $n\hat{p}$ (before) = (500)(0.52) = 260, $n\hat{q}$ (before) = (500)(0.48) = 240, $n\hat{p}$ (after) = (500)(0.54) = 270, and $n\hat{q}$ (after) = (500)(0.46) = 230 are all greater than 10, so the samples are both large enough.

Since the conditions have been satisfied, we will find a two-proportion z-interval.

$$\left(\hat{p}_A - \hat{p}_B\right) \pm z^*\sqrt{\frac{\hat{p}_A\hat{q}_A}{n_A} + \frac{\hat{p}_B\hat{q}_B}{n_B}}$$

$$= (0.54 - 0.52) \pm 1.960\sqrt{\frac{(0.54)(0.46)}{500} + \frac{(0.52)(0.48)}{500}} = (-0.042, 0.082)$$

We are 95% confident that the change in proportion of students who choose to enroll is between –4.2% and 8.2%.

b) Since 0 is contained in the interval, there is no evidence to suggest a change in the proportion of students who enroll. The program does not appear to be effective.

1. **Hurricane predictions 2012**
   a) The equation of the line of best fit for these data points is

   $\widehat{Error} = -454.448 - 8.479(year)$, where *Year* is measured in years since 1970.

   According to the linear model, the error made in predicting a hurricane's path was about 455 nautical miles, on average, in 1970. It has been declining at rate of about 8.48 nautical miles per year.

   b) H_0: There has been no change in prediction accuracy. $(\beta_1 = 0)$
   H_A: There has been a change in prediction accuracy. $(\beta_1 \neq 0)$

   c) Assuming the conditions have been met, the sampling distribution of the regression slope can be modelled by a Student's *t*-model with (36 – 2) = 34 degrees of freedom. We will use a regression slope *t*-test.
   The value of $t = -10.04$. The *P*-value ≤ 0.0001 means that the association we see in the data is unlikely to occur by chance. We reject the null hypothesis, and conclude that there is strong evidence that the prediction accuracies have in fact been changing during the time period.

   d) 71.1% of the variation in the prediction accuracy is accounted for by the linear model based on year.

3. **Movie budgets**
   a) $\widehat{Budget} - 31.387 + 0.714(Runtime)$. The model suggests that each additional minute of run time for a movie costs about $714 000.

   b) A negative intercept makes no sense, but the *P*-value of 0.07 indicates that we can't discern a difference between our estimated value and zero. The statement that a movie of zero length should cost $0 makes sense.

   c) Amounts by which movie costs differ from predictions made by this model vary, with a standard deviation of about $33 million.

   d) The standard error of the slope is 0.1541 million dollars per minute.

   e) If we constructed other models based on different samples of movies, we'd expect the slopes of the regression lines to vary, with a standard deviation of about $154 000 per minute.

5. **Movie budgets, the sequel**
   a) **Straight Enough Condition:** The scatterplot is straight enough, and the residuals plot looks unpatterned.
   Independence Assumption: The residuals plot shows no evidence of dependence.
   Does the Plot Thicken? Condition: The residuals plot shows no obvious trends in the spread.

Nearly Normal Condition, Outlier Condition: The Normal probability plot of residuals is reasonably straight.

b) Since conditions have been satisfied, the sampling distribution of the regression slope can be modelled by a Student's *t*-model with (120 – 2) = 118 degrees of freedom.

$$b_1 \pm t^*_{n-2} \times SE(b_1) = 0.714 \pm t^*_{118} \times 0.1541 \approx (0.41, 1.12)$$

We are 95% confident that the cost of making longer movies increases at a rate of between \$0.41 and \$1.12 million per minute.

7. **Hot dogs**

a) H_0: There's no association between calories and sodium content of all-beef hot dogs. $(\beta_1 = 0)$

H_A: There is an association between calories and sodium content of all-beef hot dogs. $(\beta_1 \neq 0)$

b) Assuming the conditions have been met, the sampling distribution of the regression slope can be modelled by a Student's *t*-model with (13 – 2) = 11 degrees of freedom. We will use a regression slope *t*-test. The equation of the line of best fit for these data points is:

$$\widehat{Sodium} = 90.9783 + 2.29959(Calories).$$

The value of *t* = 4.10. The *P*-value of 0.0018 means that the association we see in the data is very unlikely to occur by chance alone. We reject the null hypothesis, and conclude that there is evidence of a linear association between the number of calories in all-beef hotdogs and their sodium content. Because of the positive slope, there is evidence that hot dogs with more calories generally have higher sodium contents.

9. **Second frank**

a) Among all-beef hot dogs with the same number of calories, the sodium content varies, with a standard deviation of 59.66 mg.

b) The standard error of the slope of the regression line is 0.5607 milligrams of sodium per calorie.

c) If we tested many other samples of all-beef hot dogs, the slopes of the resulting regression lines would be expected to vary, with a standard deviation of about 0.56 mg of sodium per calorie.

11. **Last dog**

$$b_1 \pm t^*_{n-2} \times SE(b_1) = 2.29959 \pm (2.201) \times 0.5607 \approx (1.07, 3.53)$$

We are 95% confident that for every additional calorie, all-beef hot dogs have, on average, between 1.03 and 3.57 mg more sodium.

13. **Marriage age 2008**
    a) H_0: The difference in age between men and women at first marriage has not been decreasing since 1921. $(\beta_1 = 0)$

       H_A: The difference in age between men and women at first marriage has been decreasing since 1921. $(\beta_1 < 0)$

    b) **Straight Enough Condition:** The scatterplot is straight enough, but the residuals plot seems to show a curved pattern. A transformation on the response may be necessary.

       Independence Assumption: We are examining a relationship over time, so there is reason to be cautious.

       Does the Plot Thicken? Condition: The residuals plot is not consistent over time.

       Nearly Normal Condition, Outlier Condition: The histogram is not particularly unimodal and symmetric and it is skewed to the right.

    c) Assuming the conditions have been satisfied, the sampling distribution of the regression slope can be modelled by a Student's t-model with $(88 - 2) = 86$ degrees of freedom. We will use a regression slope t-test. The equation of the line of best fit for these data points is:

       $$\overline{(Men - Women)} = 49.9879 - 0.021\,(year).$$

       The value of $t = -44.02$. The P-value of less than 0.0001 (even though this is the value for a two-tailed test, it is still very small) means that the association we see in the data is unlikely to occur by chance. We reject the null hypothesis, and conclude that there is strong evidence of a negative linear relationship between difference in age at first marriage and year. The difference in marriage age between men and women appears to be decreasing over time.

15. **Marriage age 2008, again**

    $$b_1 \pm t^*_{n-2} \times SE(b_1) = -0.021036 \pm 1.988 \times 0.0004778 \approx (-0.0220, -0.0200)$$

    We are 95% confident that the mean difference in age between men and women at first marriage decreases by between 0.020 and 0.022 years in age for each year that passes.

17. **Fuel economy**
    a) H_0: There is no linear relationship between the weight of a car and its mileage. $(\beta_1 = 0)$

       H_A: There is a linear relationship between the weight of a car and its mileage. $(\beta_1 \neq 0)$

    b) **Straight Enough Condition:** The scatterplot is straight enough to try a linear model.

       Independence Assumption: The residuals plot is scattered.

Does the Plot Thicken? Condition: The residuals plot indicates some possible "thickening" as the predicted values increases, but it's probably not enough to worry about.

Nearly Normal Condition, Outlier Condition: The histogram of residuals is unimodal and symmetric, with one possible outlier. With the large sample size, it is okay to proceed.

c) Since conditions have been satisfied, the sampling distribution of the regression slope can be modelled by a Student's *t*-model with (50 – 2) = 48 degrees of freedom. We will use a regression slope *t*-test. The equation of the line of best fit for these data points is: $\widehat{MPG} = 48.7393 - 8.21362(Weight)$, where *Weight* is measured in thousands of pounds.

The value of $t = -12.2$. The *P*-value of less than 0.0001 means that the association we see in the data is unlikely to occur by chance. We reject the null hypothesis, and conclude that there is strong evidence of a linear relationship between weight of a car and its mileage. Cars that weigh more tend to have lower gas mileage.

19. **Fuel economy, part II**

a) Since conditions have been satisfied in Exercise 17, the sampling distribution of the regression slope can be modelled by a Student's *t*-model with (50 – 2) = 48 degrees of freedom. (Use $t_{45}^* = 2.014$ from the table.) We will use a regression slope *t*-interval, with 95% confidence.

$$b_1 \pm t_{n-2}^* \times SE(b_1) = -8.21362 \pm (2.014) \times 0.6738 \approx (-9.57, -6.86)$$

b) We are 95% confident that the mean mileage of cars decreases by between 6.86 and 9.57 miles per gallon for each additional 1000 pounds of weight.

21. **Fuel economy, part III**

a) The regression equation predicts that cars that weigh 2500 pounds will have a mean fuel efficiency of $48.7393 - 8.21362(2.5) = 28.20525$ miles per gallon.

$$\hat{y}_v \pm t_{n-2}^* \sqrt{SE^2(b_1) \cdot (x_v - \bar{x})^2 + \frac{s_e^2}{n}}$$

$$= 28.20525 \pm (2.014)\sqrt{0.6738^2 \cdot (2.5 - 2.8878)^2 + \frac{2.413^2}{50}}$$

$$\approx (27.34, 29.07)$$

We are 95% confident that cars weighing 2500 pounds will have mean fuel efficiency between 27.34 and 29.07 miles per gallon.

b) The regression equation predicts that cars that weigh 3450 pounds will have a mean fuel efficiency of $48.7393 - 8.21362(3.45) = 20.402311$ miles per gallon.

$$\hat{y}_v \pm t^*_{n-2}\sqrt{SE^2(b_1)\cdot(x_v-\bar{x})^2+\frac{s_e^2}{n}+s_e^2}$$

$$= 20.402311 \pm (2.014)\sqrt{0.6738^2\cdot(3.45-2.8878)^2+\frac{2.413^2}{50}+2.413^2}$$

$$\approx (15.44,\ 25.37)$$

We are 95% confident that a car weighing 3450 pounds will have fuel efficiency between 15.44 and 25.37 miles per gallon.

23. Cereal

a) H_0: There is no linear relationship between the number of calories and the sodium content of cereals. $(\beta_1 = 0)$

H_A: There is a linear relationship between the number of calories and the sodium content of cereals. $(\beta_1 \neq 0)$

Since these data were judged acceptable for inference, the sampling distribution of the regression slope can be modelled by a Student's t-model with $(77 - 2) = 75$ degrees of freedom. We will use a regression slope t-test. The equation of the line of best fit for these data points is: $\widehat{Sodium} = 21.4143 + 1.29357\,(Calories)$.

The value of $t = 2.73$. The P-value of 0.0079 means that the association we see in the data is unlikely to occur by chance. We reject the null hypothesis, and conclude that there is strong evidence of a linear relationship between the number of calories and sodium content of cereals. Cereals with higher numbers of calories tend to have higher sodium contents.

b) Only 9% of the variability in sodium content can be explained by the number of calories. The residual standard deviation is 80.49 mg, which is pretty large when you consider that the range of sodium content is only 320 mg. Although there is strong evidence of a linear association, it is too weak to be of much use. Predictions would tend to be very imprecise.

25. Another bowl

Straight Enough Condition: The scatterplot is not straight.
Independence Assumption: The residuals plot shows a curved pattern.
Does the Plot Thicken? Condition: The spread of the residuals is not consistent. The residuals plot "thickens" as the predicted values increase.
Nearly Normal Condition, Outlier Condition: The histogram of residuals is skewed to the right, with an outlier.
These data are not appropriate for inference.

27. Acid rain

a) H_0: There is no linear relationship between BCI and pH. $(\beta_1 = 0)$

H_A: There is a linear relationship between BCI and pH. $(\beta_1 \neq 0)$

b) Assuming the conditions for inference are satisfied, the sampling distribution of the regression slope can be modelled by a Student's *t*-model with (163 – 2) = 161 degrees of freedom. We will use a regression slope *t*-test. The equation of the line of best fit for these data points is: $\widehat{BCI} = 2733.37 - 197.694(pH)$. The value of t is calculated as follows:

$$t = \frac{b_1 - \beta_1}{SE(b_1)}$$

$$t = \frac{-197.694 - 0}{25.57}$$

$$t \approx -7.73$$

The corresponding *P*-value (two-sided!) is essentially 0.

c) With such a small p-value, this means that the association we see in the data is unlikely to occur by chance. We reject the null hypothesis, and conclude that there is strong evidence of a linear relationship between BCI and pH. Streams with higher pH tend to have lower BCI.

29. Ozone

a) H_0: There is no linear relationship between population and ozone level. $(\beta_1 = 0)$

H_A: There is a positive linear relationship between population and ozone level. $(\beta_1 > 0)$

Assuming the conditions for inference are satisfied, the sampling distribution of the regression slope can be modelled by a Student's *t*-model with (16 – 2) = 14 degrees of freedom. We will use a regression slope *t*-test. The equation of the line of best fit for these data points is:

$\widehat{Ozone} = 18.892 + 6.650(Population)$

where ozone level is measured in parts per million and population is measured in millions.

The value of $t \approx 3.48$.

The *P*-value of 0.0018 means that the association we see in the data is unlikely to occur by chance. We reject the null hypothesis, and conclude that there is strong evidence of a

$$t = \frac{b_1 - \beta_1}{SE(b_1)}$$

$$t = \frac{6.650 - 0}{1.910}$$

$$t \approx 3.48$$

positive linear relationship between ozone level and population. Cities with larger populations tend to have higher ozone levels.

b) City population is a good predictor of ozone level. Population explains 84% of the variability in ozone level and *s* is just over 5 parts per million.

31. **Ozone, again**
    a) $b_1 \pm t^*_{n-2} \times SE(b_1) = 6.65 \pm (1.761) \times 1.910 \approx (3.29, 10.01)$

       We are 90% confident that each additional million people will increase mean ozone levels by between 3.29 and 10.01 parts per million.

    b) The regression equation predicts that cities with a population of 600 000 people will have ozone levels of 18.892 + 6.650(0.6) = 22.882 parts per million.

    $$\hat{y}_v \pm t^*_{n-2} \sqrt{SE^2(b_1) \cdot (x_v - \bar{x})^2 + \frac{s_e^2}{n}}$$

    $$= 22.882 \pm (1.761)\sqrt{1.91^2 \cdot (0.6 - 1.7)^2 + \frac{5.454^2}{16}}$$

    $$\approx (18.47, 27.29)$$

    We are 90% confident that the mean ozone level for cities with populations of 600 000 will be between 18.47 and 27.29 parts per million.

33. **Start the car!**
    a) Since the number of degrees of freedom is 33 – 2 = 31, 33 batteries were tested.
    b) **Straight Enough Condition:** The scatterplot is roughly straight, but very scattered.

       Independence Assumption: The residuals plot shows no pattern.

       Does the Plot Thicken? Condition: The spread of the residuals is consistent.

       Nearly Normal Condition: The Normal probability plot of residuals is reasonably straight.

    c) H_0: There is no linear relationship between cost and power. $(\beta_1 = 0)$

       H_A: There is a positive linear relationship between cost and power. $(\beta_1 > 0)$

       Since the conditions for inference are satisfied, the sampling distribution of the regression slope can be modelled by a Student's t-model with (33 – 2) = 31 degrees of freedom. We will use a regression slope t-test. The equation of the line of best fit for these data points is:

       $$\widehat{Power} = 384.594 + 4.14649(Cost)$$

       with power measured in cold cranking amps, and cost measured in dollars. The value of $t \approx 3.23$. Since is a one-sided test, the P-value is 0.0029/2=0.0015, which means that the association we see in the data is unlikely to occur by chance. We reject the null hypothesis, and conclude that there is strong evidence of a positive linear relationship between cost and power. Batteries that cost more tend to have more power.

    d) Since $R^2 = 25.2\%$, only 25.2% of the variability in power can be accounted for by cost. The residual standard deviation is 116 amps. That's pretty large, considering the range battery power is only about 400 amps. Although there is strong evidence of a linear association, it is too weak to be of much use. Predictions would tend to be very imprecise.

    e) The equation of the line of best fit for these data points is:

$$\widehat{Power} = 384.594 + 4.14649(Cost)$$

with power measured in cold cranking amps, and cost measured in dollars.

f) There are 31 degrees of freedom, so use $t^*_{30} = 1.697$ as a conservative estimate from the table.

$$b_1 \pm t^*_{n-2} \times SE(b_1) = 4.14649 \pm (1.697) \times 1.282 \approx (1.97, 6.32)$$

g) We are 90% confident that the mean power increases by between 1.97 and 6.32 cold cranking amps for each additional dollar in cost.

35. Tablet computers

a) Since there are 23 – 2 = 21 degrees of freedom, there were 23 tablet computers tested.

b) **Straight enough condition:** The scatterplot is roughly straight, but scattered.
Independence assumption: The residuals plot shows no patten.
Does the plot thicken? Condition: The spread of the residuals is consistent
Nearly Normal condition: The Normal probability plot of residuals is reasonably straight.

c) H_0: There is no linear relationship between maximum brightness and battery life. ($\beta_1 = 0$)

H_A: There is a positive linear relationship between maximum brightness and battery life. ($\beta_1 > 0$)

Since the conditions for inference are satisfied, the sampling distribution of the regression slope can be modeled by a Student's t-model with (23 – 2) = 21 degrees of freedom. We will use a regression slope t-test. The equation of the line of best fit for these data points is: $\widehat{Hours} = 2.8467 + 0.01408(ScreenBrightness)$, battery life measure in hours and screen brightness measured in cd/m². The value of $t \approx 2.85$. The P-value of 0.00955 means that the association we see in the data is unlikely to occur by chance. We reject the null hypothesis, and conclude that there is strong evidence of a positive linear relationship between battery life and screen brightness. Tablets with greater screen brightness tend to have longer battery life.

d) Since $R^2 = 27.9\%$ only 27.9% of the variability in battery life can be accounted for by screen brightness. The residual standard deviation is 1.913 hours. That's pretty large, considering the range of battery life is only about 9 hours. Although there is strong evidence of a linear association, it is too weak to be of much use. Predictions would tend to be very imprecise.

e) The equation of the line of best fit for these data points is:
$\widehat{Hours} = 2.8467 + 0.01408(ScreenBrightness)$, battery life measure in hours and screen brightness measured in cd/m²

f) There are 21 degrees of freedom, so use t^*_{21}

$$b1 \pm t^*_{n-2} \times SE(b_1) = 0.01408 \pm (1.721) \times 0.004937 \sim (0.0055, 0.0226)$$

g) We are 90% confident that the mean battery life increases by between 0.0055 and 0.0226 hours for each additional cd/m² of screen brightness.

37. Body fat

a) H_0: There is no linear relationship between waist size and percent body fat. $(\beta_1 = 0)$

H_A: There is a linear relationship between waist size and percent body fat. $(\beta_1 \neq 0)$

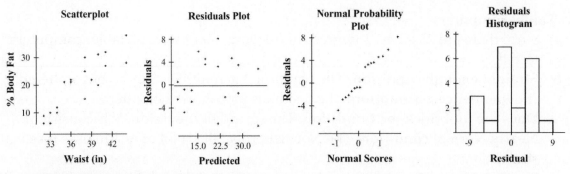

Straight Enough Condition: The scatterplot is straight enough to try linear regression.

Independence Assumption: The residuals plot shows no pattern.

Does the Plot Thicken? Condition: The spread of the residuals is consistent.

Nearly Normal Condition, Outlier Condition: The Normal probability plot of residuals is straight, and the histogram of the residuals is unimodal and symmetric with no outliers.

Since the conditions for inference are satisfied, the sampling distribution of the regression slope can be modelled by a Student's *t*-model with (20 – 2) = 18 degrees of freedom. We will use a regression slope *t*-test.

Dependent variable is: **Body Fat %**
No Selector
R squared = 78.7%   R squared (adjusted) = 77.5%
s = 4.540 with 20 - 2 = 18 degrees of freedom

Source	Sum of Squares	df	Mean Square	F-ratio
Regression	1366.79	1	1366.79	66.3
Residual	370.960	18	20.6089	

Variable	Coefficient	s.e. of Coeff	t-ratio	prob
Constant	-62.5573	10.16	-6.16	$ 0.0001
Waist (in)	2.22152	0.2728	8.14	$ 0.0001

The equation of the line of best fit for these data points is:

$$\widehat{\%BodyFat} = -62.5573 + 2.22152(Waist)$$

The value of $t \approx 8.14$. The *P*-value of essentially 0 means that the association we see in the data is unlikely to occur by chance. We reject the null hypothesis, and conclude that there is strong evidence of a linear relationship between waist size and percent body fat. People with larger waists tend to have a higher percentage of body fat.

b) The regression equation predicts that people with 40-inch waists will have $-62.5573 + 2.22152(40) = 26.3035\%$ body fat. The average waist size of the people sampled was approximately 37.05 inches.

$$\hat{y}_V \pm t^*_{n-2}\sqrt{SE^2(b_1).(x_V-\bar{x})^2 + \frac{s_e^2}{n}}$$

$$= 26.3035 \pm 2.101\sqrt{0.2728^2.(40-37.05)^2 + \frac{4.54^2}{20}}$$

$$\approx (23.58, 29.03)$$

We are 95% confident that the mean percent body fat for people with 40-inch waists is between 23.58% and 29.03%.

39. **Grades**

   a) The regression output is to the right. The model is:

   $$\widehat{Midterm_2} = 12.005 + 0.721(Midterm_1)$$

   Dependent variable is: **Midterm 2**
   No Selector
   R squared = 19.9%   R squared (adjusted) = 18.6%
   s = 16.78 with 64 - 2 = 62 degrees of freedom

Source	Sum of Squares	df	Mean Square	F-ratio
Regression	4337.14	1	4337.14	15.4
Residual	17459.5	62	281.604	

Variable	Coefficient	s.e. of Coeff	t-ratio	prob
Constant	12.0054	15.96	0.752	0.4546
Midterm 1	0.720990	0.1837	3.92	0.0002

   b) **Straight Enough Condition:** The scatterplot shows a weak, positive relationship between Midterm 2 score and Midterm 1 score. There are several outliers, but removing them only makes the relationship slightly stronger. The relationship is straight enough to try linear regression.

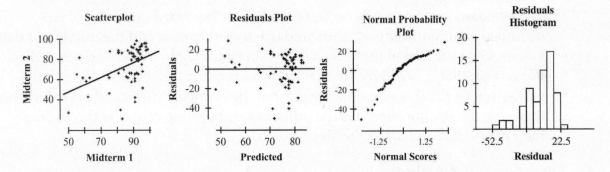

   Independence Assumption: The residuals plot shows no pattern.
   Does the Plot Thicken? Condition: The spread of the residuals is consistent.
   Nearly Normal Condition, Outlier Condition: The histogram of the residuals is unimodal, slightly skewed, with several possible outliers. The Normal probability plot shows some slight curvature.
   Since we had some difficulty with the conditions for inference, we should be cautious in making conclusions from these data. The small *P*-value of 0.0002 for the slope would indicate that the slope is statistically distinguishable from zero, but the R^2 value of 0.199 suggests that the overall relationship is weak. Midterm 1 isn't a very useful predictor of Midterm 2.

   c) The student's reasoning is not valid. The R^2 value is only 0.199 and the value of *s* is 16.8 points. Although correlation between Midterm 1 and Midterm 2 may be statistically significant, it isn't of much practical use in predicting Midterm 2 scores. It's just too weak.

41. Swimming Ontario

a) We want to know whether the times to swim across Lake Ontario have changed over time. We know the time of each swim in minutes and the year of the swim.
H_0: There is no linear relationship between year and swim time. $(\beta_1 = 0)$
H_A: Swim time and year are linearly associated. $(\beta_1 \neq 0)$

b) The Straight Enough Condition fails. There is an outlier.

c) Since the conditions are not satisfied, we cannot continue with the regression analysis.

d) With the outlier removed, the relationship is straight enough. Since some swimmers performed the feat more than once, the times may not be independent. There is no residuals plot given, but it does not appear that there is any thickening in the scatterplot. The Normal probability plot looks reasonably straight, so the Nearly Normal Condition is satisfied. The conditions appear to be met.

e) The regression model is $\widehat{Time} = -28399.9 + 14.9198(Year)$. Swim times have been increasing by about 14.9 minutes per year. The P-value 0f 0.0158 is small enough to reject the null hypothesis. There is strong evidence that the swim times have been increasing.

f) The standard error depends on three things: the standard deviation of the residuals around the line, σ, the standard deviation of x, and the number of data values. Only the first of these is changed substantially by removing the outlier. We can see that s_e was 447.3 with the outlier present, and only 277.2 with the outlier removed. This accounts for most of the change in the standard error, and therefore the P-value. Because the outlier was at a value of x near the middle of the x range, it didn't affect the slope much.

43. Education and mortality

a) **Straight Enough Condition:** The scatterplot is straight enough to try linear regression.
Independence Assumption: The residuals plot shows no pattern. If these cities are representative of other cities, we can generalize our results.
Does the Plot Thicken? Condition: The spread of the residuals is consistent.
Nearly Normal Condition, Outlier Condition: The histogram of the residuals is unimodal and symmetric with no outliers.

b) H_0: There is no linear relationship between education and mortality. $(\beta_1 = 0)$
H_A: There is a linear relationship between education and mortality. $(\beta_1 \neq 0)$

Since the conditions for inference are satisfied, the sampling distribution of the regression slope can be modelled by a Student's t-model with $(58 - 2) = 56$

degrees of freedom. We will use a regression slope *t*-test. The equation of the line of best fit for these data points is:

$\widehat{Mortality} = 1493.26 - 49.9202\,(Education)$. The value of $t \approx -6.24$.

The *P*-value of essentially 0 means that the association we see in the data is unlikely to occur by chance. We reject the null hypothesis, and conclude that there is strong evidence of a linear relationship between the level of education in a city and its mortality rate. Cities with lower education levels tend to have higher mortality rates.

$$t = \frac{b_1 - \beta_1}{SE(b_1)}$$

$$t = \frac{-49.9202 - 0}{8.000}$$

$$t \approx -6.24$$

c) We cannot conclude that getting more education is likely to prolong your life. Association does not imply causation. There may be lurking variables involved.

d) For 95% confidence, $t^*_{56} \approx 2.00327$.

$$b_1 \pm t^*_{n-2} \times SE(b_1) = -49.9202 \pm (2.003) \times 8.000 \approx (-65.95, -33.89)$$

e) We are 95% confident that the mean number of deaths per 100 000 people decreases by between 33.89 and 65.95 deaths for an increase of one year in average education level.

f) The regression equation predicts that cities with an adult population with an average of 12 years of school will have a mortality rate of $1493.26 - 49.9202(12) = 894.2176$ deaths per 100 000. The average education level was 11.0328 years.

$$\hat{y}_V \pm t^*_{n-2} \sqrt{SE^2(b_1).(x_V - \bar{x})^2 + \frac{s_e^2}{n}}$$

$$= 894.2176 \pm 2.003 \sqrt{8.00^2.(12 - 11.0328)^2 + \frac{47.92^2}{58}}$$

$$\approx (874.239, 914.196)$$

We are 95% confident that the mean mortality rate for cities with an average of 12 years of schooling is between 874.239 and 914.196 deaths per 100 000 residents.

45. **Seasonal spending**

a) **Straight Enough Condition:** The scatterplot is straight enough to try linear regression.

Independence Assumption/Randomization Condition: One cardholder's spending should not affect another's spending. The residuals plot shows no pattern. These 99 cardholders are a random sample of all cardholders.

Does the Plot Thicken? Condition: The residuals plot shows some increased spread for larger values of December charges.

Nearly Normal Condition, Outlier Condition: The histogram of residuals is unimodal and symmetric with two high outliers.

We should proceed cautiously. There are some issues with the conditions for regression.

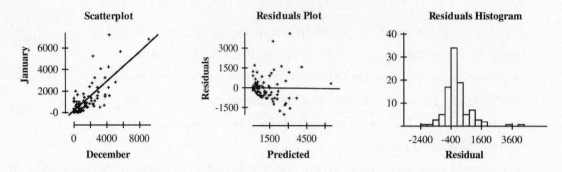

The regression model is: $\widehat{January} = 120.73 + 0.6995(December)$.

b) The regression equation predicts that cardholders who charged $2000 in December will charge $120.73 + 0.6995(2000) = \$1519.73$ in January, on average.

c) Cardholders charged an average of $1336.03 in December.

$$\hat{y}_v \pm t_{n-2}^* \sqrt{SE^2(b_1) \cdot (x_v - \overline{x})^2 + \frac{s_e^2}{n}}$$

$$= 1519.73 \pm (1.9847) \sqrt{0.0562^2 \cdot (2000 - 1336.03)^2 + \frac{874.5^2}{99}} \approx (1330.24, 1709.24)$$

We are 95% confident that the average January charges for a cardholder that charged $2000 in December will be between $1330.24 and $1709.24.

d) We are 95% confident that a cardholder who charged $2000 in December charged between $290.76 and $669.76 less than $2000 in January, on average.

e) The residuals show increasing spread, so the confidence intervals may not be valid. We should be very cautious when attempting to interpret them too literally.

47. CPU Performance

a) Yes, three points that are outliers and have high leverage.

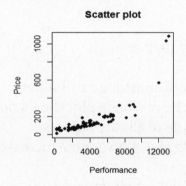

b)

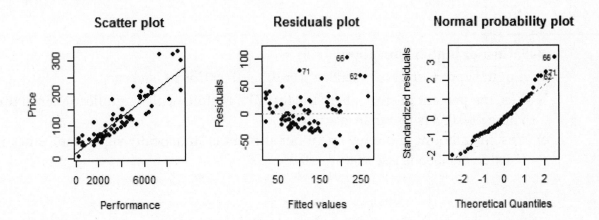

Straight Enough Condition: The scatterplot is straight enough to try linear regression.

Independence Assumption/Randomization Condition: One computer processor price should not affect another's price. These 78 processors are a random sample of all processors.

Does the Plot Thicken? Condition: The residuals plot shows some increased spread for larger values of fitted values.

Nearly Normal Condition, Outlier Condition: The Normal probability plot of residuals is reasonably straight.

Conditions seem satisfied and the following shows the results of the fit:

```
Coefficients:
              Estimate Std. Error t value Pr(>|t|)
(Intercept) 14.459564   6.935074   2.085   0.0404 *
Performance  0.027608   0.001583  17.444   <2e-16 ***
---
Signif. codes:  0 '***' 0.001 '**' 0.01 '*' 0.05 '.' 0.1 ' ' 1

Residual standard error: 31.87 on 76 degrees of freedom
Multiple R-squared: 0.8002,    Adjusted R-squared: 0.7975
F-statistic: 304.3 on 1 and 76 DF,  p-value: < 2.2e-16
```

c) The regression equation predicts the price of a new processor with a performance score of 8000 will cost 14.459 + 0.0276(8000) = 235.3221. Provided that the mean performance score is 3742.1410, the 95% prediction interval is

$$\hat{y}_V \pm t^*_{n-2}\sqrt{SE^2(b_1).(x_V - \overline{x})^2 + \frac{s_e^2}{n} + s_e^2} \approx (170.0495, 300.5948).$$

d) The regression equation predicts the price of a new processor with a performance score of 2000 will cost 14.459 + 0.0276(2000) = 69.67521. Provided that the mean performance score is 3742.1410, the 95% prediction interval is

$$\hat{y}_V \pm t^*_{n-2} \sqrt{SE^2(b_1).(x_V - \bar{x})^2 + \frac{s_e^2}{n} + s_e^2} \approx (5.561626, 133.7888).$$

e) The squared standard error of b1 is so small ($0.001583^2 = 0.0000025$) comparing to s_e^2.

*49. Cost of higher education

a) $\overline{Logit(\text{Type})} = -13.1461 + 0.08455 Top10\% + 0.000259\$ / student$

b) Yes, the percent of students in the top 10% is statistically significant, since the P-value of 0.033 is less than $\alpha = 0.05$.

c) Yes, the amount of money spent per student is statistically significant, since the P-value of 0.003 is less than $\alpha = 0.05$.

Chapter 25: Analysis of Variance

1. **Popcorn**
   a) H_0: The mean number of unpopped kernels is the same for all four brands of popcorn. $\left(\mu_1 = \mu_2 = \mu_3 = \mu_4\right)$
      H_A: The mean number of unpopped kernels is not the same for all four brands of popcorn.
   b) MS_T has $k - 1 = 4 - 1 = 3$ degrees of freedom.
      MS_E has $N - k = 16 - 4 = 12$ degrees of freedom.
   c) The F-statistic is 13.56 with 3 and 12 degrees of freedom, resulting in a P-value of 0.00037. We reject the null hypothesis and conclude that there is strong evidence that the mean number of unpopped kernels is not the same for all four brands of popcorn.
   d) To check the Similar Variance Condition, look at side-by-side boxplots of the treatment groups to see whether they have similar spreads. To check the Nearly Normal Condition, look to see if a normal probability plot of the residuals is straight, look to see that a histogram of the residuals is nearly normal, and look to see if a residuals plot shows no pattern and no systematic change in spread.

3. **Gas mileage**
   a) H_0: The mean gas mileage is the same for each muffler. $\left(\mu_1 = \mu_2 = \mu_3\right)$
      H_A: The mean gas mileage for each muffler is not the same.
   b) MS_T has $k - 1 = 3 - 1 = 2$ degrees of freedom.
      MS_E has $N - k = 24 - 3 = 21$ degrees of freedom.
   c) The F-statistic is 2.35 with 2 and 21 degrees of freedom, resulting in a P-value of 0.1199. We fail to reject the null hypothesis and conclude that there is no evidence to suggest that gas mileage associated with any single muffler is different than the others.
   d) To check the Similar Variance Condition, look at side-by-side boxplots of the treatment groups to see whether they have similar spreads. To check the Nearly Normal Condition, look to see if a normal probability plot of the residuals is straight, look to see that a histogram of the residuals is nearly normal, and look to see if a residuals plot shows no pattern and no systematic change in spread.
   e) By failing to notice that one of the mufflers resulted in significantly different gas mileage, you have committed a Type II error.

5. **Cat hair**
   a) H_0: The mean weights of cat hair remaining in the cage at the end of the day are the same for all three groups of cat receiving wet, dry, or no food.
      $\left(\mu_1 = \mu_2 = \mu_3\right)$

      H_A: The mean weights of cat hair remaining in the cage at the end of the day are not all the same.

b) MS_T has $k - 1 = 3 - 1 = 2$ degrees of freedom.
MS_E has $N - k = 60 - 3 = 57$ degrees of freedom.

c) The F-statistic is 2.58 with 2 and 57 degrees of freedom, resulting in a P-value of 0.0846. We fail to reject the null hypothesis and conclude that there is no evidence that the mean weights of cat hair for the three groups are not all the same.

d) The runs were randomized. We may need to be cautious in generalizing the results to all cats since each breed could react differently.

7. **Activating baking yeast**
a) H_0: The mean activation time is the same for all four recipes. $\left(\mu_1 = \mu_2 = \mu_3 = \mu_4\right)$
H_A: The mean activation times for each recipe are not all the same.
b) The F-statistic is 44.7392 with 3 and 12 degrees of freedom, resulting in a P-value less than 0.0001. We reject the null hypothesis and conclude that there is strong evidence that the mean activation times for each recipe are not all the same.
c) Yes, it would be appropriate to follow up with multiple comparisons because we have rejected the null hypothesis.

9. **Eye and hair colour** An analysis of variance is not appropriate because eye colour is a categorical variable. The students could consider a chi-square test of independence.

11. **Wines revisited**
a) H_0: The mean case price is the same at all three locations. $\left(\mu_C = \mu_K = \mu_S\right)$
H_A: The mean case prices for each location are not all the same.
b) The Similar Variance condition is not met, because the boxplots show distributions with radically different spreads, with an outlier in the Keuka group.

13. **Hearing**
a) H_0: The mean hearing score is the same for all four lists. $\left(\mu_1 = \mu_2 = \mu_3 = \mu_4\right)$
H_A: The mean hearing scores for each list are not all the same.
b) The F-statistic is 4.9192 with 3 and 92 degrees of freedom, resulting in a P-value equal to 0.0033. We reject the null hypothesis and conclude that there is strong evidence that the mean hearing scores for each list are not all the same.
c) Yes, it would be appropriate to follow up with multiple comparisons because we have rejected the null hypothesis.

15. Smokestack scrubbers

a) MS_T has $k - 1 = 4 - 1 = 3$ degrees of freedom.

$$MS_T = \frac{SS_T}{df_T} = \frac{81.2}{3} \approx 27.067$$

MS_E has $N - k = 20 - 4 = 16$ degrees of freedom.

$$MS_E = \frac{SS_E}{df_E} = \frac{30.8}{16} = 1.925$$

b) $F\text{-statistic} = \dfrac{MS_T}{MS_E} = \dfrac{27.067}{1.925} \approx 14.0606$

c) With a P-value equal to 0.00000949, there is very strong evidence that the mean particulate emissions for each smokestack scrubber are not all the same.

d) We have assumed that the experimental runs were performed in random order, that the variances of the treatment groups are equal, and that the errors are Normal.

e) To check the Similar Variance Condition, look at side-by-side boxplots of the treatment groups to see whether they have similar spreads. To check the Nearly Normal Condition, look to see if a normal probability plot of the residuals is straight, look to see that a histogram of the residuals is nearly normal, and look to see if a residuals plot shows no pattern and no systematic change in spread.

f) $s_p = \sqrt{MS_E} = \sqrt{1.925} \approx 1.387 \, \text{ppb}$

17. Auto noise filters

a) H_0: The mean noise level is the same for both types of filters. $(\mu_1 = \mu_2)$

H_A: The mean noise levels are different. $(\mu_1 \neq \mu_2)$

b) The F-statistic is 0.7673 with 1 and 33 degrees of freedom, resulting in a P-value equal to 0.3874. We fail to reject the null hypothesis and conclude that there is not enough evidence to suggest that the mean noise levels are different for each type of filter.

c) The Similar Variance Assumption seems reasonable. The spreads of the two distributions look similar. We cannot check the Nearly Normal Condition because we don't have a residuals plot.

d) H_0: The mean noise level is the same for both types of filters.

$(\mu_1 = \mu_2 \text{ or } \mu_1 - \mu_2 = 0)$

H_A: The mean noise level is different for each type of filters.

$(\mu_1 \neq \mu_2 \text{ or } \mu_1 - \mu_2 \neq 0)$

Independent Groups Assumption: Noise levels for the different types of filters should be independent.

Randomization Condition: Assume that this study was conducted using random allocation of the noise filters.

10% Condition: 18 and 17 are less than 10% of all automobiles.

Nearly Normal Condition: This might cause some difficulty. The distribution of the noise levels for the new device is skewed. However, the sample sizes are somewhat large, so the CLT will help to minimize any difficulty.

Since the conditions are satisfied, it is appropriate to model the sampling distribution of the difference in means with a Student's t-model, with 33 degrees of freedom ($n_1 + n_2 - 2 = 33$ for a pooled t-test). We will perform a two-sample t-test.

The pooled sample variance is $s_p^2 = \dfrac{\left((18-1)(3.22166)^2 + (17-1)(2.43708)^2\right)}{(18-1)+(17-1)} \approx 8.2265$.

The sampling distribution model has mean 0, with standard error:

$$SE_{pooled}(\bar{y}_1 - \bar{y}_2) = \sqrt{\dfrac{8.2265}{18} + \dfrac{8.2265}{17}} \approx 0.97002.$$

The observed difference between the mean scores is 81.5556 – 80.7059 = .8497.

$$t = \dfrac{(\bar{y}_1 - \bar{y}_2) - (0)}{SE_{pooled}(\bar{y}_1 - \bar{y}_2)} \approx \dfrac{0.8497 - 0}{0.97002} \approx 0.87596$$

With $t = 0.87596$ and 33 degrees of freedom, the 2-sided P-value is 0.3874. Since the P-value is high, we fail to reject the null hypothesis. There is no evidence that the mean noise levels for the two filters are different.

The P-value for the 2-sample pooled t-test = 0.3874, which is the same as the P-value for the analysis of variance test. The F-statistic, 0.7673, is approximately the same as $t^2 = (0.87596)^2 \approx 0.7673$. Also, the pooled estimate of the variance, 8.2265, is approximately equal to MS_E, the mean squared error. This is because an analysis of variance for two samples is equivalent to a 2-sample pooled t-test.

19. Fertilizers

a) H₀: The mean height of beans is the same for each of the 10 fertilizers.
   $\left(\mu_A = \mu_B = \ldots = \mu_J\right)$
   Hₐ: The mean heights of beans are not all the same for each of the 10 fertilizers.

b) The F-statistic is 1.1882 with 9 and 110 degrees of freedom, resulting in a P-value equal to 0.3097. We fail to reject the null hypothesis and conclude that there is not enough evidence to suggest that the mean bean heights are not all the same for the 10 fertilizers.

c) This does not match our findings in part b). Because the lab partner performed so many t-tests, he may have committed several Type I errors. (Type I error is the probability of rejecting the null hypothesis when there is actually no difference between the fertilizers.) Some tests would be expected to result in Type I error due to chance alone. The overall Type I error rate is higher for multiple t-tests than for an analysis of variance, or a multiple comparisons method. Because we failed to reject the null hypothesis in the analysis of variance, we should not do multiple comparison tests.

21. Cereals redux

a) H_0: The mean protein content is the same for each of the three shelves. $(\mu_1 = \mu_2 = \mu_3)$

H_A: The mean protein contents are not all the same for each of the three shelves.

b) The *F*-statistic is 5.8445 with 2 and 74 degrees of freedom, resulting in a P-value equal to 0.0044. We reject the null hypothesis and conclude that there is strong evidence to suggest that the mean protein content of cereal on at least one shelf differs from that on other shelves.

c) We cannot conclude that cereals on shelf 2 have a lower mean protein content than cereals on shelf 3, or that cereals on shelf 2 have a lower mean protein content than cereals on shelf 1. We can conclude only that the mean protein contents are not all equal.

d) We can conclude that the mean protein content on shelf 2 is significantly different from the mean protein contents on shelf 3. In fact, there is evidence that the mean protein content on shelf 3 is greater than that of shelf 2. The other pairwise comparisons are not significant at $\alpha = 0.05$.

23. Analgesics

a) H_0: The mean pain level reported is the same for each of the three drugs. $(\mu_A = \mu_B = \mu_C)$

H_A: The mean pain levels reported are not all the same for the three drugs.

Analysis of Variance

Source	DF	Sum of Squares	Mean Square	F-ratio	P-Value
Drug	2	28.2222	14.1111	11.9062	0.0003
Error	24	28.4444	1.1852		
Total	26	56.6666			

b) The *F*-statistic is 11.9062 with 2 and 24 degrees of freedom, resulting in a P-value equal to 0.0003. We reject the null hypothesis and conclude that there is strong evidence that the mean pain level reported is different in at least one of the three drugs.

c) **Randomization Condition**: The volunteers were randomly allocated to treatment groups.

Similar Variance Condition: The boxplots show similar spreads for the distributions of pain levels for the different drugs, although the outlier for drug B may cause some concern.

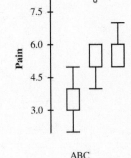

Nearly Normal Condition: The normal probability plot of residuals is reasonably straight.

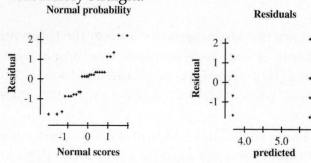

d) A Bonferroni test shows that drug A's mean pain level as reported is significantly below the other two, but that drug B's and C's means are indistinguishable at the $\alpha = 0.05$ level.

25. **Popsicle sticks, redux** This is unacceptable as we looked at the results before forming our hypothesis. This follow-up test would only be permissible if we had planned it before conducting the experiment. Also, this is a meaningless test for the same reasons as in the previous question.

27. **Popsicle sticks, dessert**
    a) **Equal variance assumption:** The distributions of the residuals show similar spreads for the four glue types.
    b) **Outliers:** Some outliers are found in the hot and white glue type.
    c) **Normality of residuals:** The normal probability plot of residuals shows some small deviation from the identity line from both ends.
    d) Conditions seem satisfied and we could proceed with our usual inferences.

Chapter 26: Multifactor Analysis of Variance

1. **Popcorn revisited**
   a) H_0: The effect due to *Power* level is the same for each level. ($\gamma_{Low} = \gamma_{Med} = \gamma_{High}$)
      H_A: Not all of the *Power* levels have the same effect on popcorn popping.
      H_0: The effect due to popping *Time* is the same for each time. ($\tau_3 = \tau_4 = \tau_5$)
      H_A: Not all of the popping *Times* have the same effect on popcorn popping.
   b) The *Power* sum of squares has 3 – 1 = 2 degrees of freedom.
      The *Time* sum of squares has 3 – 1 = 2 degrees of freedom.
      The error sum of squares has (9 – 1) – 2 – 2 = 4 degrees of freedom.
   c) Because the experiment did not include replication, there are no degrees of freedom left for the interaction term. The interaction term would require 2(2) = 4 degrees of freedom, leaving none for the error term, making any tests impossible.

3. **Popcorn again**
   a) The *F*-statistic for *Power* is 13.56 with 2 and 4 degrees of freedom, resulting in a P-value equal to 0.0165. The *F*-statistic for *Time* is 9.36 with 2 and 4 degrees of freedom, resulting in a P-value equal to 0.0310.
   b) With a P-value equal to 0.0165, we reject the null hypothesis that *Power* has no effect, and conclude that the mean number of uncooked kernels is not equal across all three *Power* levels. With a P-value equal to 0.0310, we reject the null hypothesis that *Time* has no effect and conclude that the mean number of uncooked kernels is not equal across all three *Time* levels.
   c) **Randomization Condition**: The bags of popcorn should be randomly assigned to treatments.
      Similar Variance Condition: The side-by-side boxplots should have similar spreads. The residuals plot should show no pattern and no systematic change in spread.
      Nearly Normal Condition: The Normal probability plot of the residuals should be straight, and the histogram of the residuals should be unimodal and symmetric.

5. **Crash analysis**
   a) H_0: The effect on head injury severity is the same for both seats. ($\gamma_D = \gamma_P$)
      H_A: The effects on head injury severity are different for the two seats.
      H_0: The size of the vehicle has no effect on head injury severity.
      ($\tau_1 = \tau_2 = \tau_3 = \tau_4 = \tau_5 = \tau_6$)
      H_A: The size of the vehicle does have an effect on head injury severity.
   b) **Randomization Condition**: Assume that the cars are representative of all cars.
      Additive Enough Condition: The interaction plot is reasonably parallel.
      Similar Variance Condition: The side-by-side boxplots have similar spreads.

The residuals plot shows no pattern and no systematic change in spread.
Nearly Normal Condition: The Normal probability plot of the residuals should be straight, and the histogram of the residuals should be unimodal and symmetric.

c) With a P-value equal to 0.838, the interaction term between *Seat* and vehicle *Size* is not significant. With a P-value less than 0.0001 for both *Seat* and *Size*, we reject both null hypotheses and conclude that both *Seat* and *Size* affect the severity of head injury. By looking at one of the partial boxplots we see that the mean head injury severity is higher for the driver's side. The effect of the driver's *Seat* seems to be roughly the same for all six cars.

7.   Baldness and heart disease

a) A two-factor ANOVA must have a quantitative response variable. Here the response is whether they exhibited baldness or not, which is a categorical variable. A two-factor ANOVA is not appropriate.

b) We could use a chi-square analysis to test whether baldness and heart disease are independent.

9.   Baldness and heart disease again

a) A chi-square test of independence gives a chi-square statistic of 14.510 with $(4 - 1)(2 - 1) = 3$ degrees of freedom, which correspond to a P-value equal to 0.0023. We reject the hypothesis that baldness and heart disease are independent.

b) No, the fact that these are not independent does not mean that one causes the other. There could be a lurking variable (such as age) that influences both.

11.   Basketball shots

a) H_0: The time of day has no effect on the number of shots made. $(\gamma_M = \gamma_N)$
H_A: The time of day does have an effect on the number of shots made.
H_0: The type of shoe has no effect on the number of shots made. $(\tau_F = \tau_O)$
H_A: The type of shoe does have an effect on the number of shots made.

b) The partial boxplots show little effect of either *Time of day* or *Shoes* on the number of shots made.

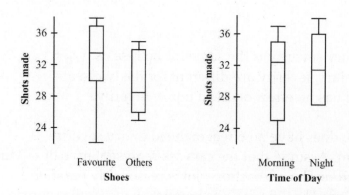

Randomization Condition: We assume that the number of shots made were independent from one treatment condition to the next.

Similar Variance Condition: The side-by-side boxplots have similar spreads. The residuals plot shows no pattern and no systematic change in spread.

Nearly Normal Condition: The Normal probability plot of the residuals is straight.

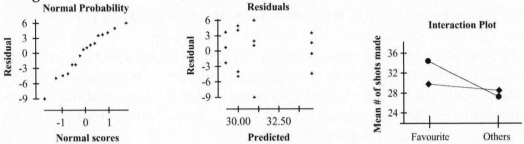

The interaction plot shows a possible interaction effect. It looks as though the favourite shoes may make more of a difference at night. Here is the ANOVA table:

	Df	Sum Sq	Mean Sq	F value	Pr(>F)
Shoes	1	39.06	39.06	1.777	0.207
Time	1	7.56	7.56	0.344	0.568
Shoes:Time	1	18.06	18.06	0.822	0.382
Residuals	12	263.75	21.98		

We fail to reject the null hypothesis that there is no interaction effect. In fact, none of the effects appears to be significant. It looks as though she cannot conclude that either *Shoes* or *Time of day* affect her mean free throw percentage.

13. **Sprouts again**

a) H_0: The temperature level has no effect on the number of sprouts. ($\gamma_{32} = \gamma_{34} = \gamma_{36}$)
   H_A: The temperature level does have an effect on the number of sprouts.
   H_0: The salinity level has no effect on the number of sprouts. ($\tau_0 = \tau_4 = \tau_8 = \tau_{12}$)
   H_A: The salinity level does have an effect on the number of sprouts.

b) **Randomization Condition**: We assume that the sprouts were representative of all sprouts.

   Similar Variance Condition: There appears to be more spread in the number of sprouts for the lower salinity levels. This is cause for some concern, but most likely does not affect the conclusions.

   Nearly Normal Condition: The Normal probability plot of the residuals should be straight.

   The partial boxplots show that *Salinity* level appears to have an effect on the number of bean sprouts while *Temperature* does not. The ANOVA supports this. With a P-value less than 0.0001 for *Salinity*, there is strong evidence that the salinity level affects the number of bean sprouts. However, it appears that neither the interaction term nor *Temperature* has a significant effect on the number of bean sprouts. The interaction term has a P-value equal to 0.7549 and *Temperature* has a P-value equal to 0.3779.

15. Gas additives

H_0: The car type has no effect on gas mileage. ($\gamma_H = \gamma_M = \gamma_S$)

H_A: The car type does have an effect on gas mileage.

H_0: The mean gas mileage is the same for both additives. ($\tau_G = \tau_R$)

H_A: The mean gas mileage is different for the additives.

The ANOVA table shows that both *Car Type* and *Additive* affect gas mileage, with P-values less than 0.0001. There is a significant interaction effect as well that makes interpretation of the main effects problematic. However, the residual plot shows a strong increase in variance, which makes the whole analysis suspect. The Similar Variance Condition appears to be violated.

17. Gas additives again

After the re-expression of the response, gas mileage, the Normal probability plot of residuals looks straight and the residuals plot shows constant spread over the predicted values.

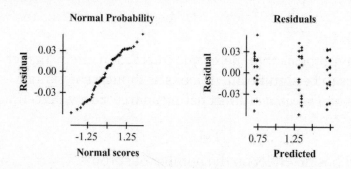

Analysis of Variance					
Source	DF	Sum of Squares	Mean Square	F-ratio	P-Value
Type	2	10.1254	5.06268	5923.1	<0.0001
Additive	1	0.026092	0.0260915	30.526	<0.0001
Interaction	2	7.57E-05	3.78E-05	0.044265	0.9567
Error	54	0.046156	8.55E-04		
Total	59	10.1977			

After the re-expression, the ANOVA table only shows the main effects to be significant, while the interaction term is not. We can conclude that both the *Car type* and *Additive* have an effect on mileage and that the effects are constant (in log(*mpg*)) over the values of the various levels of the other factor.

19. **Batteries again**
    a) H_0: The mean times are the same under both environments. $(\gamma_C = \gamma_{RT})$
       H_A: The mean times are different under the two environments.
       H_0: The brand of battery has no effect on time. $(\tau_A = \tau_B = \tau_C = \tau_D)$
       H_A: The brand of battery does have an effect on time.
    b) From the partial boxplots it appears that *Environment* does have an effect on time, but it is unclear whether *Brand* has an effect on time.
    c) Yes, the ANOVA does match our intuition based on the boxplots. The *Brand* effect has a P-value equal to 0.099. While this is not significant at the $\alpha = 0.05$ level, it is significant at the $\alpha = 0.10$ level. As it appeared on the boxplot, the *Environment* is clearly significant with a P-value less than 0.0001.
    d) There is also an interaction, however, which makes the statement about *Brands* problematic. Not all *Brands* are affected by the environment in the same way. Brand C, which works best in the cold, performs the worst at room temperature.
    e) I would be uncomfortable recommending brand C because it performs the worst of the four at room temperature.

21. **Batteries once more** In this one-way ANOVA, we can see that the means vary across treatments. (However, boxplots with only two observations are not appropriate.) By looking closely, it seems obvious that the four flashlights at room temperature lasted much longer than the ones in the cold. It is much harder to see whether the means of the four brands are different, or whether they differ by the same amounts across both environmental conditions. The two-way ANOVA with interaction makes these distinctions clear.

Chapter 27: Multiple Regression

1. **Interpretations**
   a) There are two problems with this interpretation. First, the other predictors are not mentioned. Second, the prediction should be stated in terms of a mean, not a precise value.
   b) This is a correct interpretation.
   c) This interpretation attempts to predict in the wrong direction. This model cannot predict *Lotsize* from *Price*.
   d) R^2 concerns the fraction of variability accounted for by the regression model, not the fraction of data values.

3. **Predicting final exams**
   a) $\widehat{Final} = -6.72 + 0.2560(Test1) + 0.3912(Test2) + 0.9015(Test3)$
   b) $R^2 = 77.7\%$, which means that 77.7% of the variation in *Final* grade is accounted for by the multiple regression model.
   c) According to the multiple regression model, each additional point on *Test3* is associated with an average increase of 0.9015 points on the final, for students with given *Test1* and *Test2* scores.
   d) Test scores are probably collinear. If we are only concerned about predicting the final exam score, *Test1* may not add much to the regression. However, we would expect it to be associated with the final exam score.

5. **Home prices**
   a) $\widehat{Price} = -152037 + 9530(Baths) + 1398.87(Sqft)$
   b) $R^2 = 71.1\%$, which means that 71.1% of the variation in asking price is accounted for by the multiple regression model.
   c) According to the multiple regression model, the asking price increases, on average, by about $139.87 for each additional square foot, for homes with the same number of bathrooms.
   d) The number of bathrooms is probably correlated with the size of the house, even after considering the square footage of the bathroom itself. This correlation may account for the coefficient of *Baths* not being discernibly different from 0. Moreover, the regression model does not predict what will happen when a house is modified, for example, by converting existing space into a bathroom.

7. **Predicting finals II**
   Straight Enough Condition: The plot of residuals versus predicted values looks curved, rising in the middle, and falling on both ends. This is a potential difficulty.
   Randomization Condition: It is reasonable to think of this class as a representative sample of all classes.
   Nearly Normal Condition: The Normal probability plot and the histogram of residuals suggest that the highest five residuals are extraordinarily high.

Does the Plot Thicken? Condition: The spread is not consistent over the range of predicted values.

These data may benefit from a re-expression.

9. **Receptionist performance.**

a) $\widehat{Salary} = 9.788 + 0.11(Service) + 0.053(Education) + 0.071(Score) + 0.004(Speed) + 0.065(Dictation)$

b) $\widehat{Salary} = 9.788 + 0.11(120) + 0.053(9) + 0.071(50) + 0.004(60) + 0.065(30) \approx \$29\ 200$

c) H_0: When including the other potential predictors, typing speed does not increase our ability to predict salary. $(\beta_4 = 0)$

H_A: When including the other potential predictors, typing speed increases or decreases our ability to predict salary. $(\beta_4 \neq 0)$

With $t = 0.013$, and $30 - 6 = 24$ degrees of freedom, the p-value equals 0.9897. Since the p-value is so large, we fail to reject the null hypothesis. There is no evidence to suggest that the coefficient for typing *Speed* is anything other than 0.

d) Omitting typing *Speed* would simplify the model, and probably result in a model that was almost as good. Other predictors might also be omitted, but we cannot make that decision from the information given.

e) *Age* may be collinear with other predictors in the model. In particular, it is likely to be highly associated with months of *Service*.

11. **Body fat revisited**

a) H_0: There is no linear relationship between weight and percent body fat. $(\beta_W = 0)$

H_A: There is a linear relationship between weight and percent body fat. $(\beta_W \neq 0)$

With $t = 12.4$, and $250 - 2 = 248$ degrees of freedom, the p-value is less than 0.0001. Since the p-value is so small, we reject the null hypothesis. There is strong evidence of a linear relationship between *Weight* and *%Body fat*.

b) According to the linear model, each pound of *Weight* is associated, on average, with a 0.189% increase in *%Body fat*.

c) After removing the linear effects of *Waist size* and *Height*, each pound of *Weight* is associated, on average, with a decrease of 0.10% in *%Body fat*. The change in coefficient and sign is a result of including the other predictors. We expect *Weight* to be correlated with both *Waist size* and *Height*. It may be collinear with them.

d) The p-value of 0.1567 says that if the coefficient of *Height* in this model is truly 0, we would expect to observe a sample regression coefficient at least as far from 0 as the one we have here about 15.67% of the time.

13. **Body fat again**
    a) H_0: There is no linear relationship between chest size and percent body fat.
       $(\beta_C = 0)$
       H_A: There is a linear relationship between chest size and percent body fat.
       $(\beta_C \neq 0)$
       With $t = 15.5$, and $250 - 2 = 248$ degrees of freedom, the p-value is less than
       0.0001. Since the p-value is so small, we reject the null hypothesis. There is strong
       evidence of a linear relationship between *Chest size* and *%Body fat*.
    b) According to the linear model, each additional inch in *Chest size* is associated
       with an average increase of 0.71272% in *%Body rat*.
    c) After allowing for the linear effects of *Waist size* and *Height*, the multiple
       regression model predicts an average decrease of 0.233531% in *%Body fat* for
       each additional inch in *Chest size*.
    d) Each of the variables appears to contribute to the model. There does not appear
       to be an advantage to removing any of them.

15. **Fifty states 2009**
    a) The only model that seems to do poorly is the one that omits *Murder*. The other
       three are hard to choose among.
       $$\widehat{Lifeexp} = 74.781 - 0.39266(Murder) + 0.007670(HSgrad) + 0.00010773(Income)$$
       $$R^2 = 68.7\%$$
       $$\widehat{Lifeexp} = 77.332 - 0.45525(Murder) + 0.01538(HSgrad) + 5.060(Illiteracy)$$
       $$R^2 = 57.6\%$$
       $$\widehat{Lifeexp} = 75.3472 - 0.42699(Murder) + 0.00010727(Income) + 1.335(Illiteracy)$$
       $$R^2 = 68.5\%$$
    b) Each of the models has at least one coefficient with a large P-value. This
       predictor variable could be omitted to simplify the model without degrading it
       too much.
    c) No. Regression models cannot be interpreted that way. Association is not the
       same thing as causation.
    d) Plots of the residuals highlight some states as possible outliers. You may want
       to consider setting them aside to see if the model changes.

17. **Burger King revisited**
    a) With an $R^2 = 100\%$, the model should make excellent predictions.
    b) The value of s, 3.140 calories, is very small compared to the initial standard
       variation of *Calories*. This means that the model fits the data quite well, leaving
       very little variation unaccounted for.
    c) No, the residuals are not all 0. Indeed, we know that their standard deviation is s
       = 3.140 calories. They are very small compared with the original values. The true
       value of R^2 was rounded up to 100%.

Chapter 28: Multiple Regression Wisdom

1. **Climate change 2009**
   a) The distribution of Studentized residuals might be bimodal. We should check for evidence of two groups in the data.
   b) Answers may vary. It is also known that carbon dioxide levels changed in character in the late 1970s, and began a more rapid increase than had been seen earlier in the 20th century. Perhaps fitting two different regression models to the data up to 1977 and to the data after 1977 will show a different result of the effect of *Year* on *Global temperature anomaly*. This is an example of Simpson's Paradox.

3. **Healthy breakfast**
   a) The slope of a partial regression plot is the coefficient of the corresponding predictor — in this case, – 1.020.
   b) Quaker oatmeal makes the slope more strongly negative. It appears to have substantial influence on this slope.
   c) Not surprisingly, omitting Quaker oatmeal changes the coefficient of *Fibre*. It is now positive (although not significantly different from 0). This second regression model has a higher R^2, suggesting that it fits the data better. Without the influential point, the second regression is probably the better model.
   d) The coefficient of *Fibre* is not discernibly different from 0. We have no evidence that it contributes significantly to calories.

5. **A nutritious breakfast**
   a) After allowing for the effects of *Sodium* and *Sugars*, the model predicts a decrease of 0.019 calories for each additional gram of *Potassium*.
   b) Those points pull the slope of the relationship down. Omitting them should increase the value of the coefficient of *Potassium*. It would likely become positive, since the remaining points show a positive slope in the partial regression plot.
   c) These appear to be influential points. They have both high leverage and large residuals, and the partial regression plot shows their influence.
   d) If our goal is to understand the relationships among these variables, then it might be best to omit these cereals because they seem to behave as if they are from a separate subgroup.

7. **Traffic delays**
   a) This set of indicators uses *Medium* size as its base. The coefficients of *Small*, *Large*, and *Very large* estimate the average change in amount of *Delay/person* relative to the amount of *Delay/person* for *Medium* size cities found for each of the

other three sizes. If there were an indicator variable for *Medium* as well, the four indicators would be collinear, so the coefficients could not be estimated.

b) On average, the total *Delay/person* is 5.01 hours longer for people in *Large* cities relative to *Medium* cities, after allowing for the effects of *Arterial road speed* and *Highway road speed*.

9.   More traffic

a)   An assumption required for indicator variables to be useful is that the regression models fit for the different groups identified by the indicators are parallel, i.e., they have the same slope. These lines are not parallel.

b)   The coefficient of *Am × Sml* adjusts the slope of the regression model fit for the small cities. We would say that the slope of *Delay/person* on *Arterial mph* for *Small* cities (after allowing for the linear effects of the other variables in the model) is –2.60848 + 3.81461 = 1.20613.

c)   The regression model seems to do a good job. The R^2 shows that 80.7% of the variability in *Delay/person* is accounted for by the model. Most of the p-values for the coefficients are small. The coefficients concerning the very large cities have larger p-values, but that may be due to having a relatively small number of such cities. It may still be wise to keep those predictors in the model.

11.   Influential traffic? Colorado Springs has high leverage, but it does not have a particularly high Studentized residual. It appears that the point has influence but does not exert it. Removing this case from the regression may not result in a large change in the model, so the case is probably not influential.

13. Popsicle sticks

a)   It is the expected change in stress for a unit change in area, for a large overlap joint, holding everything else constant.

b)   There are way more medium joints than small joints. Since the relationship is very weak for medium joints, the P-value is 'swamped' by them.

c)   0.7773

d)   (–10.6376 + 0.7773)*10 = -98.6

Part VII Review: Inference When Variables Are Related

1. **Tableware**
   a) Since there are 57 degrees of freedom, there were 59 different products in the analysis.
   b) 84.5% of the variation in retail price is explained by the polishing time.
   c) Assuming the conditions have been met, the sampling distribution of the regression slope can be modelled by a Student's t-model with $(59 - 2) = 57$ degrees of freedom. We will use a regression slope t-interval.
      For 95% confidence, use $t_{57}^* \approx 2.0025$, or estimate from the table $t_{50}^* \approx 2.009$.
      $$b_1 \pm t_{n-2}^* \times SE(b_1) = 2.49244 \pm (2.0025) \times 0.1416 \approx (2.21,\ 2.78)$$
   d) We are 95% confident that the average price increases between $2.21 and $2.78 for each additional minute of polishing time.

3. **Mutual funds**
   a) **Paired Data Assumption**: These data are paired by mutual fund.
      Randomization Condition: Assume that these funds are representative of all large cap mutual funds.
      Nearly Normal Condition: The histogram of differences is unimodal and symmetric.

      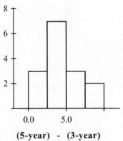

      Since the conditions are satisfied, the sampling distribution of the difference can be modelled with a Student's t-model with $15 - 1 = 14$ degrees of freedom. We will find a paired t-interval, with 95% confidence.
      $$\bar{d} \pm t_{n-1}^* \left(\frac{s_d}{\sqrt{n}} \right) = 4.54 \pm t_{14}^* \left(\frac{2.50508}{\sqrt{15}} \right) \approx (3.15, 5.93)$$

      Provided that these mutual funds are representative of all large cap mutual funds, we are 95% confident that, on average, 5-year yields are between 3.15% and 5.93% higher than 3-year yields.
   b) H_0: There is no linear relationship between 3-year and 5-year rates of return. $(\beta_1 = 0)$
      H_A: There is a linear relationship between 3-year and 5-year rates of return. $(\beta_1 \neq 0)$

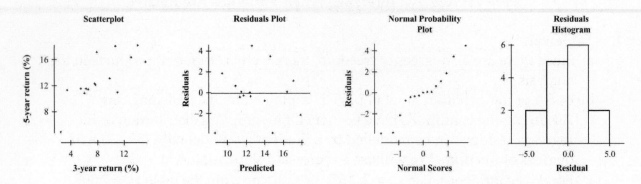

Straight Enough Condition: The scatterplot is straight enough to try linear regression.

Independence Assumption: The residuals plot shows no pattern.

Does the Plot Thicken? Condition: The spread of the residuals is consistent.

Nearly Normal Condition: The Normal probability plot of residuals isn't very straight. However, the histogram of residuals is unimodal and symmetric. With a sample size of 15, it is probably okay to proceed.

Since the conditions for inference are satisfied, the sampling distribution of the regression slope can be modelled by a Student's t-model with $(15 - 2) = 13$ degrees of freedom. We will use a regression slope t-test.

The equation of the line of best fit for these data points is:

$$(\widehat{5Year}) = 6.92904 + 0.719157(3year).$$

Dependent variable is: **5-year**
No Selector
R squared = 58.4%   R squared (adjusted) = 55.2%
s = 2.360 with 15 - 2 = 13 degrees of freedom

Source	Sum of Squares	df	Mean Square	F-ratio
Regression	101.477	1	101.477	18.2
Residual	72.3804	13	5.56773	

Variable	Coefficient	s.e. of Coeff	t-ratio	prob
Constant	6.92904	1.557	4.45	0.0007
3-year	0.719157	0.1685	4.27	0.0009

The value of $t \approx 4.27$. The P-value of 0.0009 means that the association we see in the data is unlikely to occur by chance. We reject the null hypothesis, and conclude that there is strong evidence of a linear relationship between the rates of return for 3-year and 5-year periods. Provided that these mutual funds are representative of all large cap mutual funds, mutual funds with higher 3-year returns tend to have higher 5-year returns.

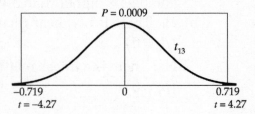

5. **Golf**
   a) She should have performed a one-way ANOVA and an F-test.
   b) H_0: The mean distance from the target is the same for all four clubs.
      $(\mu_1 = \mu_2 = \mu_3 = \mu_4)$
      H_A: The mean distances from the target are not all the same.

c) With a *p*-value equal to 0.0245, we reject the null hypothesis and conclude that the mean distances from the target are not the same for all four clubs.

d) **Randomization Condition**: The clubs used were randomized.
Similar Variance Condition: The boxplots should show similar spread.
Nearly Normal Condition: The histogram of the residuals should be unimodal and symmetric.

e) She might want to perform a multiple comparison test to determine which club is best.

7. **North and hard**

a) *Derby* is an indicator variable.

b) According to the model, the mortality rate north of Derby is, on average, 158.9 deaths per 100 000 higher than south of Derby, after allowing for the linear effects of the water hardness.

c) The regression with the indicator variable appears to be a better regression. The indicator has a highly significant (*p*-value equal to 0.0001) coefficient, and the R^2 for the overall model has increased from 43% to 56.2%.

9. **Bowling**

a) H_0: The mean number of pins knocked down is the same for all three weights.
$(\gamma_{Low} = \gamma_{Med} = \gamma_{High})$
H_A: The mean number of pins knocked down is not all the same.
H_0: The mean number of pins knocked down is the same whether walking or standing. $(\tau_S = \tau_W)$
H_A: The mean number of pins knocked down is not the same for the two approaches.

b) The *Weight* sum of squares has 3 – 1 = 2 degrees of freedom.
The *Approach* sum of squares has 2 – 1 = 1 degrees of freedom.
The error sum of squares has (24 – 1) – 2 – 1 = 20 degrees of freedom.

c) If the interaction plot shows any evidence of not being parallel, she should fit an interaction term, using 2(1) = 2 degrees of freedom.

11. **Resume fraud**

To estimate the true percentage of people who have misrepresented their backgrounds to within ± 5%, the company would have to perform about 406 random checks.

$$ME = z^* \sqrt{\frac{\hat{p}\hat{q}}{n}}$$

$$0.05 = 2.326\sqrt{\frac{(0.25)(0.75)}{n}}$$

$$n = \frac{(2.326)^2(0.25)(0.75)}{(0.05)^2}$$

$$n \approx 406 \text{ random checks}$$

13. **Nuclear power**
    a) Both scatterplots are straight enough. Larger plants are more expensive. Plants got more expensive over time.
    b) $\widehat{cost} = 111.741 + 0.423831(mwatts)$.

    According to the linear model, the cost of nuclear plants increased by \$42 383, on average, for each Megawatt of power.
    c) **Straight Enough Condition**: The plot of residuals versus predicted values looks reasonably straight. **Independence Assumption**: It is reasonable to think of these nuclear power plants as representative of all nuclear power plants. **Nearly Normal Condition**: The Normal probability plot looks straight enough.
    Does the Plot Thicken? Condition: The residuals plot shows reasonably constant spread.
    d) H_0: The power plant capacity (*Mwatts*) is not linearly associated with the cost of the plant. $(\beta_{Mwatts} = 0)$
    H_A: The power plant capacity (*Mwatts*) is linearly associated with the cost of the plant. $(\beta_{Mwatts} \neq 0)$
    With $t = 2.93$, and 30 degrees of freedom, the *p*-value equals 0.0064. Since the *p*-value is so small, we reject the null hypothesis; *Mwatts* is linearly associated with the cost of the plant.
    e) From the regression model we can find the cost of a 1000 *Mwatt* plant: $111.741 + 0.42383(1000) \approx 535.57$ which is \$535 570 000.
    f) The R^2 means that 22.3% of the variation in *cost* of nuclear plants can be accounted for by the linear relationship between *Cost* and the size of the plant, measured in megawatts (*Mwatts*). A scatterplot of residuals against *Date* shows a strong linear patter, so it looks like *Date* could account for more of the variation.
    g) The coefficients of *Mwatts* changed very little from the simple linear regression to the multiple regression.
    h) Because the coefficient changed little when we added *Date* to the model, we can expect that *Date* and *Mwatts* are relatively uncorrelated. In fact, their correlation is 0.02.

15. Sleep

a) H_0: The mean hours slept are the same for both sexes. $(\gamma_F = \gamma_M)$
   H_A: The mean hours slept are not the same for both sexes.
   H_0: The mean hours slept are the same for all years. $(\tau_{Fr} = \tau_{So} = \tau_{Jr} = \tau_{Sr})$
   H_A: The mean hours slept are not the same for all years.

b) With all three p-values so large, none of the effects appear to be significant.

c) There are a few outliers. Based on the residuals plot, the Similar Variance Condition appears to be met, but we do not have a Normal probability plot of residuals. The main concern is that we may not have enough power to detect differences between groups. In particular it may be that upper year women (juniors and seniors) sleep more than other groups, but we would need more data to tell.

17. Pregnancy

a) H_0: The proportion of live births is the same for women under the age of 38 as it is for women over the age of 38. $\left(p_{<38} = p_{\geq 38} \text{ or } p_{<38} - p_{\geq 38} = 0\right)$

   H_A: The proportion of live births is different for women under the age of 38 than for women over the age of 38. $\left(p_{<38} \neq p_{\geq 38} \text{ or } p_{<38} - p_{\geq 38} \neq 0\right)$

   Randomization Condition: Assume that the women studied are representative of all women.

   10% Condition: 157 and 89 are both less than 10% of all women.

   Independent Samples Condition: The groups are not associated.

   Success/Failure Condition: $n\hat{p}$ (under 38) = 42, $n\hat{q}$ (under 38) = 115, $n\hat{p}$ (38 and over) = 7, and $n\hat{q}$ (38 and over) = 82 are not all greater than 10, since the observed number of live births is only 7. However, if we check the pooled value, $n\hat{p}_{pooled}$ (38 and over) = (89)(0.191) = 17, all of the samples are large enough.

   Since the conditions have been satisfied, we will model the sampling distribution of the difference in proportion with a Normal model with mean 0 and standard deviation estimated by

   $$SE_{pooled}\left(\hat{p}_{<38} - \hat{p}_{\geq 38}\right) = \sqrt{\frac{\hat{p}_{pooled}\hat{q}_{pooled}}{n_{<38}} + \frac{\hat{p}_{pooled}\hat{q}_{pooled}}{n_{\geq 38}}} = \sqrt{\frac{\left(\frac{49}{246}\right)\left(\frac{197}{246}\right)}{157} + \frac{\left(\frac{49}{246}\right)\left(\frac{197}{246}\right)}{89}} \approx 0.0530 .$$

   The observed difference between the proportions is:
   0.2675 – 0.0787 = 0.1888.

   Since the P-value = 0.0004 is low, we reject the null hypothesis. There is strong evidence to suggest a difference in the proportion of live births for women under 38 and women 38 and over at this clinic.

   In fact, the evidence suggests that women under 38 have a higher proportion of live births.

   $z = \dfrac{0.1888 - 0}{0.0530}$

   $z \approx 3.56$

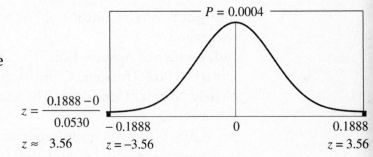

b) H_0: Age and birth rate are independent.

H_A: There is an association between age and birth rate

Counted Data Condition: The data are counts.

Randomization Condition: Assume that these women are representative of all women.

Expected Cell Frequency Condition: The expected counts are all greater than 5.

	Live birth *(Obs / Exp)*	No live birth *(Obs / Exp)*
Under 38	42 / 31.272	115 / 125.73
38 and over	7 / 17.728	82 / 71.27

Under these conditions, the sampling distribution of the test statistic is χ^2 on 1 degree of freedom. We will use a chi-square test for independence.

$$\chi^2 = \sum_{all\ cells} \frac{(Obs - Exp)^2}{Exp} \approx 12.70, \text{ and the P-value} \approx 0.0004.$$

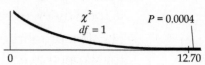

Since the P-value ≈ 0.0004 is low, we reject the null hypothesis. There is strong evidence of an association between age and birth rate. Younger mothers tend to have higher birth rates.

c) A two-proportion z-test and a chi-square test for independence with 1 degree of freedom are equivalent. $z^2 = (3.563944)^2 = 12.70 = \chi^2$. The P-values are both the same.

19. Old Faithful

a) There is a moderate, linear, positive association between duration of the previous eruption and interval between eruptions for Old Faithful. Relatively long eruptions appear to be associated with relatively long intervals until the next eruption.

b) H_0: There is no linear relationship between duration of the eruption and interval until the next eruption. $(\beta_1 = 0)$

H_A: There is a linear relationship between duration of the eruption and interval until the next eruption. $(\beta_1 \neq 0)$

c) **Straight Enough Condition**: The scatterplot is straight enough to try linear regression.

Independence Assumption: The residuals plot shows no pattern.

Does the Plot Thicken? Condition: The spread of the residuals is consistent.

Nearly Normal Condition: The histogram of residuals is unimodal and symmetric.

Since the conditions for inference are satisfied, the sampling distribution of the regression slope can be modelled by a Student's t-model with (222 – 2) = 220 degrees of freedom. We will use a regression slope *t*-test. The equation of the line of best fit for these data points is: $\widehat{Interval} = 33.9668 + 10.3582(Duration)$.

d) The value of $t \approx 27.1$. The P-value of essentially 0 means that the association we see in the data is unlikely to occur by chance. We reject the null hypothesis, and conclude that there is strong evidence of a linear relationship between duration and interval. Relatively long eruptions tend to be associated with relatively long intervals until the next eruption.

e) The regression equation predicts that an eruption with duration of 2 minutes will have an interval until the next eruption of $33.9668 + 10.3582(2) = 54.6832$ minutes. $(t_{220}^{*} \approx 1.9708)$

$$\hat{y}_v \pm t_{n-2}^{*} \sqrt{SE^2(b_1) \cdot (x_v - \bar{x})^2 + \frac{s_e^2}{n}}$$

$$= 54.6832 \pm (1.9708) \sqrt{0.3822^2 \cdot (2 - 3.57613)^2 + \frac{6.159^2}{222}}$$

$$\approx (53.24, 56.12)$$

We are 95% confident that, after a 2-minute eruption, the mean length of time until the next eruption will be between 53.24 and 56.12 minutes.

f) The regression equation predicts that an eruption with duration of 4 minutes will have an interval until the next eruption of $33.9668 + 10.3582(4) = 75.3996$ minutes. $(t_{220}^{*} \approx 1.9708)$

$$\hat{y}_v \pm t_{n-2}^{*} \sqrt{SE^2(b_1) \cdot (x_v - \bar{x})^2 + \frac{s_e^2}{n} + s_e^2}$$

$$= 75.3996 \pm (1.9708) \sqrt{0.3822^2 \cdot (4 - 3.57613)^2 + \frac{6.159^2}{222} + 6.159^2}$$

$$\approx (63.23, 87.57)$$

We are 95% confident that the length of time until the next eruption will be between 63.23 and 87.57 minutes, following a 4-minute eruption.

21. Is Old Faithful getting older?

a) $\widehat{interval} = 35.2463 + 10.4348(duration) - 0.126316(day)$

b) H$_0$: After allowing for the effects of duration, *Interval* does not change over time. $(\beta_{Day} = 0)$

H$_A$: After allowing for the effects of duration, *Interval* does change over time. $(\beta_{Day} \neq 0)$

With a *p*-value equal to 0.0166, we reject the null hypothesis. It appears there is a change over time, after we account for the *Duration* of the previous eruption.

c) The coefficient for *Day* is not about the two variable association, but about the relationship between *Interval* and *Day* after allowing for the linear effects of *Duration*.

d) The amount of change is only about –0.126316 (60) = –7.58 seconds per day. This doesn't seem particularly meaningful, although we expect a change of about 46 minutes per year.

23. Preemies

H_0: The proportion of "preemies" who are of "subnormal height" as adults is the same as the proportion of normal birth weight babies who are.
$$(p_P = p_N \text{ or } p_P - p_N = 0)$$
H_A: The proportion of "preemies" who are of "subnormal height" as adults is greater than the proportion of normal birth weight babies who are.
$$(p_P > p_N \text{ or } p_P - p_N > 0)$$
Randomization Condition: Assume that these children are representative of all children.
10% Condition: 242 and 233 are both less than 10% of all children.
Independent Samples Condition: The groups are not associated.
Success/Failure Condition: $n\hat{p}$ (preemies) = 24, $n\hat{q}$ (preemies) = 218, $n\hat{p}$ (normal) = 12, and $n\hat{q}$ (normal) = 221 are all greater than 10, so the samples are both large enough.

Since the conditions have been satisfied, we will model the sampling distribution of the difference in proportion with a Normal model with mean 0 and standard deviation estimated by

$$SE_{pooled}(\hat{p}_P - \hat{p}_N) = \sqrt{\frac{\hat{p}_{pooled}\hat{q}_{pooled}}{n_P} + \frac{\hat{p}_{pooled}\hat{q}_{pooled}}{n_N}} = \sqrt{\frac{\left(\frac{36}{475}\right)\left(\frac{439}{475}\right)}{242} + \frac{\left(\frac{36}{475}\right)\left(\frac{439}{475}\right)}{233}} \approx 0.02429.$$

The observed difference between the proportions is:
0.09917 – 0.05150 = 0.04767.

Since the P-value = 0.0249 is low, we reject the null hypothesis. There is moderate evidence that "preemies" are more likely to be of "subnormal height" as adults than children of normal birth weight.

$$z = \frac{0.04767 - 0}{0.02429}$$

$$z \approx 1.96$$

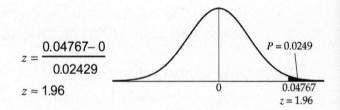

25. More teen traffic

a) This regression model has a much higher R^2. We should look at the residuals plot to make sure that there is no pattern and that the residuals are smaller than in the regression in the previous exercise.

b) Female teen traffic deaths are estimated to have increased at the rate of 20.4 per year, after allowing for the linear effects of male traffic deaths.

c) The number of female teen traffic deaths has probably been decreasing over time. In the second regression, male and female traffic deaths are likely to be collinear.

27. Typing again

a) H_0: Typing speed is the same when the television is on or when it is off. ($\gamma_{On} = \gamma_{Off}$)

H_A: Typing speeds are not the same when the television is on or when it is off.

H_0: Typing speed is the same when the music is on or when it is off. ($\tau_{On} = \tau_{Off}$)

H_A: Typing speeds are not the same when the music is on or when it is off.

b) Based on the partial boxplots, both effects appear to be small and probably not statistically significant.

c) The interaction plot is not parallel so the Additive Enough Condition is not met. Therefore, an interaction term should be fit.

d) The interaction term may be real. It appears that the effect of television is stronger with the music off than on.

e) All three effects have large p-values. *Music* has a p-value equal to 0.4312, *Television* has a p-value equal to 0.2960, and the interaction effect has a p-value equal to 0.5986. None of these effects appear to be significant.

f) None of the effects seem strong. He seems to type just as well with the music and/or the television on.

g) $s = \sqrt{\dfrac{197.5}{28}} = 2.66$ words per minute. This seems consistent with the size of the variation shown in the partial boxplots.

h) Turning the television and/or the music on will not increase his typing speed, nor will it decrease it.

i) The Normal probability plot of residuals should be straight. The residuals plot shows constant spread across the predicted values. The Similar Variance Condition appears to be met. For the levels of the factors that he used, it seems that to the level of experimental error present, neither the music nor the television affects his typing speed. If he wanted to see smaller effects, he would have to increase his sample size.

29.   **Births**

a) $\widehat{Births} = 4422.98 - 15.1898(Age) - 1.8983(Year)$

b)   According to the model, births seem to be declining, on average, at a rate of 1.89 births per 1000 women each year, after allowing for differences across the age of women. According to the model, from 1990 to 1999, birth rate decreased by about 17 births per 1000 women.

c)   The scatterplot shows both clumping and a curved relationship. We might want to re-express *Births* or add a quadratic term to the model. The clumping is due to having data for each year of the decade of the 90s for the age bracket of women. It probably does not indicate a failure of the linearity assumption.

31. **Learning math**

a)   H₀: The mean score of Accelerated Math students is the same as the mean score of traditional students. $\left(\mu_A = \mu_T \text{ or } \mu_A - \mu_T = 0\right)$

H_A: The mean score of Accelerated Math students is different from the mean score of traditional students. $\left(\mu_A \neq \mu_T \text{ or } \mu_A - \mu_T \neq 0\right)$

Independent Groups Assumption: Scores of students from different classes should be independent.

Randomization Condition: Although not specifically stated, classes in this experiment were probably randomly assigned to learn either Accelerated Math or traditional curricula..

Nearly Normal Condition: We don't have the actual data, so we can't check the distribution of the sample. However, the samples are large. The Central Limit Theorem allows us to proceed.

Since the conditions are satisfied, it is appropriate to model the sampling distribution of the difference in means with a Student's *t*-model, with 459.24 degrees of freedom (from the approximation formula).

We will perform a two-sample *t*-test. The sampling distribution model has mean 0, with standard error: $SE(\bar{y}_A - \bar{y}_T) = \sqrt{\dfrac{84.29^2}{231} + \dfrac{74.68^2}{245}} \approx 7.3158$.

The observed difference between the mean scores is 560.01 – 549.65 = 10.36

Since the P-value = 0.1574, we fail to reject the null hypothesis. There is no evidence that the Accelerated Math students have a different mean score on the pretest than the traditional students.

$$t = \frac{(\bar{y}_A - \bar{y}_T) - (0)}{SE(\bar{y}_A - \bar{y}_T)}$$

$$t \approx \frac{10.36}{7.3158}$$

$$t \approx 1.42$$

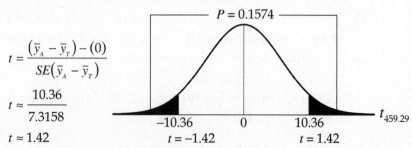

b)   H₀: Accelerated Math students do not show significant improvement in test scores. The mean individual gain for Accelerated Math is zero. $\left(\mu_d = 0\right)$

H_A: Accelerated Math students show significant improvement in test scores. The mean individual gain for Accelerated Math is greater than zero. $(\mu_d > 0)$

Paired Data Assumption: The data are paired by student.

Randomization Condition: Although not specifically stated, classes in this experiment were probably randomly assigned to learn either Accelerated Math or traditional curricula.

Nearly Normal Condition: We don't have the actual data, so we cannot look at a graphical display, but since the sample is large, it is safe to proceed.

The Accelerated Math students had a mean individual gain of $\bar{d} = 77.53$ points and a standard deviation of 78.01 points. Since the conditions for inference are satisfied, we can model the sampling distribution of the mean individual gain with a Student's t model, with $231 - 1 = 230$ degrees of freedom, $t_{230}\left(0, \dfrac{78.01}{\sqrt{231}}\right)$.

We will perform a paired t-test.

$$t = \frac{\bar{d} - 0}{\dfrac{s_d}{\sqrt{n}}}$$

Since the P-value is essentially 0, we reject the null hypothesis. There is strong evidence that the mean individual gain is greater than zero. The Accelerated Math students showed significant improvement.

$$t = \frac{77.53 - 0}{\dfrac{78.01}{\sqrt{231}}}$$

$$t \approx 15.11$$

c) H_0: Students taught using traditional methods do not show significant improvement in test scores. The mean individual gain for traditional methods is zero. $(\mu_d = 0)$

H_A: Students taught using traditional methods show significant improvement in test scores. The mean individual gain for traditional methods is greater than zero. $(\mu_d > 0)$

Paired Data Assumption: The data are paired by student.

Randomization Condition: Although not specifically stated, classes in this experiment were probably randomly assigned to learn either Accelerated Math or traditional curricula.

Nearly Normal Condition: We don't have the actual data, so we cannot look at a graphical display, but since the sample is large, it is safe to proceed.

The students taught using traditional methods had a mean individual gain of $\bar{d} = 39.11$ points and a standard deviation of 66.25 points. Since the conditions for inference are satisfied, we can model the sampling distribution of the mean individual gain with a Student's t model, with $245 - 1 = 244$ degrees of freedom,

$$t = \frac{\bar{d} - 0}{\dfrac{s_d}{\sqrt{n}}}$$

$$t = \frac{39.11 - 0}{\dfrac{66.25}{\sqrt{245}}}$$

$$t \approx 9.24$$

$$t_{244}\left(0, \frac{66.25}{\sqrt{245}}\right).$$

We will perform a paired *t*-test.

Since the P-value is essentially 0, we reject the null hypothesis. There is strong evidence that the mean individual gain is greater than zero. The students taught using traditional methods showed significant improvement.

d) H₀: The mean individual gain of Accelerated Math students is the same as the mean individual gain of traditional students. $\left(\mu_{dA} = \mu_{dT} \text{ or } \mu_{dA} - \mu_{dT} = 0\right)$

Hₐ: The mean individual gain of Accelerated Math students is greater than the mean individual gain of traditional students. $\left(\mu_{dA} > \mu_{dT} \text{ or } \mu_{dA} - \mu_{dT} > 0\right)$

Independent Groups Assumption: Individual gains of students from different classes should be independent.

Randomization Condition: Although not specifically stated, classes in this experiment were probably randomly assigned to learn either Accelerated Math or traditional curricula.

Nearly Normal Condition: We don't have the actual data, so we can't check the distribution of the sample. However, the samples are large. The Central Limit Theorem allows us to proceed.

Since the conditions are satisfied, it is appropriate to model the sampling distribution of the difference in means with a Student's *t*-model, with 452.10 degrees of freedom (from the approximation formula).

We will perform a two-sample *t*-test. The sampling distribution model has mean 0, with standard error: $SE(\bar{d}_A - \bar{d}_T) = \sqrt{\dfrac{78.01^2}{231} + \dfrac{66.25^2}{245}} \approx 6.6527$.

The observed difference between the mean scores is 77.53 – 39.11 = 38.42

Since the P-value is less than 0.0001, we reject the null hypothesis. There is strong evidence that the Accelerated Math students have an individual gain that is significantly higher than the individual gain of the students taught using traditional methods.

$$t = \frac{(\bar{d}_A - \bar{d}_T) - 0}{SE(\bar{d}_A - \bar{d}_T)}$$

$$t = \frac{38.42 - 0}{6.6527}$$

$$t \approx 5.78$$

33. Dairy sales

a) Since the CEO is interested in the association between cottage cheese sales and ice cream sales, the regression analysis is appropriate.

b) There is a moderate, linear, positive association between cottage cheese and ice cream sales. For each additional million pounds of cottage cheese sold, an average of 1.19 million pounds of ice cream are sold.

c) The regression will not help here. A paired *t*-test will tell us whether there is an average difference in sales.

d) There is evidence that the company sells more cottage cheese than ice cream, on average.

e) In part a), we are assuming that the relationship is linear, that errors are independent with constant variation, and that the distribution of errors is Normal.

   In part c), we are assuming that the observations are independent and that the distribution of the differences is Normal. This may not be a valid assumption, since the histogram of differences looks bimodal.

f) The equation of the regression line is

   $\widehat{IceCream} = -26.5306 + 1.19334(CottageCheese)$. In a month in which 82 million pounds of cottage cheese are sold we expect to sell:

   $\widehat{IceCream} = -26.5306 + 1.19334(82) = 71.32$ million pounds of ice cream.

g) Assuming the conditions for inference are satisfied, the sampling distribution of the regression slope can be modelled by a Student's *t*-model with (12 – 2) = 10 degrees of freedom. We will use a regression slope *t*-interval, with 95% confidence.

   $b_1 \pm t^*_{n-2} \times SE(b_1) = 1.19334 \pm (2.228) \times 0.4936 \approx (0.09, 2.29)$

h) We are 95% confident that the mean number of pounds of ice cream sold increases by between 0.09 and 2.29 pounds for each additional pound of cottage cheese sold.

35. Javelin

a) H_0: Type of javelin has no effect on the distance of her throw. $(\gamma_P = \gamma_S)$
   H_A: Type of javelin does have an effect on the distance of her throw.
   H_0: Preparation has no effect on the distance of her throw. $(\tau_W = \tau_C)$
   H_A: Preparation does have an effect on the distance of her throw.

b) The partial boxplots show that both factors seem to have an effect on the distance of her throw.
   The interaction plot is parallel. The Additive Enough Condition is met. No interaction term is needed.
   Randomization Condition: The experiment was performed in random order.
   Similar Variance Condition: The side-by-side boxplots have similar spreads. The residuals plot shows no pattern and no systematic change in spread.
   Nearly Normal Condition: The Normal probability plot of the residuals should be straight.

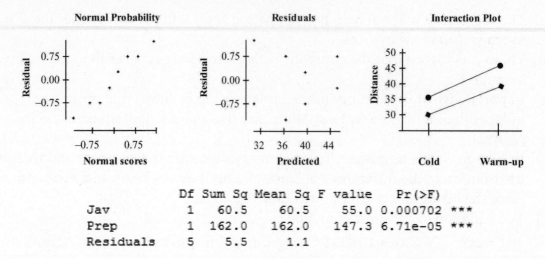

	Df	Sum Sq	Mean Sq	F value	Pr(>F)	
Jav	1	60.5	60.5	55.0	0.000702	***
Prep	1	162.0	162.0	147.3	6.71e-05	***
Residuals	5	5.5	1.1			

With very small *p*-values, both effects are significant. It appears that depending on the cost, the premium javelin may be worth it—it increases the distance about 5.5 metres, on average. Warming up increases distance about 9 metres, on average. She should always warm up and consider using the premium javelin.

37. **Infliximab**
H_0: The remission rates are the same for the three groups.
H_A: The remission rates are different for the three groups.
Counted Data Condition: The data are counts.
Randomization Condition: Assume that these patients are representative of all patients.
Expected Cell Frequency Condition: The expected counts are all greater than 5.

	Placebo (Obs / Exp)	5 mg (Obs / Exp)	10 mg (Obs / Exp)
Remission	23 / 38.418	44 / 39.466	50 / 39.116
No Remission	87 / 71.582	69 / 73.534	62 / 72.884

Under these conditions, the sampling distribution of the test statistic is χ^2 on 2 degrees of freedom. We will use a chi-square test for homogeneity.

$$\chi^2 = \sum_{all\ cells} \frac{(Obs - Exp)^2}{Exp} \approx 14.96,$$

and the P-value ≈ 0.0006.
Since the P-value ≈ 0.0006 is low, we reject the null hypothesis. There is strong evidence that the remission rates are different in the three groups. Patients receiving 10 mg of Infliximab have higher remission rates than the other groups. These data indicate that continued treatment with Infliximab is of value to Crohn's disease patients who exhibit a positive initial response to the drug.

39. **TV and athletics**
    a) H_0: The mean hours of television watched is the same for all three groups.
       $$\left(\mu_1 = \mu_2 = \mu_3\right)$$
       H_A: The mean hours of television watched are not the same for all three groups.
    b) The variance for the *none* group appears to be slightly smaller, and there are outliers in all three groups. We do not have a Normal probability plot of the residuals, but we suspect that the data may not be Normal enough.
    c) The ANOVA *F*-test indicates that athletic participation is significant with a *p*-value equal to 0.0167. We conclude that the number of television hours watched is not the same for all three types of athletic participation.
    d) It seems that the differences are evident even when the outliers are removed. The conclusion seems valid.

41. **Weight loss**
    Randomization Condition: The respondents were randomly selected from among the clients of the weight loss clinic.

    Nearly Normal Condition: The histogram of the number of pounds lost for each respondent is unimodal and symmetric, with no outliers.
    The clients in the sample had a mean weight loss of 9.15 pounds, with a standard deviation of 1.94733 pounds. Since the conditions have been satisfied, construct a one-sample *t*-interval, with 20 – 1 = 19 degrees of freedom, at 95% confidence.
    $$\bar{y} \pm t^*_{n-1}\left(\frac{s}{\sqrt{n}}\right) = 9.15 \pm t^*_{19}\left(\frac{1.94733}{\sqrt{20}}\right) \approx (8.24, 10.06)$$
    We are 95% confident that the mean weight loss experienced by clients of this clinic is between 8.24 and 10.06 pounds. Since 10 pounds is contained within the interval, the claim that the program will allow clients to lose 10 pounds in a month is plausible. Answers may vary, depending on the chosen level of confidence.

43. **Education vs. income**
    a) **Straight Enough Condition**: The scatterplot is straight enough to try linear regression.
       Independence Assumption: The residuals plot shows no pattern.
       Does the Plot Thicken? Condition: The spread of the residuals is consistent.
       Nearly Normal Condition: The Normal probability plot is reasonably straight.
       Since the conditions for inference are satisfied, the sampling distribution of the regression slope can be modelled by a Student's *t*-model with (57 – 2) = 55 degrees of freedom. We will use a regression slope *t*-test. The equation of the line of best fit for these data points is: $\widehat{Income} = 5970.05 + 2444.79(Education)$.
    b) The value of $t \approx 5.19$. The P-value of less than 0.0001 means that the association we see in the data is unlikely to occur by chance. We reject the null hypothesis, and conclude that there is strong evidence of a linear relationship between

education level and income. Cities in which median education level is relatively high have relatively high median incomes.

c) If the data were plotted for individuals, the association would appear to be weaker. Individuals vary more than averages.

d) $b_1 \pm t^*_{n-2} \times SE(b_1) = 2444.79 \pm (2.004) \times 471.2 \approx (1500, 3389)$

We are 95% confident that each additional year of median education level in a city is associated with an increase of between \$1500 and \$3389 in median income.

e) The regression equation predicts that a city with a median education level of 11 years of school will have a median income of $5970.05 + 2444.79(11) = \$32862.74$ ($t^*_{55} \approx 1.6730$)

$$\hat{y}_v \pm t^*_{n-2} \sqrt{SE^2(b_1) \cdot (x_v - \bar{x})^2 + \frac{s_e^2}{n}}$$

$$= 32862.74 \pm (1.6730) \sqrt{471.2^2 \cdot (11 - 10.9509)^2 + \frac{2991^2}{57}}$$

$$\approx (32199, 33527)$$

We are 90% confident that cities with 11 years for median education level will have an average income of between \$32 199 and \$33 527.

Chapter 29: Rank-Based Nonparametric Tests

1. Course grades again

If we assume that the rank sum has a normal distribution, its mean is

$$\frac{n_1(n_1+n_2+1)}{2}=\frac{5\times(5+6+1)}{2}=30$$

and the variance is $\dfrac{n_1 n_2(n_1+n_2+1)}{12}=\dfrac{5\times6\times(5+6+1)}{12}=30.$

$z=\dfrac{24-30}{\sqrt{30}}=-1.10$ and the p-value is $2\times0.1357=0.2714$ (for a non-directional test).

In the chapter, we found that the p-value for the Wilcoxon rank sum test is greater than 0.10.

3. Music students

The ranks are shown below:

Joe(14) Bini(13) **Jack(12) Bob(11) Shuang(10)** John(9) Sally(8) **Mary(7)** Mo(6)
Celina(5) Zheng (4) **Aditi (3)** Rodrigo(2) Astrud(1)

The sum of ranks of the students from Lawrence School of Music is 14 + 12 + 11 + 10 + 7 + 5 + 3 = 62 and the sum of ranks of the students in the other school is 13 + 9 + 8 + 6 + 4 + 2 + 1 = 43.

From the table, for sample sizes of 7 and 7, the value closest to 43 is 39, and the probability to the left of 39 is approximately 0.05. Hence the P-value for our observed sum of 43 is greater than 0.05 × 2, after doubling the lower tail area for our two-tailed test. The difference is still not significant at the 5% level.

5. Downloading again

Kruskal-Wallis Test: Time(sec) versus Time of day

Kruskal-Wallis Test on Time(sec)

Time of day	N	Median	Avg Rank	Z
Early (7a.m.)	16	91.50	10.6	–4.87
Evening (5p.m.)	16	264.50	38.4	4.88
Latenight (12a.m.)	16	198.50	24.5	–0.01
Overall	48		24.5	

H = 31.64 DF = 2 P = 0.000
H = 31.65 DF = 2 P = 0.000 (adjusted for ties)

$$H = \frac{12}{N(N+1)} \sum \frac{T_i^2}{n_i} - 3(N+1) = 31.6$$

$\chi^2_{2,0.05} = 5.991$. $H > \chi^2_{2,0.05}$ and so reject H_0: all populations have the same distributions.

The results from the one-way ANOVA is copied below, and based on that we reject the null hypothesis of equal population means.

One-way ANOVA: Time(sec) versus Time of day

Source	DF	SS	MS	F	P
Time of day	2	204641	102320	46.03	0.000
Error	45	100020	2223		
Total	47	304661			

S = 47.15 R-Sq = 67.17% R-Sq(adj) = 65.71%

```
                                      Individual 95% CIs For Mean Based on
                                      Pooled StDev
Level              N    Mean   StDev  -----+---------+---------+---------+----
Early(7a.m.)       16  113.38  47.65  (---*---)
Evening(5p.m.)     16  273.31  52.19                                (---*---)
Latenight(12a.m.)  16  193.06  40.90                   (---*---)
                                      -----+---------+---------+---------+----
                                         120       180       240       300
```

Pooled StDev = 47.15

7. **Extreme values** The results from the Kruskal-Wallis test and one-way ANOVA are given below. The one-way ANOVA shows a greater difference. *F*-value changes from 11.91 to 43.97 as the value of the Kruskal-Wallis statistics (H) only changes from 13.74 to 17.43.

Kruskal-Wallis Test: pain2 versus Drug

```
Kruskal-Wallis Test on pain2

Drug       N Median Ave Rank      Z
A          9 2.000       5.0  -4.17
B          9 6.000      18.0   1.85
C          9 6.000      19.0   2.31
Overall   27 14.0

H = 17.43 DF = 2 P = 0.000
H = 17.98 DF = 2 P = 0.000 (adjusted for ties)
```

One-way ANOVA: pain2 versus Drug

Source	DF	SS	MS	F	P
Drug	2	104.22	52.11	43.97	0.000
Error	24	28.44	1.19		
Total	26	132.67			

S = 1.089 R-Sq = 78.56% R-Sq(adj) = 76.77%

```
                               Individual 95% CIs For Mean Based on
                               Pooled StDev
Level  N   Mean   StDev   ----+---------+---------+---------+-----
A      9   1.667  0.866   (----*----)
B      9   5.778  1.481                            (----*----)
C      9   5.889  0.782                            (----*----)
                          ----+---------+---------+---------+-----
                            1.5       3.0       4.5       6.0
```

Pooled StDev = 1.089

9. **Washing again**
   Friedman Test: Score versus Temp blocked by Cycle

   S = 10.88 DF = 3 P = 0.012
   S = 11.15 DF = 3 P = 0.011 (adjusted for ties)

```
                  Sum
            Est   of
Temp      N Median Ranks
Cold-cold 4 3.469  4.0
Cold-warm 4 4.419  8.0
Hot-hot   4 6.806 14.5
Warm-hot  4 6.331 13.5
```

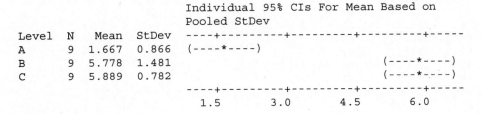

$$S = \frac{12}{bt(t+1)}\sum_i T_i^2 - 3b(t+1) = \frac{12}{4\times4(4+1)}\left(4^2+8^2+14.5^2+13.5^2\right) - 3\times4(4+1) = 10.88$$

$\chi^2_{3,0.05} = 7.815$. $H > \chi^2_{2,0.05}$ and so reject H$_0$: all populations have the same distributions.

The results from the ANOVA is copied below and based on that we reject the null hypothesis of equal population means.

Two-way ANOVA: Score versus Temp, Cycle

```
Source DF     SS      MS     F     P
Temp    3 33.2519 11.0840 23.47 0.000
Cycle   3  7.1969  2.3990  5.08 0.025
Error   9  4.2506  0.4723
Total  15 44.6994

S = 0.6872 R-Sq = 90.49% R-Sq(adj) = 84.15%
```

11. Women's 1500-m skate

a) There are 7 negative differences and 10 positive differences, and the p-value of the sign test is 0.6291. The P-value is large and so there's insufficient evidence to declare any difference.

```
Pair InnerTime OuterTime difference
 1     129.24       *          *
 2     125.75    122.34      3.41
 3     121.63    122.12     -0.49
 4     122.24    123.35     -1.11
 5     120.85    120.45      0.40
 6     122.19    123.07     -0.88
 7     122.15    122.75     -0.60
 8     122.16    121.22      0.94
 9     121.85    119.96      1.89
10     121.17    121.03      0.14
11     124.77    118.87      5.90
12     118.76    121.85     -3.09
13     119.74    120.13     -0.39
14     121.60    120.15      1.45
15     119.33    116.74      2.59
16     119.30    119.15      0.15
17     117.31    115.27      2.04
18     116.90    120.77     -3.87
```

Sign Test for Median: difference

```
Sign test of median = 0.00000 versus not = 0.00000

                 N N* Below Equal Above     P Median
difference 17 1      7      0    10 0.6291 0.1500
```

b) Wilcoxon Statistic = 93.0 and the p-value is 0.449. The P-value is large and so there's insufficient evidence to declare any difference.

Wilcoxon Signed Rank Test: Difference

```
Test of median = 0.000000 versus median not = 0.000000

                    N
                  for    Wilcoxon            Estimated
            N    Test   Statistic      P      Median
Difference  17    17         93.0  0.449      0.4650
```

c) The p-values for the paired *t*-test, Wilcoxon's signed rank test and sign test, are 0.391, 0.449, and 0.6291.

Paired T-Test and CI: InnerTime, OuterTime

```
Paired T for InnerTime - OuterTime

            N    Mean    StDev    SE Mean
```

Chapter 30: The Bootstrap

1. Resampling I

```
Sample
  1        2.0   2.0   2.0
  2        5.0   5.0   5.0
  3        8.0   8.0   8.0

  4        2.0   2.0   5.0
  5        2.0   2.0   8.0
  6        5.0   5.0   2.0
  7        5.0   5.0   8.0
  8        8.0   8.0   2.0
  9        8.0   8.0   5.0

 10        2.0   5.0   8.0
```

Samples with all three identical measurements, like {2.0, 2.0, 2.0}, are least likely since there is only one way such a sample can arise (with probability = 1/3 × 1/3 × 1/3). Other samples have higher probabilities since they can arise in multiple ways, e.g., selecting 2.0, 2.0, 5.0, or 2.0, 5.0, 2.0, or 5.0, 2.0, 2.0 gives rise to the same distinct sample.

3. Resampling III

```
Sample                                        mean
   1     (  2    2    6   10   )              5
   2     (  2    2    6   14   )              6
   3     (  2    2   10   14   )              7

   4     (  6    6    2   10   )              6
   5     (  6    6    2   14   )              7
   6     (  6    6   10   14   )              9

   7     (  10   10    2    6   )             7
   8     (  10   10    2   14   )             9
   9     (  10   10    6   14   )            10

  10     (  14   14    2    6   )             9
  11     (  14   14    2   10   )            10
  12     (  14   14    6   10   )            11
```

The standard deviation of these sample means is the standard error and is equal to 1.907.
The standard deviation of the four values (2.0, 6.0, 10.0, 14.0) is

$\sigma = \sqrt{\frac{\sum_i (x_i - \bar{x})^2}{n}} = 4.47$ and the theoretical formula gives

$$SE(\bar{X}) = \frac{\sigma}{\sqrt{n}} = \frac{4.47}{\sqrt{4}} = 2.235.$$

5. More popcorn The answers will differ, depending on the resamples drawn. The means of the 200 resamples are shown below:

Mean

7.825	4.700	7.075	6.125	5.775	6.050	7.175	5.725	3.550
7.850	7.675	4.650	7.400	5.200	6.475	7.475	9.075	5.975
6.400	7.650	6.150	5.650	7.500	4.875	7.650	6.475	6.275
6.750	7.075	6.150	6.600	4.450	6.150	8.150	5.775	5.800
4.775	6.225	5.675	6.550	6.825	5.550	6.800	6.775	5.875
7.300	6.675	4.375	4.850	7.350	4.900	5.350	5.775	6.575
6.675	5.625	7.325	5.700	6.650	5.925	6.625	6.250	4.800
8.325	8.925	3.875	7.575	7.000	6.150	7.175	4.625	8.375
5.950	6.500	6.400	7.400	6.425	5.475	3.900	6.650	8.725
7.475	7.150	8.675	8.350	6.300	5.725	5.875	5.850	9.950
6.575	7.775	7.225	5.500	6.900	6.575	6.600	6.625	7.825
8.325	6.900	6.450	7.975	5.400	5.800	6.600	7.025	5.250
8.925	7.725	6.750	8.150	3.750	7.000	6.325	6.650	6.600
7.625	6.875	8.325	6.150	7.425	7.750	5.800	7.725	8.075
5.700	6.525	6.075	7.075	5.450	6.800	6.125	4.700	6.300
7.050	5.700	6.250	6.125	7.150	5.675	5.650	5.825	5.275
6.450	5.000	6.575	5.800	6.600	7.200	4.650	4.950	5.800
7.450	6.500	6.550	6.725	8.675	5.650	5.975	6.150	6.600
7.075	7.450	7.200	7.975	7.000	8.825	8.050	5.875	6.775
5.750	5.825	6.375	4.375	8.025	5.800	7.675	6.750	8.500
5.625	7.350	7.250	5.000	8.000	5.400	7.775	6.500	9.275
6.700	7.550	4.875	5.700	5.175	6.675	6.450	7.250	5.775
6.350	7.425							

Mean and the standard deviation of the 200 sample means:

Descriptive Statistics: Mean

Variable	N	Mean	StDev
Mean	200	6.5449	1.1357

Bootstrap estimate of the standard error of the mean percentage of unpopped kernels = 1.1357.

The standard deviation (population) of the eight values in the data set is

$$\sigma = \sqrt{\frac{\sum_i (x_i - \bar{x})^2}{n}} = 3.4 \text{ and the theoretical formula gives } SE(\bar{X}) = \frac{\sigma}{\sqrt{n}} = \frac{3.4}{\sqrt{8}} = 1.2.$$

Note: The above simulation can be done using most statistical software. The following MINITAB macro was used in the answer.

gmacro
Bootstrap
let k1 = 8 # This is the sample size. The data must be in c1.
let k4 = 200 # The number of samples

```
do k5 = 1:k4
sample k1 c1 c2;
replace.
let c5(k5) = mean(c2)
enddo
name c5 'Mean'
Describe 'Mean';
Mean;
StDeviation;
N.
Histogram 'Mean';
Grid 2;
MGrid 2;
Grid 1;
Mgrid 1;
Bar.
endmacro
```

7. **Chips of chocolate**
   a) The answers will differ, depending on the resamples drawn. The medians of the 200 resamples are shown below:

Median

1214.0	1207.0	1231.5	1244.0	1214.0	1258.0	1258.0	1258.0
1229.0	1216.5	1200.0	1291.5	1258.0	1219.0	1216.5	1214.0
1231.5	1244.0	1258.0	1244.0	1258.0	1270.0	1167.5	1195.5
1251.0	1195.5	1264.0	1244.0	1200.0	1244.0	1251.0	1195.5
1258.0	1251.0	1251.0	1214.0	1219.0	1258.0	1195.5	1200.0
1219.0	1214.0	1191.0	1257.0	1238.5	1244.0	1214.0	1244.0
1209.5	1219.0	1282.5	1161.5	1219.0	1258.0	1229.0	1219.0
1216.5	1251.0	1269.5	1163.0	1282.5	1214.0	1219.0	1216.5
1325.0	1200.0	1219.0	1214.0	1295.0	1269.5	1282.5	1135.0
1231.5	1219.0	1219.0	1195.5	1191.0	1244.0	1207.0	1219.0
1222.0	1195.5	1264.0	1307.5	1231.5	1264.0	1244.0	1195.5
1231.5	1257.0	1258.0	1257.0	1251.0	1207.0	1216.5	1209.5
1174.5	1276.5	1229.0	1163.0	1229.0	1161.5	1282.5	1297.5
1231.5	1231.5	1207.0	1191.0	1257.0	1251.0	1310.0	1222.0
1207.0	1214.0	1251.0	1236.0	1231.5	1231.5	1270.0	1219.0
1244.0	1219.0	1244.0	1229.0	1238.5	1238.5	1257.0	1270.0
1251.0	1251.0	1244.0	1214.0	1251.0	1264.0	1257.0	1200.0
1276.5	1251.0	1219.0	1195.5	1282.5	1238.5	1251.0	1207.0
1231.5	1244.0	1251.0	1195.5	1207.0	1200.0	1205.0	1251.0
1264.0	1219.0	1216.5	1251.0	1264.0	1191.0	1191.0	1200.0

1216.5	1251.0	1229.0	1264.0	1216.5	1264.0	1251.0	1214.0
1258.0	1251.0	1238.5	1251.0	1167.5	1251.0	1257.0	1264.0
1231.5	1244.0	1251.0	1216.5	1244.0	1214.0	1244.0	1229.0
1251.0	1238.5	1264.0	1251.0	1219.0	1269.5	1251.0	1258.0
1209.5	1207.0	1195.5	1200.0	1200.0	1257.0	1295.0	1229.0

Mean and the standard deviation of the 200 sample medians:

Descriptive Statistics: Median

Variable	N	Mean	StDev
Median	200	1233.4	30.4

Bootstrap estimate of the standard error of the median = 30.4.

b) The answers will differ, depending on the resamples drawn. The means of the 200 resamples are shown below:

Mean

1215.31	1245.31	1198.00	1229.88	1264.88	1253.13	1236.06
1239.94	1243.19	1246.25	1244.50	1224.19	1251.81	1217.88
1269.44	1262.88	1228.19	1240.31	1256.00	1222.69	1250.38
1260.75	1249.50	1263.31	1261.75	1227.00	1217.50	1242.63
1221.38	1245.06	1238.88	1276.19	1226.25	1274.19	1286.75
1252.75	1247.81	1246.63	1213.13	1207.50	1267.88	1187.00
1244.38	1270.56	1220.13	1223.06	1222.56	1256.69	1264.56
1258.75	1257.31	1269.94	1260.75	1249.19	1250.44	1236.38
1181.50	1227.69	1202.31	1228.06	1256.94	1254.88	1227.81
1213.06	1252.88	1274.25	1253.69	1217.81	1255.13	1257.44
1206.19	1234.25	1199.13	1217.81	1274.94	1284.94	1257.00
1271.31	1231.69	1268.50	1255.44	1238.19	1253.63	1246.25
1249.13	1241.94	1238.88	1260.19	1230.44	1224.19	1229.69
1252.13	1254.00	1278.38	1235.50	1228.19	1242.94	1274.50
1253.56	1251.94	1252.19	1262.50	1238.31	1229.56	1227.25
1260.63	1244.19	1227.69	1217.69	1223.31	1253.75	1271.75
1227.00	1285.00	1208.44	1263.13	1222.94	1240.94	1223.94
1213.25	1274.13	1206.75	1229.88	1240.38	1249.06	1247.94
1194.25	1223.81	1223.81	1224.31	1247.38	1272.94	1236.25
1234.88	1259.44	1229.75	1195.81	1238.13	1261.94	1252.50
1222.00	1231.13	1248.44	1197.50	1240.69	1262.81	1221.75
1230.31	1200.56	1252.56	1255.06	1234.75	1237.31	1294.00
1285.63	1231.31	1278.94	1229.25	1247.69	1240.75	1232.19
1235.00	1195.94	1273.19	1255.75	1237.38	1236.56	1265.13
1209.00	1221.44	1266.25	1270.31	1242.31	1255.75	1180.81
1251.31	1230.56	1260.88	1222.69	1225.13	1231.88	1210.31
1219.00	1208.19	1217.00	1239.88	1263.50	1183.56	1270.38

| 1240.38 | 1230.75 | 1184.13 | 1264.19 | 1266.63 | 1238.88 | 1236.31 |
| 1257.75 | 1242.00 | 1236.44 | 1279.00 | | | |

Mean and the standard deviation of the 200 sample means:
Descriptive Statistics: Mean

Variable	N	Mean	StDev
Mean	200	1241.0	22.7

Bootstrap estimate of the standard error of the mean = 22.7.
Mean has less standard error than the median, so mean is a better estimator.

c) The formula for the standard error of the sample median is somewhat complicated and so bootstrap methods can be useful in estimating the standard error of the sample median. For the sample mean, a simple formula is available for the standard error, so it is not necessary to use bootstrap methods in part (b). It can be calculated as follows:
The standard deviation (population) of the eight values in the data set is

$$\sigma = \sqrt{\frac{\sum_i (x_i - \bar{x})^2}{n}} = 91.3 \text{ and the theoretical formula gives}$$

$$SE(\bar{X}) = \frac{\sigma}{\sqrt{n}} = \frac{91.3}{\sqrt{16}} = 22.8.$$

9. **More chocolate chips** The answers will differ, depending on the resamples drawn. The trimmed means of the 200 resamples are shown below:

Trimmed Mean

1214.14	1235.71	1175.93	1222.64	1238.86	1225.93	1226.79
1192.00	1205.29	1236.43	1207.79	1238.71	1255.36	1288.07
1226.14	1212.14	1258.36	1239.50	1265.71	1234.86	1229.71
1266.29	1240.07	1222.86	1242.50	1304.57	1217.86	1190.00
1229.14	1230.71	1237.07	1239.93	1251.71	1264.93	1277.50
1245.86	1267.64	1231.64	1192.79	1234.79	1233.29	1240.71
1254.43	1242.00	1229.29	1226.71	1255.43	1207.71	1233.07
1266.07	1217.50	1193.93	1245.43	1258.57	1247.36	1158.79
1238.57	1250.21	1220.29	1211.14	1194.86	1232.71	1214.00
1249.43	1234.07	1224.86	1271.21	1225.14	1232.79	1249.71
1243.71	1199.79	1229.14	1213.57	1214.14	1255.86	1214.71
1226.79	1194.64	1232.64	1243.21	1218.29	1245.64	1253.21
1250.36	1218.21	1244.64	1229.57	1225.43	1251.64	1220.93
1248.57	1240.36	1255.71	1191.43	1254.29	1241.64	1219.36
1227.86	1260.57	1240.50	1246.00	1223.29	1261.93	1238.93

1272.36	1237.07	1211.57	1262.14	1201.14	1201.36	1203.93
1235.21	1259.43	1228.07	1236.43	1263.79	1198.07	1215.00
1251.43	1230.93	1241.29	1210.57	1208.43	1228.43	1213.21
1285.29	1231.14	1257.50	1286.14	1222.29	1281.36	1261.43
1221.36	1260.21	1248.21	1207.00	1220.36	1227.86	1262.36
1247.50	1230.50	1239.79	1248.43	1221.50	1212.21	1209.64
1256.86	1193.21	1224.93	1211.93	1241.29	1275.43	1225.86
1208.71	1255.50	1232.07	1224.43	1206.71	1232.64	1227.21
1261.14	1248.14	1240.43	1185.93	1243.50	1243.43	1206.36
1230.86	1247.36	1224.86	1221.36	1265.93	1296.93	1192.36
1242.71	1238.00	1236.07	1188.43	1216.29	1237.07	1241.79
1277.64	1244.79	1219.43	1192.07	1230.14	1197.21	1260.14
1225.14	1158.86	1257.64	1214.43	1229.00	1248.57	1271.00
1228.36	1242.21	1263.64	1234.21			

Mean and the standard deviation of the 200 sample trimmed means:

Descriptive Statistics: Trimmed Mean

Variable	N	Mean	StDev
Trimmed Mean	200	1233.8	24.1

Bootstrap estimate of the standard error of the mean = 24.1.

Since there is no easy formula for the standard error, the Bootstrap methods are very useful in estimating it.

Note: The above simulation can be done using most statistical software. The following MINITAB macro was used in the answer:

```
gmacro
Bootstrap
#==============================================================
====================
# Please update info in this section
let k1 = 16      # This is the sample size (n)
let k4 = 200  # The number of samples (B)
let k10 = 5      # The percentage to be trimmed out from each end
#==============================================================
====================
Let k21 = round((k10/100)*k1)  # The number of values to be trimmed out from
each end
Let k22 = k1 - k21  # The values after k22 will be removed when calculating the
mean
do k5 = 1:k4
sample k1 c1 c2;  # C2 contains the bootstap sample
replace.
# The section below calculates the k10 percent trimmed mean of the
```

```
# bootstrap sample in c2
name c3 "sorted"  # Labels column C3 for sorted data
sort c2 'sorted'; # sorts data in C2
by c2.
Let k25 = 0 # Initializing a variable for the summation
do k30 = k21+1:k22
let k25 = k25 + c3(k30)
enddo
let k35=k25/(k1-2*k21) # This is the k10 percent Trimmed mean of the data in
column c2
let c5(k5) = k35
enddo
name c5 'Trimmed Mean'
Describe 'Trimmed Mean';
Mean;
StDeviation;
Minimum;
Maximum;
N.
Histogram 'Trimmed Mean';
Grid 2;
MGrid 2;
Grid 1;
Mgrid 1;
Bar.
Endmacro
```

11. **Lots of chocolate chips** The standard errors with $B = 50, 200$, and 500 are shown below:

Descriptive Statistics: Mean

Variable	N	StDev
Mean	50	20.1

Descriptive Statistics: Mean

Variable	N	StDev
Mean	200	23.1

Descriptive Statistics: Mean

Variable	N	StDev
Mean	500	22.9

There are no big differences in the estimates of the standard error after about 200 resamples. Individual results will vary.

13. **Reading scores** The answers will differ, depending on the resamples drawn. 200 resamples were drawn with replacement from each sample and their sample means and the differences between those sample means were calculated. These differences are shown below:

Difference in Means

-2.7071	-12.5101	-6.2727	-3.6010	-12.9545	-10.2677	-13.8081
-14.2576	-13.2828	-11.4949	-19.3131	-15.3535	-5.7172	-14.3030
-17.3131	-7.5000	-5.8182	-9.9899	-2.8889	-16.9848	-8.0758
-11.4040	-7.8182	-9.2576	-8.3586	-8.1869	-2.6768	-9.3838
-13.9343	-10.6212	-0.7980	-12.6818	-7.5859	-15.0253	-6.7525
-9.0859	-4.4192	-10.9848	-13.4091	-14.4646	-10.7374	-9.2121
-6.1566	-12.2323	-14.1616	0.7071	-6.8030	-18.9747	-8.0909
-11.4697	-14.3081	-11.2121	-12.6364	-4.0051	-15.7121	-5.6919
-8.7576	-8.4798	-16.1263	-8.5303	-3.6162	-11.4444	-13.0606
-9.4343	-16.7475	-3.2071	-11.2273	-8.7424	-11.5303	-9.4646
-9.5101	-9.6970	-10.2626	-7.5202	-12.1010	-11.3081	-1.7475
-13.3434	-11.8838	-13.6919	-11.2424	-10.2273	-12.5808	-12.9343
-15.5051	-6.3081	-14.0758	-5.8889	-13.8990	-13.6768	-15.7374
-3.8434	-14.2980	-5.9545	-1.1313	-12.6970	-10.3232	-10.8232
-3.4444	-10.8182	-13.5101	-3.4798	-16.4040	-5.6313	-16.3333
-11.8232	-8.4293	-20.1061	-17.5808	-6.0354	-11.2879	-13.2879
-11.3081	-9.8990	-12.8889	-6.7525	-16.0152	-13.7121	-7.3283
-11.2677	-13.5707	-7.9343	-11.8889	-4.5000	-8.2222	-17.2525
-10.3990	-10.4394	-15.1364	1.9899	-10.6111	-12.2828	-8.2677
-9.1313	-14.9444	-16.4596	-13.7879	-14.7424	-8.6263	-10.3636
-15.8081	-12.8737	-14.1869	-10.4495	-6.9899	-9.3485	-5.8384
-4.8636	-9.5000	-13.8889	-2.5101	-10.8838	-6.8687	-13.7929
-9.8232	-16.1667	-12.7172	-7.7828	-10.1364	-9.6818	-8.1919
-18.1212	-6.1970	-9.7778	-11.3990	-16.6616	-7.4899	-13.5505
-8.2374	-9.8737	-6.4091	-9.3788	-3.6111	-8.6212	-16.8535
-7.9495	-7.8788	-8.3232	-8.9192	-11.4040	-13.0354	-7.1010
-12.1313	-7.5101	-9.1970	-13.8333	-9.2374	-3.8687	-9.7778
-2.4242	-17.6667	-13.3485	-17.6970	-7.2980	-11.0859	-7.7626
-6.8586	-15.2929	-0.0556	-9.6010			

Mean and the standard deviation of these 200 differences:
Descriptive Statistics: Difference in Means

Variable	N	Mean	StDev
Difference in Me	200	-10.281	4.229

Bootstrap estimate of the standard error of the difference of the two sample means = 4.229.

The means and the standard deviations of the two original samples are:

```
               N   Mean  StDev  SE Mean
Control        22  41.8  16.6      3.5
NewActivities  18  51.7  11.7      2.8
```

The standard error of the difference between the two means calculated using the formula = $\sqrt{3.5^2 + 2.8^2} = 4.48$.

Note: The above simulation can be done using most statistical software. The following MINITAB macro was used in the answer.

```
gmacro
Bootstrap
let k1 = 22 # This is the size of the first sample
let k2 = 18 # This is the size of the second sample
let k4 = 200 # The number of samples
do k5 = 1:k4
sample k1 c1 c4;
replace.
let c7(k5) = mean(c4)
enddo
do k6 = 1:k4
sample k2 c2 c5;
replace.
let c8(k6) = mean(c5)
enddo
let c10 = c7 - c8
name c10 'Difference in Means'
Describe 'Difference in Means';
Mean;
StDeviation;
Minimum;
Maximum;
N.
Histogram 'Difference in Means';
Grid 2;
MGrid 2;
Grid 1;
Mgrid 1;
Bar.
Endmacro
```

15. Washing hands

a) The answers will differ, depending on the resamples drawn. 200 resamples were drawn with replacement from each sample and their sample means and the differences between those sample means were calculated. These differences are shown below:

Difference in Means

-78.875	-79.125	-84.750	-88.625	-78.250	-68.875	-80.750
-85.750	-97.625	-81.375	-74.875	-77.500	-68.750	-75.875
-84.625	-99.875	-89.875	-74.250	-78.375	-50.000	-78.375
-77.250	-89.875	-64.125	-77.125	-72.375	-66.875	-96.500
-74.500	-85.625	-55.250	-74.250	-89.000	-82.250	-58.375
-86.500	-107.000	-71.375	-58.500	-64.750	-80.875	-86.000
-68.500	-94.375	-61.375	-84.750	-97.125	-73.250	-75.125
-92.000	-65.875	-87.375	-65.625	-73.500	-71.375	-82.250
-99.750	-97.250	-105.875	-90.375	-68.000	-71.750	-66.250
-90.000	-90.500	-103.000	-66.875	-92.500	-71.500	-107.250
-67.375	-82.500	-56.875	-64.375	-69.750	-78.125	-80.125
-73.500	-89.500	-113.000	-76.500	-91.875	-87.000	-92.625
-72.125	-61.500	-53.750	-73.250	-71.750	-85.500	-67.125
-68.500	-53.375	-62.750	-68.500	-91.625	-57.500	-81.250
-60.750	-66.125	-71.625	-62.500	-100.875	-108.250	-85.000
-68.250	-84.625	-76.625	-74.750	-80.625	-80.750	-67.125
-88.750	-76.000	-78.250	-84.750	-41.375	-69.250	-79.750
-71.875	-78.750	-93.875	-97.750	-99.875	-70.000	-64.625
-78.125	-81.250	-92.000	-61.750	-76.250	-60.125	-82.250
-84.125	-95.500	-78.125	-92.125	-88.250	-80.125	-80.625
-94.125	-85.125	-65.750	-73.000	-103.625	-93.250	-66.375
-84.250	-64.500	-96.375	-69.000	-59.625	-61.250	-80.125
-94.375	-101.500	-62.625	-70.750	-72.375	-108.000	-62.750
-93.750	-90.125	-79.375	-83.875	-91.750	-83.750	-84.625
-65.250	-78.375	-60.625	-97.000	-90.500	-91.500	-88.000
-41.125	-70.000	-75.000	-87.375	-91.125	-96.500	-87.875
-83.750	-85.375	-84.000	-90.375	-75.625	-74.625	-60.750

Mean and the standard deviation of these 200 differences:

Descriptive Statistics: Difference in Means

Variable	N	Mean	StDev
Difference in Means	200	-79.051	13.227

Bootstrap estimate of the standard error of the difference of the two sample mean = 13.227

b) The histogram of the above 200 differences is shown below. Based on this we see that it is unlikely to have a difference of zero between the two means.

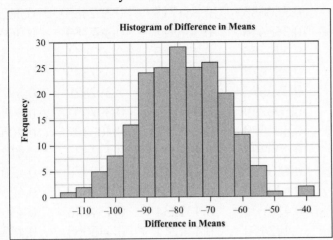

c) Assuming normality, we can test for the equality of the two means using a two-sample *t*-test as shown below. The test indicates a significant difference between the two means. The standard error is the difference between the two sample means = $\sqrt{9.4^2 + 11.0^2} = 14.5$.

Two-Sample *t*-Test and CI: Alcohol, Water

Two-sample *t* for Alcohol vs Water

N	Mean	StDev	SE Mean
Alcohol 8	37.5	26.6	9.4
Water	8 117.0	31.1	11

Difference = mu (Alcohol) - mu (Water)
Estimate for difference: -79.5000
95% CI for difference: (-110.7561, -48.2439)
t-Test of difference = 0 (vs not =): T-Value = -5.49 P-Value = 0.000 DF = 13

17. **Movie budgets again** The answers will differ, depending on the resamples drawn. The sample correlations of the 200 resamples are shown below (arranged in increasing order):

Correlation

0.200598	0.203693	0.204524	0.207919	0.215304	0.235844	0.237436
0.238880	0.244068	0.244185	0.257506	0.257902	0.259889	0.260567
0.276175	0.276681	0.284213	0.288857	0.289554	0.290694	0.291995
0.293896	0.294071	0.299159	0.303086	0.308969	0.309029	0.310699
0.314050	0.314882	0.317267	0.318481	0.319164	0.320704	0.320732
0.321724	0.322511	0.326242	0.327452	0.333717	0.334874	0.335372

0.335381	0.335547	0.336336	0.336496	0.339413	0.340369	0.340912
0.341166	0.347025	0.347889	0.348145	0.355494	0.356217	0.356886
0.357040	0.357523	0.359749	0.360636	0.361713	0.362485	0.362552
0.364712	0.365204	0.365483	0.365646	0.366315	0.367589	0.368848
0.369505	0.369814	0.370334	0.370755	0.373089	0.373641	0.374363
0.375042	0.377443	0.378099	0.381677	0.383962	0.387595	0.388788
0.389049	0.390227	0.391184	0.392672	0.392751	0.392829	0.392989
0.393648	0.394665	0.396201	0.396239	0.397026	0.397457	0.398197
0.398434	0.399689	0.399905	0.400579	0.403271	0.403397	0.405644
0.406203	0.407440	0.408306	0.408397	0.413348	0.413432	0.414477
0.417260	0.417652	0.418136	0.418289	0.418745	0.419970	0.420130
0.421059	0.421371	0.421837	0.422472	0.425090	0.426623	0.426945
0.427333	0.427606	0.429286	0.430954	0.431234	0.434076	0.436650
0.437073	0.437815	0.441259	0.442450	0.444264	0.445058	0.445478
0.446602	0.446658	0.448464	0.448745	0.449599	0.453086	0.453126
0.453946	0.454934	0.455412	0.456402	0.457730	0.458857	0.459299
0.464396	0.464589	0.464644	0.465341	0.468707	0.471396	0.471593
0.472326	0.472346	0.474062	0.475949	0.476441	0.476713	0.476988
0.477399	0.478365	0.480248	0.484047	0.486055	0.488018	0.491710
0.496624	0.497230	0.505020	0.505194	0.505447	0.506256	0.510945
0.512534	0.517102	0.520673	0.527540	0.532018	0.534094	0.537388
0.543511	0.554632	0.555805	0.565348	0.567340	0.575615	0.584636
0.587042	0.594028	0.603128	0.630306			

Mean and the standard deviation of the 200 sample correlations:

Descriptive Statistics: Correlation

Variable	N	Mean	StDev
Correlation	200	0.40093	0.08501

Bootstrap estimate of the standard error of the sample correlation = 0.08501.
For the 95% confidence interval, just pick the 5th values from the beginning and the end of the ordered correlations (i.e., approximate 2.5th and 97.5th percentiles). For this simulation it is approximately (0.215, 0.585). This interval does not include 0 and so based on this simulation, it looks unlikely for the correlation to be zero.

19. **CensusAtSchool again** The answers will differ, depending on the resamples drawn. First, let's use the data for girls. The sample correlations of the 200 resamples are shown below (arranged in increasing order):

Correlation

-0.321010	-0.275845	-0.271973	-0.261744	-0.260817	-0.259400
-0.259379	-0.258802	-0.255268	-0.248051	-0.243988	-0.238521

-0.235636	-0.222072	-0.215914	-0.211232	-0.210164	-0.208636
-0.199853	-0.198295	-0.198176	-0.197089	-0.190203	-0.188319
-0.187085	-0.184813	-0.174361	-0.160582	-0.153964	-0.146144
-0.139733	-0.137177	-0.136438	-0.133837	-0.127474	-0.126103
-0.125033	-0.119803	-0.113999	-0.110191	-0.108768	-0.106701
-0.098932	-0.097855	-0.095221	-0.093627	-0.093526	-0.089290
-0.089174	-0.088060	-0.087128	-0.086593	-0.082884	-0.080907
-0.079873	-0.079679	-0.078529	-0.075911	-0.072921	-0.071370
-0.069974	-0.067689	-0.067596	-0.066988	-0.066319	-0.064885
-0.063946	-0.061063	-0.059526	-0.053220	-0.049746	-0.046907
-0.043603	-0.040824	-0.037671	-0.036998	-0.035953	-0.035439
-0.030301	-0.029476	-0.027310	-0.026724	-0.017858	-0.017856
-0.017781	-0.016938	-0.014821	-0.013000	-0.012991	-0.012891
-0.001774	-0.000744	0.001490	0.004282	0.005490	0.011196
0.013219	0.014151	0.020654	0.024523	0.025532	0.025935
0.026738	0.028162	0.028313	0.030939	0.032546	0.032579
0.033360	0.033395	0.035334	0.039032	0.040978	0.041382
0.043915	0.044167	0.044317	0.046596	0.050111	0.050114
0.050543	0.050557	0.051210	0.053277	0.055379	0.055784
0.057610	0.058157	0.059672	0.060228	0.062768	0.063330
0.066720	0.068692	0.069586	0.071245	0.072224	0.074427
0.074434	0.076648	0.077000	0.084452	0.090321	0.095748
0.101561	0.105715	0.106341	0.107063	0.111600	0.122567
0.126200	0.127850	0.128500	0.130718	0.137376	0.137529
0.138116	0.144142	0.144433	0.150663	0.151233	0.151799
0.152964	0.153198	0.153364	0.153968	0.155570	0.158814
0.159701	0.160349	0.160804	0.161471	0.164167	0.164735
0.168804	0.172396	0.174599	0.178065	0.179041	0.184083
0.187504	0.187697	0.189127	0.189726	0.195518	0.207653
0.217265	0.232921	0.246649	0.264007	0.289946	0.292456
0.293703	0.321656	0.325494	0.336030	0.362482	0.381601
0.406137	0.455074				

Mean and the standard deviation of the 200 sample correlations:

Descriptive Statistics: Correlation

Variable	N	Mean	StDev
Correlation	200	0.0121	0.1501

Bootstrap estimate of the standard error of the sample correlation = 0.1501.
For the 95% confidence interval, just pick the 5th values from the beginning and the
end of the ordered correlations (i.e., approximate 2.5[th] and 97.5[th] percentiles). For

this simulation it is approximately (-0.261, 0.336). This interval includes 0, and this indicates that it is possible for the correlation to be zero.

Now let's look at the data for boys. The sample correlations of the 200 resamples are shown below (arranged in increasing order):

Correlation

0.076594	0.107678	0.125385	0.140372	0.155502	0.160641	0.172120
0.172132	0.174294	0.174588	0.177429	0.182077	0.214746	0.215070
0.218464	0.221345	0.225599	0.240627	0.244953	0.247703	0.250812
0.254627	0.255681	0.271429	0.273017	0.278207	0.278938	0.279310
0.282017	0.290109	0.290581	0.292909	0.301122	0.303799	0.306616
0.310674	0.312524	0.313305	0.316432	0.317332	0.320252	0.321984
0.322839	0.329484	0.331207	0.331209	0.332068	0.333575	0.334149
0.334986	0.336402	0.336782	0.337067	0.338324	0.339721	0.346060
0.346460	0.346617	0.348044	0.349744	0.350283	0.351098	0.355799
0.356485	0.356816	0.358150	0.359293	0.359542	0.359709	0.360302
0.367204	0.368202	0.368209	0.370324	0.371143	0.371183	0.373382
0.374312	0.376598	0.378019	0.381652	0.383867	0.384290	0.385205
0.385621	0.387691	0.387809	0.388980	0.389173	0.391049	0.393676
0.395479	0.397773	0.397908	0.399205	0.400247	0.401568	0.402527
0.402804	0.403806	0.405901	0.409505	0.409676	0.409910	0.419915
0.425402	0.427731	0.428168	0.429053	0.431885	0.432701	0.435306
0.435441	0.435753	0.437879	0.438817	0.439452	0.440544	0.440976
0.442923	0.442937	0.444322	0.444385	0.445133	0.445464	0.445654
0.448416	0.448727	0.449361	0.449573	0.450670	0.453238	0.454556
0.455494	0.457354	0.457592	0.459474	0.459728	0.460971	0.461082
0.462003	0.462496	0.463298	0.467316	0.469853	0.470442	0.478627
0.480062	0.482836	0.482921	0.483915	0.486714	0.489304	0.495848
0.496183	0.496862	0.499062	0.501704	0.501815	0.503066	0.504244
0.505002	0.506247	0.510371	0.511047	0.511141	0.519738	0.520178
0.524034	0.525570	0.526142	0.529668	0.530248	0.530294	0.531708
0.533834	0.542166	0.545207	0.545345	0.547910	0.549403	0.551635
0.555601	0.558277	0.564112	0.569155	0.574222	0.575517	0.584477
0.590003	0.590216	0.592314	0.604588	0.612498	0.637047	0.643558
0.655124	0.656766	0.669667	0.676449			

Mean and the standard deviation of the 200 sample correlations:

Descriptive Statistics: Correlation

Variable	N	Mean	StDev
Correlation	200	0.40543	0.11695

Bootstrap estimate of the standard error of the sample correlation = 0.11695.
For the 95% confidence interval, just pick the 5th values from the beginning and the end of the ordered correlations (i.e., approximate 2.5th and 97.5th percentiles). For this simulation it is approximately (0.156, 0.644). This interval does not include 0 and so based on this simulation, it looks unlikely for the correlation to be zero.

21. **Mean** The answers will differ due to randomness. The sample of 30 observations generated from a Normal distribution with a known mean (of, say, $\mu = 65$), a known standard deviation (of, say, $\sigma = 5$), and mean and the standard deviation of the sample are shown below:

59.1787	60.3044	67.0518	65.4541	66.6844	64.1374	53.5239
63.8788	63.1310	61.0043	61.5790	64.4940	69.0911	64.5024
65.0824	69.6807	56.6600	67.2456	55.8048	64.2704	58.5570
72.8720	70.2878	65.8018	65.4818	61.4000	59.0011	66.4736
70.4672	60.4118					

Variable	N	Mean	SE Mean	StDev
C1	30	63.784	0.842	4.612

The means of the 200 resamples from the above sample and the histogram of those 200 sample means are shown below:

Mean

63.7447	63.8069	65.1318	64.1033	63.1662	63.7740	63.4635
62.7708	61.9800	63.1597	64.3513	64.0902	62.0807	64.5234
62.8716	63.6124	63.8453	62.2446	62.8600	63.1033	63.4459
63.7811	61.5864	65.5432	63.8388	64.7823	63.1369	61.8433
64.5491	65.5470	65.1019	62.6649	64.2730	63.1670	65.5454
62.9408	64.0521	64.8397	63.2670	64.0826	64.7394	64.7204
63.9626	62.8048	62.3764	64.5700	64.6598	63.3908	63.0051
62.6054	62.9576	63.7873	63.5749	64.4591	63.2810	62.1458
64.0168	62.9784	61.6540	62.9439	64.8909	63.3646	63.5682
63.2562	63.0667	63.0963	64.7465	63.4385	63.9771	64.4517
64.6042	64.2514	64.4759	64.0430	64.5107	63.6507	62.9070
62.6761	63.3080	63.2322	64.8306	63.2788	63.9941	63.7376
63.5327	64.2255	65.7598	62.9145	63.2147	63.4901	63.5470

62.5478	63.8546	63.6364	63.9743	63.8648	63.0994	63.8441
63.8297	63.5441	64.7062	64.2176	65.2195	64.6484	63.7047
64.1748	63.9021	62.4065	63.5739	63.1801	63.8946	62.6778
63.9002	64.9348	63.9331	64.0164	64.2924	64.1491	64.3211
64.8336	63.8866	62.7561	63.9456	63.7989	62.4128	62.2528
63.6070	63.5246	63.5601	62.1951	64.7309	63.7079	64.0152
63.1925	63.6502	64.0483	64.2712	63.8390	62.2928	61.6062
62.9311	62.8511	64.2809	62.0376	63.6406	64.2565	64.0274
63.6495	64.5283	63.0867	61.9325	63.5002	63.8589	63.4350
65.2124	63.2876	63.9733	64.7694	62.5573	65.2039	65.5147
62.9923	64.6568	64.0628	62.6455	62.6041	65.2800	64.6694
63.8130	63.5205	65.0073	64.9451	64.8911	63.3968	63.8602
63.5086	65.3508	63.4538	63.2068	63.8054	63.4675	63.3915
62.8429	63.5545	63.5763	64.1105	64.0116	63.7011	63.7861
62.7053	65.1161	63.2066	62.7636	63.7947	64.3211	64.4501
64.0046	62.7599	62.8489	63.1579			

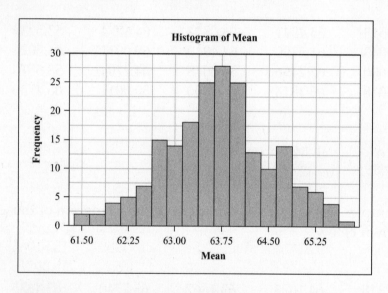

The mean and the standard deviation (the bootstrap standard error) of the 200 means:

Descriptive Statistics: Mean

Variable	N	Mean	StDev
Mean	200	63.696	0.857

The distribution is symmetric centered around 63.75 (close to the mean of the sample from the normal distribution).

23. **Standard deviation** The answers will differ due to randomness. The sample of 30 observations generated from a normal distribution with a known mean (of, say, $\mu =$

65) and a known standard deviation (of, say, $\sigma = 5$) and mean and the standard deviation of the sample are shown below:

C1

62.2614	64.2326	58.1349	67.8349	69.2504	63.4110	59.0211
63.5680	67.9223	74.6798	60.7009	66.4197	72.8893	50.5429
62.2437	70.6465	59.2842	68.9763	56.7706	65.4978	65.1057
66.2726	61.4787	64.9551	67.2118	60.5744	54.7737	68.4119
67.2500	62.5024					

Descriptive Statistics: C1

Variable	N	Mean	StDev
C1	30	64.094	5.295

The standard deviations of the 200 resamples from the above sample and the histogram of those 200 sample standard deviations are shown below:

Std deviation

4.66001	4.97557	5.01786	5.50359	5.66153	6.28903	3.99984
3.77663	5.00275	5.33451	5.42411	4.45115	6.66293	5.58117
4.54422	4.45296	5.17018	5.43343	5.67459	5.63363	4.88812
6.01936	5.38827	4.54415	5.91733	6.09181	5.22481	5.79918
5.57581	6.43161	5.14508	6.16960	4.83818	5.41626	3.96438
6.16723	4.51435	4.44443	4.80409	5.38541	5.31486	4.94219
5.32203	5.40650	5.39299	5.76170	4.56696	6.16282	4.96266
6.35409	5.25756	4.53397	5.18265	5.26029	5.53973	4.98305
4.97266	5.51348	5.25206	4.25202	4.79806	5.49566	4.48149
4.32959	4.26105	4.38888	4.84733	4.95296	4.16755	5.38639
5.18504	4.90396	5.41484	4.67684	5.54291	5.45316	4.89019
4.67559	5.42218	6.40685	5.81436	4.36872	4.73909	5.31081
5.97795	5.74172	5.04325	5.61432	5.50440	4.36547	5.30276
4.67350	5.86654	4.12740	5.25279	4.96729	4.90542	3.99175
5.95744	5.70628	5.45084	5.88647	4.56510	4.82318	5.63808
4.24854	6.26507	4.88949	6.06017	3.83108	5.42714	5.65611
3.88603	4.03755	4.50240	5.85583	4.61996	4.37661	5.08605
4.29964	6.30767	4.85154	5.53200	4.87651	6.12621	5.45537
4.49753	4.50189	4.73739	5.73770	4.94640	5.74848	6.36516
4.83004	4.95062	5.35772	4.84219	5.45511	5.83329	4.94890
4.95356	5.08019	4.44861	5.53432	6.38075	5.78856	4.95368
4.52841	4.29279	5.26769	5.65561	4.56361	4.78608	5.30467
5.69703	5.49891	4.46331	4.16347	4.45143	5.40321	4.70341
3.99124	5.09801	5.52631	4.17098	6.60630	5.40862	6.57939
7.22280	5.50610	4.12690	4.48439	5.01642	5.74896	5.73342
6.17881	4.02427	4.57888	5.19126	5.05119	5.01790	5.01334

4.54014	4.04914	5.23537	5.15926	5.00286	4.41312	4.44039
5.88761	5.03478	3.49237	5.69494	4.42845	5.26108	4.98231
4.79918	5.15061	5.81604	4.67808			

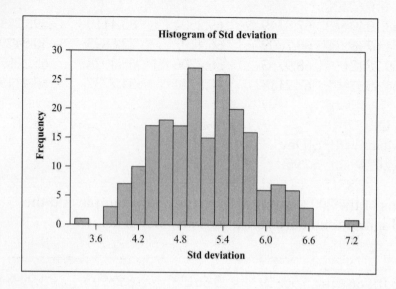

The mean and the standard deviation (the bootstrap standard error) of the 200 sample standard deviations:

Descriptive Statistics: Std deviation

Variable	N	Mean	StDev
Std deviation	200	5.1370	0.6634

The histogram is somewhat symmetric, with centre around 5.3 and ranging approximately from 3 to 7.

25. **Correlation** The answers will differ due to randomness. The medians of the 200 resamples from the above sample and the histogram of those 200 sample medians are shown below:

Correlation

0.785698	0.613946	0.586843	0.501893	0.526401	0.396375
0.474427	0.241789	0.759066	0.644223	0.487315	0.519607
0.447036	0.471923	0.590963	0.522964	0.604221	0.546688
0.494286	0.774408	0.448096	0.678720	0.519220	0.486575
0.614990	0.309607	0.501002	0.647937	0.771385	0.596495
0.473838	0.557004	0.624819	0.640079	0.642788	0.458855
0.566706	0.267041	0.710863	0.591193	0.614932	0.654733
0.559902	0.530933	0.679616	0.648657	0.373612	0.348047
0.822826	0.605153	0.595275	0.657847	0.410380	0.666054
0.502975	0.626914	0.557472	0.527009	0.510862	0.534441

0.540771	0.576620	0.686379	0.560456	0.814902	0.747498
0.765250	0.315355	0.576484	0.584094	0.663106	0.448412
0.661316	0.686589	0.722054	0.141658	0.610311	0.706787
0.576382	0.513419	0.624309	0.580795	0.416723	0.471252
0.470901	0.586005	0.346633	0.620381	0.695170	0.641054
0.358568	0.524399	0.623996	0.364895	0.603395	0.642135
0.624157	0.601091	0.490992	0.663136	0.337762	0.561285
0.716795	0.709292	0.599795	0.664173	0.401727	0.758430
0.436710	0.569211	0.453085	0.373676	0.676526	0.610883
0.725133	0.431512	0.720192	0.669769	0.720809	0.431717
0.669493	0.413389	0.667052	0.711596	0.592035	0.636997
0.558879	0.780707	0.760007	0.588851	0.609481	0.605180
0.676924	0.382512	0.685123	0.611094	0.657267	0.435200
0.663943	0.628098	0.477997	0.568765	0.543731	0.540185
0.602348	0.719019	0.546733	0.745190	0.617606	0.686872
0.717183	-0.022223	0.524841	0.690547	0.565033	0.369898
0.159915	0.541112	0.681576	0.543320	0.081134	0.702172
0.546986	0.676251	0.260466	0.559879	0.516077	0.613583
0.573403	0.697825	0.636131	0.593778	0.309336	0.615788
0.591174	0.731522	0.500375	0.633493	0.702300	0.584927
0.738964	0.436784	0.481917	0.557847	0.803252	0.807568
0.443769	0.491287	0.669460	0.560219	0.560573	0.703965
0.523546	0.567730	0.297033	0.691778	0.613286	0.541919
0.466034	0.467393				

The mean and the standard deviation (the bootstrap standard error) of the 200 sample correlations:

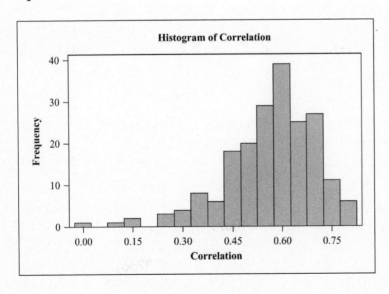

Descriptive Statistics: Correlation

Variable	N	Mean	StDev
Correlation	200	0.56823	0.13731

The histogram shows a left-skewed (negatively-skewed) shape with a mode close to 0.6.